Geometric Function Theory in Several Complex Variables

TRANSLATIONS OF MATHEMATICAL MONOGRAPHS

VOLUME **80**

Geometric Function Theory in Several Complex Variables

JUNJIRO NOGUCHI
TAKUSHIRO OCHIAI

American Mathematical Society · Providence · Rhode Island

幾何学的関数論

KIKAGAKUTEKI KANSU-RON (Geometric Function Theory in Several Complex Variables)
by Junjiro Noguchi & Takushiro Ochiai

Originally published in Japanese by Iwanami Shoten, Publishers, Tokyo in 1984

Translated from the Japanese by Junjiro Noguchi

1980 *Mathematics Subject Classification* (1985 *Revision*). Primary 32H30; Secondary 32F05.

ABSTRACT. This expanded edition in English gives a self-contained account of recent developments in geometric function theory in several complex variables and a fundamental treatise on the theory of positive currents, plurisubharmonic functions, and meromorphic mappings, culminating on the value-distribution theory for holomorphic curves.
Bibliography: 125 titles.

Library of Congress Cataloging-in-Publication Data

Noguchi, Junjirō, 1948–
[Kikagakuteki kansūron. English]
Geometric function theory in several complex variables/Junjiro Noguchi and Takushiro Ochiai: translated by Junjiro Noguchi.
p. cm.—(Translations of mathematical monographs; 80)
Translation of: Kikagakuteki kansūron.
ISBN 0-8218-4533-0
1. Geometric function theory. 2. Functions of several complex variables. I. Ochiai, Takushiro. II. Title. III. Series.
QA360.N64 1990 90-546
515′.94—dc20 CIP

This volume was printed from a TROFF file prepared by the author.
10 9 8 7 6 5 4 3 2 1 95 94 93 92 91 90

Foreword

This book is an expanded and largely rewritten version by J. Noguchi of the original Japanese edition by Noguchi with T. Ochiai ([89]). The purpose of this book is two fold. The first one is to give a self-contained and coherent account of recent developments in geometric function theory in several complex variables to those readers who have already learned the basics of complex function theory and the very elementary part of the theory of differential and complex manifolds. The second is to present, in a self-contained manner, sufficient fundamental accounts of the theory of positive currents and plurisubharmonic functions, and of the notion of meromorphic mappings, which are nowadays indispensable in the analytic and geometric theories of complex functions in several variables.

The elementary complex function theory in one variable consists, roughly speaking, of the following three themes:

I The contents from the definition of holomorphic functions to Cauchy's integral formula and the applications:

II Riemann's mapping theorem and the construction of Riemann surfaces via analytic continuation:

III The value distribution of meromorphic functions.

In the course, we have learned such theorems as Montel's theorem on normal families and Picard's little and big theorems. R. Nevanlinna evolved those theorems to the so-called Nevanlinna theory by establishing his first and second main theorems. The contents of III may be considered the basic part of the Nevanlinna theory.

In recent years, the contents of III have been generalized and extended to problems of the complex function theory in several variables and of the value distribution of holomorphic and meromorphic mappings between higher dimensional complex manifolds, and many interesting results have been obtained.

Problem of value distribution. *Let M and N be complex manifolds and $f: M \to N$ a holomorphic mapping (or more generally, meromorphic mapping). Then, investigate the properties of f and the image $f(M)$ in N.*

In this book we discuss the problem of the value distribution through geometric methods and treat the subject from the elementary level to the latest developments and topics.

Chapter I is devoted to the theory of Kobayashi pseudo-distances. S. Kobayashi defined a pseudo-distance d_M for an arbitrary complex manifold M, and called M a hyperbolic manifold if d_M is a distance. The definition is so natural that it immediately implies the decreasing principle $d_N(f(x), f(y)) \leq d_M(x, y)$ for all $x, y \in M$, which plays an essential role throughout his theory. A part of the problem of the value distribution can be reduced to those of Kobayashi pseudo-distances and be systematically discussed. Kobayashi wrote a comprehensive text book [54] and a survey paper [55]. We here define the Kobayashi pseudo-distance by making use of its infinitesimal form F_M due to Royden, and show that both coincide. Those results which have been obtained after the publication of Kobayashi's text book are described in detail. It might be helpful to read this chapter along with Chapters 4-6 of his book.

In Chapter II, we investigate the above problem of value distribution in the equidimensional case ($\dim M = \dim N$) for non-degenerate f; i.e., the differential df has maximal rank at a point. Similarly to the definition of F_M, we define the hyperbolic pseudo-volume form Ψ_M on M, which also satisfies the decreasing principle $f^*\Psi_N \leq \Psi_M$. We discuss the properties of Ψ_M and give applications.

Our next main object is to extend the Nevanlinna theory to the case of meromorphic mappings $f: \mathbf{C}^k \to M$, where M is compact. There is a long history in this subject beginning with R. Nevanlinna and represented by names such as H. Cartan, A. Bloch, H. and J. Weyl, L. Ahlfors, W. Stoll, and S. S. Chern. In the early 1970's, P. Griffiths and his co-authors gave a new insight in this subject and showed abundant connections to other fields such as differential geometry and algebraic geometry. The extended first main theorem is described in terms of holomorphic line bundles and Chern classes. As for the second main theorem, the problem, however, is more complicated. Nonetheless, P. Griffiths and his coauthors established a satisfactory second main theorem in the equidimensional case for non-degenerate f. We devote Chapter V to these works. The Poincaré-Lelong formula connecting plurisubharmonic functions with positive currents plays an important role there. Therefore we need the following items:

(i) Holomorphic line bundles and Chern classes,

(ii) Positive currents and plurisubharmonic functions,

(iii) Meromorphic mappings between complex manifolds,

(iv) Poincaré-Lelong formula.

We explain (i) at the beginning of Chapter II. Chapter III is devoted to (ii). The theory of positive currents was initiated by P. Lelong. Since the importance of (ii) has been lately increasing in various aspects of complex analysis and complex geometry, we made Chapter III worthwhile to be read independently and to be used as a reference.

We deal with (iii) in Chapter IV. We here discuss the notion of meromorphic mappings after R. Remmert. Elementary facts from the theory of complex analytic spaces are recalled without proofs (suitable references are given there). Then we give complete proofs to theorems on meromorphic mappings. Chapter IV should be used as a reference throughout the present book and may provide a standard reference on meromorphic mappings.

In Chapter VI we investigate the value distribution of holomorphic curves $f:\mathbf{C}\to M$. Except for certain special cases dealt by H. Cartan and L. Ahlfors, there is no known satisfactory second main theorem for $f:\mathbf{C}\to M$. On the other hand A. Bloch [9] stated a theorem that if M is projective algebraic and carries linearly independent holomorphic 1-forms more than the dimension of M, then the image $f(\mathbf{C})$ of f must be contained in a proper algebraic subset of M. His proof contained serious gaps and the statement was called Bloch's conjecture. We give a rigorous proof to this theorem. One notes that the theorem is connected to the problem of establishing the second main theorem for holomorphic curves (see Noguchi [77, 80, 82, 84, 85]). Here we use rather freely terminology from algebraic geometry, which is briefly explained in the second section. Readers without general basics of the algebraic geometry may find this chapter somewhat difficult, but we tried to make clear the essence of this theorem and our main idea.

In this English edition two appendices are added. In Appendix I, we explain holomorphic vector bundles and the determinant bundle. Then we calculate the Chern classes of canonical bundles of some complex submanifolds of the complex projective space $\mathbf{P}^m(\mathbf{C})$. We use these to discuss a number of examples.

Appendix II is devoted to the Weierstrass-Stoll canonical function on $\mathbf{C}^m$, which is a generalization of Weierstrass' canonical product. As an immediate consequence, we see that an effective analytic divisor D on $\mathbf{C}^m$ is algebraic if and only if the mass of $D\cap B(r)$ has polynomial growth in r, where $B(r)$ denotes a ball of $\mathbf{C}^m$ with radius r. We use this in Chapter V. Our construction is after P. Lelong and reveals a nice application of the theory of positive currents.

Since we started our research in these subjects, we have been inspired by Professor S. Kobayashi through his book, papers, lectures and many conversations. It is our pleasure to express deep gratitude to him.

We also thank Professor W. Stoll. By his invitation, J. Noguchi visited Notre Dame in 1984-1985 and gave a series of lectures on these subjects to graduate students, notes of which make up a part of this book.

Junjiro Noguchi
Takushiro Ochiai

October 1988

Table of Contents

Remarks and Notation

(i) In this book, theorems, propositions, lemmas and equations are consecutively numbered, so that for instance, (1.2.3) appears in Chapter I, §1 and after (1.2.2).

(ii) Throughout this book, manifolds are assumed to be connected unless otherwise mentioned. We follow the usual conventions in notation **N**, **Z**, **Q**, **R**, **C**, $\otimes$, $\wedge$, etc.; i.e., **N** denotes the set of natural numbers, **Z** the ring of integers, **Q** the field of rational numbers, **R** the field of real numbers, **C** the field of complex numbers, $\otimes$ the tensor product, and $\wedge$ the exterior product. For sets A and B, $A - B = \{x \in A\, ;\, x \notin B\}$. For more symbols, see Symbols at the end of the book.

CHAPTER I

Hyperbolic Manifolds

1.1 Geometry on Discs

We write **C** for the complex plane. Let $z = x + iy$ be the standard coordinate on **C**, where x and y are real variables. Set

$$dz = dx + idy, \quad d\bar{z} = dx - idy,$$

$$\frac{\partial}{\partial z} = \frac{1}{2}\left(\frac{\partial}{\partial x} - i\frac{\partial}{\partial y}\right), \quad \frac{\partial}{\partial \bar{z}} = \frac{1}{2}\left(\frac{\partial}{\partial x} + i\frac{\partial}{\partial y}\right),$$

$$dzd\bar{z} = \frac{1}{2}(dz \otimes d\bar{z} + d\bar{z} \otimes dz) = dx \otimes dx + dy \otimes dy.$$

Let U be an open subset of **C**. A **hermitian metric** g on U is a symmetric tensor which is written as

$$g = 2a(z)\, dzd\bar{z},$$

where $a(z)$ is a positive, C^∞-function on U. A symmetric tensor $h = 2b(z)\, dzd\bar{z}$ is called a **hermitian pseudo-metric** on U if

(1.1.1) (i) b is continuous, real-valued and $b \geq 0$;

(ii) Zero $(h) = \{z \in U;\, b(z) = 0\}$ is a discrete subset of U;

(iii) b is a C^∞-function on U − Zero (h).

Thus h is a hermitian metric if and only if Zero $(h) = \varnothing$. For example, let $f: U \to$ **C** be a holomorphic function with $f \not\equiv 0$. Then $h = 2|f(z)|\, dzd\bar{z}$ is a hermitian pseudo-metric on U. Let $h = 2b(z)\, dzd\bar{z}$ be a hermitian pseudo-metric on U. The **Gaussian curvature function** $K_h \colon U \to [-\infty, \infty)$ of h is defined by

$$K_h(z) = \begin{cases} -\dfrac{1}{b(z)} \dfrac{\partial^2 \log b(z)}{\partial z \partial \bar{z}} & \text{for } z \in U - \text{Zero}(h) \\ -\infty & \text{for } z \in \text{Zero}(h). \end{cases}$$

Let $w = w(z)$ be a holomorphic function on a neighborhood V of a given point of U. If w is injective (one-to-one) on V, then $dw/dz(z) \neq 0$ on V and w gives rise to a holomorphic local coordinate on V. In this case, h is also written as

$$h = 2c(w)\, dw d\bar{w} \quad \text{on } V,$$

where $c(w(z)) = b(z)|dw/dz|^{-2}$. Since $dw/dz \neq 0$, the zeros of c are identical to those of b in V. By a simple computation we see that

$$-\frac{1}{b(z)} \frac{\partial^2 \log b(z)}{\partial z \partial \bar{z}} = -\frac{1}{c(w)} \frac{\partial^2 \log c(w)}{\partial w \partial \bar{w}}$$

on $V - \text{Zero}(h)$. Therefore we have the following.

(1.1.2) Proposition. *For a hermitian pseudo-metric h, $K_h(z)$ is invariant under the changes of holomorphic local coordinates.*

We shall define a distance (metric) $d_h(z, w)$ on U with respect to h. Let $a, b \in \mathbf{R}$ be real numbers with $a < b$ and let $\phi: [a, b] \to U$ be a piecewise C^∞-curve in U. Then the length of ϕ with respect to h is given by

$$L_h(\phi) = \int_a^b \sqrt{2b(\phi(t))} \left| \frac{d\phi}{dt}(t) \right| dt.$$

For $z, w \in U$, we define the distance $d_h(z, w)$ between z and w with respect to h by

$$d_h(z, w) = \inf\{L_h(\phi)\},$$

where the infimum is taken for all piecewise C^∞-curves $\phi: [a, b] \to U$ with $\phi(a) = z$ and $\phi(b) = w$.

(1.1.3) *Example.* For $r > 0$ and $w \in \mathbf{C}$, set

$$\Delta(w; r) = \{z \in \mathbf{C};\ |z - w| < r\}, \quad \Delta(r) = \Delta(0; r).$$

We call $\Delta(w; r)$ the **disc** of radius r. Set

$$a_r(z) = \frac{2r^2}{(r^2 - |z|^2)^2}, \quad g_r = 2a_r(z)\, dz d\bar{z} \quad (z \in \Delta(r)).$$

Then g_r is a hermitian metric on $\Delta(r)$ and called the **Poincaré metric** on $\Delta(r)$. By a simple calculation we have

(1.1.4) $$K_{g_r} \equiv -1.$$

Moreover we have

(1.1.5) $$d_{g_r}(z, w) = \log \frac{r + |\alpha|}{r - |\alpha|} \quad (z, w \in \Delta(r)),$$

where $\alpha = r^2(w - z)/(r^2 - \bar{z}w)$. This d_{g_r} is called the **Poincaré distance** of $\Delta(r)$.

For two hermitian pseudo-metrics $h_i = 2b_i(z)\, dz d\bar{z}$, $i = 1, 2$ on U, we write

$$h_1 \le h_2 \text{ (resp. } h_1 < h_2)$$

if $b_1(z) \le b_2(z)$ (resp. $b_1(z) < b_2(z)$) for all $z \in U$.

(1.1.6) Lemma (Schwarz-Pick-Ahlfors). (i) *Let $h = 2b(z)\, dz d\bar{z}$ be a hermitian pseudo-metric on the disc $\Delta(r)$. Assume that $K_h \le -1$. Then $h \le g_r$.*

(ii) *Let $f : \Delta(r) \to \Delta(r)$ be a holomorphic mapping. Then $f^* g_r \le g_r$.*

Proof. (i) We first assume that b is defined on a neighborhood of the closure $\overline{\Delta(r)}$ of $\Delta(r)$. Put

$$\mu(z) = \log \frac{b(z)}{a_r(z)}.$$

Since $\mu(z) \to -\infty$ as $z \to \partial\Delta(r)$, there is a point $z_0 \in \Delta(r)$ such that

$$\mu(z_0) = \sup\{\mu(z); z \in \Delta(r)\} > -\infty.$$

Then $b(z_0) > 0$ and b is C^∞ around z_0. Noting that $4\partial^2/\partial z \partial \bar{z} = \partial^2/\partial x^2 + \partial^2/\partial y^2$, we have by (1.1.4) and the assumption, $K_h \le -1$

$$0 \ge \frac{\partial^2 \mu}{\partial z \partial \bar{z}}(z_0) = \frac{\partial^2 \log b}{\partial z \partial \bar{z}}(z_0) - \frac{\partial^2 \log a_r}{\partial z \partial \bar{z}}(z_0)$$
$$= -b(z_0) K_h(z_0) - a_r(z_0) \ge b(z_0) - a_r(z_0).$$

Hence $a_r(z_0) \ge b(z_0)$ and so $\mu(z_0) \le 0$. Therefore we see that $\mu(z) \le 0$ on $\Delta(r)$, so that $b(z) \le a_r(z)$ on $\Delta(r)$.

For the general $b(z)$ we take an increasing sequence $\{r_\nu\}_{\nu=1}^\infty$ such that $0 < r_\nu < r$ and $r_\nu \to r$ as $\nu \to \infty$. Now $b(z)$ satisfies the above assumption on $\Delta(r_\nu)$. Hence we have

$$b(z) \le a_{r_\nu}(z) \text{ for all } z \in \Delta(r_\nu).$$

Since $\{a_{r_\nu}(z)\}$ converges to $a_r(z)$, we have

$$b(z) \le a_r(z) \text{ for all } z \in \Delta(r).$$

(ii) Put $h = f^* g_r$. If $df/dz \equiv 0$, then $h \equiv 0 < g_r$. Otherwise, h is a hermitian pseudo-metric on $\Delta(r)$. Then Proposition (1.1.2) implies that $K_h(z)$ takes value -1 or $-\infty$ according to $df/dz\,(z) \neq 0$ or $= 0$. Applying (i), we have $h \le g_r$. *Q.E.D.*

We denote by $\mathrm{Aut}(\Delta(r))$ the group of all holomorphic transformations of $\Delta(r)$.

(1.1.7) Theorem. (i) $\mathrm{Aut}(\Delta(r)) = \left\{ T(z) = e^{i\theta} \dfrac{r^2(z-w)}{r^2 - \bar{w}z} ;\ w \in \Delta(r),\ \theta \in \mathbf{R} \right\}$.

(ii) *The group* $\mathrm{Aut}(\Delta(r))$ *acts transitively on* $\Delta(r)$*; i.e., for arbitrary* $z, w \in \Delta(r)$*, there exists* $T \in \mathrm{Aut}(\Delta(r))$ *such that* $T(z) = w$.

(iii) *All* $T \in \mathrm{Aut}(\Delta(r))$ *are isometries with respect to the Poincaré metric* g_r*; i.e.,* $T^* g_r = g_r$.

(iv) *For any biholomorphic mapping* $T\colon \Delta(r) \to \Delta(r')$, $T^* g_{r'} = g_r$.

Proof. Since $\mathrm{Aut}(\Delta(r)) \supset \left\{ T(z) = e^{i\theta} \dfrac{r^2(z-w)}{r^2 - \bar{w}z} ;\ w \in \Delta(r),\ \theta \in \mathbf{R} \right\}$, (ii) follows. Lemma (1.1.6), (ii) implies (iii). Putting $T_0\colon z \in \Delta(r) \to r'z/r \in \Delta(r')$, we see by definition that $T_0^* g_{r'} = g_r$. Combining this with (iii), we have (iv). Let $T \in \mathrm{Aut}(\Delta(r))$ be an arbitrary element. Put

$$S(z) = \frac{r^2(z + T^{-1}(0))}{r^2 + \overline{T^{-1}(0)}z}.$$

Then $T_1 = T \circ S \in \mathrm{Aut}(\Delta(r))$ and $T_1(0) = 0$. By (iii), $|dT_1/dz(0)| = 1$. Note that $T_1(z)/z$ is a holomorphic function on $\Delta(r)$ with values in $\Delta(1)$ and takes the value $dT_1/dz(0)$ at 0. Hence, by the maximum principle we see that $T_1(z) = e^{i\theta}z$ with $\theta \in \mathbf{R}$, so that $T = T_1 \circ S^{-1}$. *Q.E.D.*

1.2 Kobayashi Differential Metric

Let M be an m-dimensional complex manifold. Let $(U, \phi, \Delta(1)^m)$ be a holomorphic local coordinate neighborhood system around $x \in M$; i.e., U is a neighborhood of x and $\phi\colon U \to \Delta(1)^m$ is a biholomorphic mapping. Put $\phi = (z^1, \dots, z^m)$. Then $(z^1, \dots, z^m)$ is a holomorphic local coordinate system around x. We also write $(U, (z^1, \dots, z^m))$ for the holomorphic local coordinate, when we need to specify the domain where it is defined. Put $z^\alpha = x^\alpha + iy^\alpha$, where x^α and y^α are real valued. Then $(x^1, \dots, x^m, y^1, \dots, y^m)$ is a real local coordinate system around x, when M is considered as a $2m$-dimensional differentiable manifold. We write $T(M)_x$ for the tangent space of M at x. Then $T(M)_x$ is a $2m$-dimensional real vector space, and $\{(\partial/\partial x^1)_x, \dots, (\partial/\partial x^m)_x, (\partial/\partial y^1)_x, \dots, (\partial/\partial y^m)_x\}$ is a basis of $T(M)_x$. We write $T(M)_x \otimes_{\mathbf{R}} \mathbf{C}$ for the complexification of $T(M)_x$. Then $\{(\partial/\partial x^1)_x, \dots, (\partial/\partial x^m)_x, (\partial/\partial y^1)_x, \dots, (\partial/\partial y^m)_x\}$ is also a basis of the complex vector space $T(M)_x \otimes_{\mathbf{R}} \mathbf{C}$. Set

$$\frac{\partial}{\partial z^j} = \frac{1}{2}\left(\frac{\partial}{\partial x^j} - i\frac{\partial}{\partial y^j}\right), \quad 1 \le j \le m.$$

We write $\mathbf{T}(M)_x$ for the complex linear subspace $\left\{ \sum_{j=1}^{m} \xi^j (\partial / \partial z^j)_x;\ \xi^j \in \mathbf{C} \right\}$. Then $\mathbf{T}(M)_x$ is an m-dimensional complex linear subspace of $T(M)_x \otimes_{\mathbf{R}} \mathbf{C}$, which is independent of the particular choice of a holomorphic local coordinate system $(z^1, \ldots, z^m)$. We call $\mathbf{T}(M)_x$ the holomorphic tangent space of M at x. Set $\mathbf{T}(M) = \cup_{x \in M} \mathbf{T}(M)_x$ (disjoint union). We define the projection π: $\mathbf{T}(M) \to M$ by the condition $\pi(\mathbf{T}(M)_x) = x$. Then $\mathbf{T}(M)$ naturally carries the structure of a $2m$-dimensional complex manifold such that π: $\mathbf{T}(M) \to M$ is a holomorphic mapping. More precisely, let $\{z^1, \ldots, z^m\}$ be a holomorphic local coordinate system defined on an open subset U of M. Then we have

$$\pi^{-1}(U) = \left\{ \sum_{j=1}^{m} \xi^j \left(\frac{\partial}{\partial z^j}\right)_x ;\ x \in U,\ (\xi^1, \ldots, \xi^m) \in \mathbf{C}^m \right\}.$$

The mapping

$$\sum_{j=1}^{m} \xi^j \left(\frac{\partial}{\partial z^j}\right)_x \in \pi^{-1}(U) \to (z^1(x), \ldots, z^m(x), \xi^1, \ldots, \xi^m) \in \mathbf{C}^{2m}$$

is a holomorphic local coordinate system of $\mathbf{T}(M)$. We call $\mathbf{T}(M)$ the **holomorphic tangent bundle** of M. Denote by O_x the zero vector of $\mathbf{T}(M)_x$ and set $O = \cup_{x \in M} O_x$. Then O is a complex submanifold of $\mathbf{T}(M)$ and the mapping, $x \in M \to O_x \in O$ is biholomorphic. This mapping or its image O is called the zero section of $\mathbf{T}(M)$. Let N be a complex manifold and $f: M \to N$ a holomorphic mapping. We write f_* for the differential of f. For any $x \in M$ f_* is a linear mapping from $T(M)_x$ into $T(N)_{f(x)}$. Therefore f_* can be extended to a complex linear mapping from $T(M)_x \otimes_{\mathbf{R}} \mathbf{C}$ into $T(N)_{f(x)} \otimes_{\mathbf{R}} \mathbf{C}$, also denoted by f_*. Since f is holomorphic, we have $f_*(\mathbf{T}(M)_x) \subset \mathbf{T}(N)_{f(x)}$ and hence a holomorphic mapping f_*: $\mathbf{T}(M) \to \mathbf{T}(N)$, called the **holomorphic differential** of f.

Set $\mathbf{R}^+ = \{t \in \mathbf{R};\ t \ge 0\}$. A mapping $F: \mathbf{T}(M) \to \mathbf{R}^+$ is called a **differential metric** if the following conditions are satisfied:

(1.2.1) $F(O_x) = 0$, where O_x is the zero vector of $\mathbf{T}(M)_x$;

(1.2.2) for any $\xi_x \in \mathbf{T}(M)_x$ and any $a \in \mathbf{C}$, $F(a\xi_x) = |a| F(\xi_x)$.

Moreover, if F is continuous and $F(\xi_x) \ne 0$ for any $\xi_x \in \mathbf{T}(M)_x - \{O_x\}$, then F is especially called a **Finsler metric.**

Now we define a mapping F_M: $\mathbf{T}(M) \to \mathbf{R}^+$ as follows: For any $\xi_x \in \mathbf{T}(M)_x$, set

(1.2.3)
$$F_M(\xi_x) = \inf\left\{\frac{1}{r};\ \text{there exists a holomorphic mapping } f:\Delta(r)\to M \text{ such that } f(0)=x \text{ and } f_*\left(\left(\frac{\partial}{\partial z}\right)_0\right)=\xi_x\right\}.$$

Here we note that for any $s > 0$ with $F_M(\xi_x) < 1/s$, there is a holomorphic mapping $f:\Delta(s)\to M$ such that $f(0)=x$ and $f_*((\partial/\partial z)_0)=\xi_x$. By making use of the biholomorphic mapping $T: z\in\Delta(1)\to re^{i\theta}z\in\Delta(r)$, we easily have

(1.2.4)
$$F_M(\xi_x) = \inf\left\{|a|;\ \text{there exists a holomorphic mapping } f:\Delta(1)\to M \text{ such that } f(0)=x \text{ and } f_*\left(a\left(\frac{\partial}{\partial z}\right)_0\right)=\xi_x,\ a\in\mathbf{C}\right\}.$$

For the sake of simplicity, we set

$$f'(z) = f_*\left(\left(\frac{\partial}{\partial z}\right)_z\right).$$

(1.2.5) Lemma. *The mapping F_M: $\mathbf{T}(M)\to\mathbf{R}^+$ is a differential metric.*

Proof. First we show $F_M(O_x)=0$. For any $r>0$, take the constant mapping $f:\Delta(r)\to M$ such that $f(z)=x$ for $z\in\Delta(r)$. Then $f(0)=x$ and $f'(0)=O_x$. Letting $r\to\infty$, we have $F_M(O_x)=0$. Next we show that $F_M(a\xi_x)=|a|F_M(\xi_x)$. We may assume $a\neq 0$. Let $f:\Delta(1)\to M$ be a holomorphic mapping such that $f(0)=x$ and $f_*(c(\partial/\partial z)_0)=\xi_x$ with $c\in\mathbf{C}$. Since $f_*(ac(\partial/\partial z)_0)=a\xi_x$, $F_M(a\xi_x)\le|a||c|$ by (1.2.4), so that $F_M(a\xi_x)\le|a|F_M(\xi_x)$. It follows that

$$F_M(\xi_x) = F_M\left(\frac{1}{a}\cdot a\xi_x\right) \le \frac{1}{|a|}F_M(a\xi_x).$$

Therefore we see that $F_M(a\xi_x)=|a|F_M(\xi_x)$. *Q.E.D.*

We call F_M the **Kobayashi differential metric** on M.

(1.2.6) Theorem. *Let M and N be complex manifolds and $f:M\to N$ a holomorphic mapping. Then we have $f^*F_N\le F_M$; i.e., for any $\xi_x\in\mathbf{T}(M)_x$,*

$F_N(f_*(\xi_x)) \le F_M(\xi_x)$. *In particular, if f is biholomorphic,* $f^*F_N = F_M$.

Proof. Take any $\xi_x \in \mathbf{T}(M)_x$. Let $h: \Delta(r) \to M$ be a holomorphic mapping such that $h'(0) = \xi_x$. Then $f \circ h: \Delta(r) \to N$ is a holomorphic mapping such that $(f \circ h)'(0) = f_*(\xi_x)$. Hence, $F_N(f_*(\xi_x)) \le 1/r$. Since h is arbitrary, we have $F_N(f_*(\xi_x)) \le F_M(\xi_x)$. Moreover, if f is biholomorphic, we see that $F_M(\xi_x) = F_M(f_*^{-1}(f_*(\xi_x))) \le F_N(f_*(\xi_x))$. Hence we obtain $F_M(\xi_x) = F_N(f_*(\xi_x))$. *Q.E.D.*

(1.2.7) Proposition. *Let* M_j, $j = 1, 2$, *be complex manifolds. Then for any* $\xi_x + \eta_y \in \mathbf{T}(M_1)_x + \mathbf{T}(M_2)_y$ *we have*

$$F_{M_1 \times M_2}(\xi_x + \eta_y) = \max\{F_{M_1}(\xi_x), F_{M_2}(\eta_y)\}.$$

Proof. Consider the natural projections $\pi_j: M_1 \times M_2 \to M_j$ $(j = 1, 2)$. Applying Theorem (1.2.6) to the holomorphic mappings π_j, we have

$$F_{M_1 \times M_2}(\xi_x + \eta_y) \ge \max\{F_{M_1}(\xi_x), F_{M_2}(\eta_y)\}.$$

Let $f_j: \Delta(r_j) \to M_j$ $(j = 1, 2)$ be holomorphic mappings such that $f_1'(0) = \xi_x$ and $f_2'(0) = \eta_y$. Put $r = \min\{r_1, r_2\}$. Then the holomorphic mapping $f: z \in \Delta(r) \to (f_1(z), f_2(z)) \in M_1 \times M_2$ satisfies $f'(0) = \xi_x + \eta_y$. Hence $F_{M_1 \times M_2}(\xi_x + \eta_y) \le 1/r = \max\{1/r_1, 1/r_2\}$, so that

$$F_{M_1 \times M_2}(\xi_x + \eta_y) \le \max\{F_{M_1}(\xi_x), F_{M_2}(\eta_y)\}. \quad Q.E.D.$$

(1.2.8) Proposition. *Let* $\pi: \tilde{M} \to M$ *be a holomorphic unramified covering of a complex manifold M. Then* $F_{\tilde{M}} = \pi^* F_M$.

Proof. It follows from Theorem (1.2.6) that $\pi^* F_M \le F_{\tilde{M}}$. Take any $\xi_x \in \mathbf{T}(\tilde{M})_x$ with $x \in \tilde{M}$. Let $f: \Delta(r) \to M$ be a holomorphic mapping such that $f'(0) = \pi_*(\xi_x)$. Since $\Delta(r)$ is simply connected, there exists a holomorphic mapping $\tilde{f}: \Delta(r) \to \tilde{M}$ such that $\pi \circ \tilde{f} = f$ and $\tilde{f}(0) = x$. Since $\pi_*: \mathbf{T}(\tilde{M})_x \to \mathbf{T}(M)_{\pi(x)}$ is a linear isomorphism, we have $\tilde{f}'(0) = \xi_x$. Therefore $F_{\tilde{M}}(\xi_x) \le 1/r$. Since f is arbitrary, we have $F_{\tilde{M}}(\xi_x) \le F_M(\pi_*(\xi_x))$. *Q.E.D.*

(1.2.9) *Example.* We have $F_{\mathbf{C}^m} \equiv 0$. In fact, for any $r > 0$ and for any $\xi_x \in \mathbf{T}(\mathbf{C}^m)_x \cong \mathbf{C}^m$, the holomorphic mapping $f: z \in \Delta(r) \to z\xi_x + x \in \mathbf{C}^m$ satisfies $f'(0) = \xi_x$. Letting $r \to \infty$, we see that $F_{\mathbf{C}^m}(\xi_x) = 0$.

(1.2.10) *Example.* For any $\xi_z = a(\partial/\partial z)_z \in \mathbf{T}(\Delta(r))_z$, we have

$$F_{\Delta(r)}(\xi_z) = \frac{1}{\sqrt{2}}\sqrt{g_r(\xi_z, \bar{\xi}_z)} = \frac{r}{r^2 - |z|^2}|a|.$$

By Theorem (1.1.7), (iv) and Theorem (1.2.6) we may assume that $r = 1$. By Lemma (1.2.5) we may also assume that $a = 1$. Let $f: \Delta(1) \to \Delta(1)$ be a holomorphic mapping such that $f(0)=z$ and $f_*(b(\partial/\partial z)_0)=(\partial/\partial z)_z$ with $b\in \mathbf{C}$. By Lemma (1.1.6), (ii) we have $f^*g_1 \le g_1$. Hence $g_1(b(\partial/\partial z)_0, \overline{b(\partial/\partial z)_0}) \ge g_1((\partial/\partial z)_z, \overline{(\partial/\partial z)_z})$, so that $|b| \ge (1-|z|^2)^{-1}$. It follows from (1.2.4) that

$$F_{\Delta(1)}((\partial/\partial z)_z) \ge \frac{1}{1-|z|^2} = \frac{1}{\sqrt{2}}\sqrt{g_1((\partial/\partial z)_z, \overline{(\partial/\partial z)_z})}.$$

On the other hand, by Theorem (1.1.7), (ii), there is a biholomorphic mapping $T\in \mathrm{Aut}(\Delta(1))$ such that $T(0)=x$ and $T^*g_1 = g_1$. Therefore there is a number $b\in \mathbf{C}$, $b\neq 0$ such that $T_*(b(\partial/\partial z)_0) = (\partial/\partial z)_z$. By (1.2.4), $F_{\Delta(1)}((\partial/\partial z)_z) \le |b|$. Using $T^*g_1 = g_1$, we obtain

$$2|b|^2 = g_1(b(\partial/\partial z)_0, \overline{b(\partial/\partial z)_0})$$
$$= g_1((\partial/\partial z)_z, \overline{(\partial/\partial z)_z}) = \frac{2}{(1-|z|^2)^2}.$$

Therefore $F_{\Delta(1)}((\partial/\partial z)_z) = (1-|z|^2)^{-1}$.

(1.2.11) *Example*. Set $\Delta^*(r) = \Delta(r) - \{0\}$. For any $\xi_z = a(\partial/\partial z)_z \in \mathbf{T}(\Delta^*(r))$, we have

$$F_{\Delta^*(r)}(\xi_z) = \frac{|a|}{|z|\,\left|\log|z/r|^2\right|}.$$

In fact, set $H = \{z = x+iy\in \mathbf{C};\, y>0\}$. The mapping

$$\sigma\colon z\in H \to \frac{z-i}{z+i} \in \Delta(1)$$

is biholomorphic. By a simple computation we have

$$\sigma^* g_1 = \frac{dz d\bar{z}}{y^2}, \tag{1.2.12}$$

which is called the **Poincaré metric** on the upper half-plane H. Consider the unramified covering mapping $\tau\colon z\in H \to e^{2\pi i z}\in \Delta^*(1)$. Let h_r be the hermitian metric on $\Delta^*(r)$ defined by

$$h_r = \frac{4}{|z|^2(\log|z/r|^2)^2} dz d\bar{z},$$

which is called the **Poincaré metric** on $\Delta^*(r)$. It is easy to see that

$$\tau^* h_1 = \frac{dz d\bar{z}}{y^2}. \tag{1.2.13}$$

Now $\tau\circ\sigma^{-1}\colon \Delta(1) \to \Delta^*(1)$ is an unramified covering mapping, and it follows from (1.2.12) and (1.2.13) that

$$(\tau \circ \sigma^{-1})^* h_1 = g_1. \tag{1.2.14}$$

By Proposition (1.2.8) and (1.2.14) we have

$$F_{\Delta^*(1)}(\xi_z) = \frac{1}{\sqrt{2}} \sqrt{h_1(\xi_z, \bar{\xi}_z)} = \frac{|a|}{|z| \left| \log |z|^2 \right|}.$$

Put $S: z \in \Delta^*(r) \to z/r \in \Delta^*(1)$, which is a biholomorphic mapping. Then $S^* h_1 = h_r$ and $S^* F_{\Delta^*(1)} = F_{\Delta^*(r)}$ by Theorem (1.2.6). Therefore we have

$$F_{\Delta^*(r)} = \frac{1}{\sqrt{2}} \sqrt{h_r(\xi_z, \bar{\xi}_z)} = \frac{|a|}{|z| \left| \log |z/r|^2 \right|}.$$

(1.2.15) *Example.* For positive real numbers $r_1, \ldots, r_m$, put $D = \Delta(r_1) \times \cdots \times \Delta(r_m)$. Then we see by Proposition (1.2.7) and Example (1.2.10) that

$$F_D = \max_{1 \le j \le m} \{F_{\Delta(r_j)}\} : \mathbf{T}(D) \to \mathbf{R}^+$$

is continuous.

For a general complex manifold M, we do not know the continuity of F_M, but we have the following proved by Royden [97].

(1.2.16) Theorem. *Let M be a complex manifold. Then the Kobayashi differential metric F_M: $\mathbf{T}(M) \to \mathbf{R}^+$ is upper semicontinuous; i.e., for any $\xi \in \mathbf{T}(M)$ and any $\varepsilon > 0$, there is a neighborhood U of ξ in $\mathbf{T}(M)$ such that $F_M(\eta) < F_M(\xi) + \varepsilon$ for all $\eta \in U$.*

In order to prove this theorem, we need the following lemma due to Royden. Since its proof is difficult to present here, we refer the readers to Royden [98].

(1.2.17) Lemma. *Let M be a complex manifold and $h: \Delta(r) \to M$ a holomorphic mapping with $h'(0) \ne O_{h(0)}$. Then for any positive $s < r$, there is a holomorphic mapping $H: \Delta(s) \times \Delta(1)^{m-1} \to M$ such that H is biholomorphic in a neighborhood of O and $H(z^1, 0, \ldots, 0) = h(z^1)$ for all $z^1 \in \Delta(s)$. Moreover, if h is a local imbedding, then H is taken to be a local imbedding.*

Proof of Theorem (1.2.16). Take any $\xi_x \in \mathbf{T}(M)_x$ with $\xi_x \ne O_x$ and any $\varepsilon > 0$. By definition, there are $r > 0$ and a holomorphic mapping $h: \Delta(r) \to M$ such that $h(0) = x$, $h'(0) = \xi_x$ and

$$F_M(\xi_x) \le \frac{1}{r} < F_M(\xi_x) + \varepsilon.$$

Fix $0 < s < r$ so that $1/s < F_M(\xi_x) + \varepsilon$. By Lemma (1.2.17) there is a holomorphic mapping

$$H: \Delta(s) \times \Delta(1)^{m-1} \to M$$

such that H is biholomorphic in a neighborhood of O and $H(z^1, 0, ..., 0) = h(z)$ for $z^1 \in \Delta(s)$. Set $D = \Delta(s) \times \Delta(1)^{m-1}$. Note that $H(O) = x$ and $H_*((\partial / \partial z^1)_O) = \xi_x$. By Example (1.2.15) we have

$$F_D((\partial / \partial z^1)_O) = \frac{1}{s} < F_M(\xi_x) + \varepsilon. \tag{1.2.18}$$

By Example (1.2.15), F_D is continuous. Hence there is a neighborhood V of $(\partial / \partial z^1)_O$ in $\mathbf{T}(D)$ such that

$$F_D(\zeta) < F_D((\partial / \partial z^1)_O) + \varepsilon \tag{1.2.19}$$

for all $\zeta \in V$. Since H is biholomorphic around O, we can take V so that $U = H_*(V)$ is a neighborhood of ξ_x in $\mathbf{T}(M)$ and the mapping $H_*: V \to U$ is biholomorphic. Take any $\eta \in U$. Then there exists $\zeta \in V$ such that $H_*(\zeta) = \eta$. It follows from Theorem (1.2.6), (1.2.18) and (1.2.19) that

$$\begin{aligned} F_M(\eta) = F_M(H_*(\zeta)) \le F_D(\zeta) &< F_D((\partial / \partial z^1)_O) + \varepsilon \\ &< F_M(\xi_x) + 2\varepsilon. \end{aligned}$$

Thus we see that F_M is upper semicontinuous at $\xi_x \neq O_x$. To show that F_M is also upper semicontinuous at O_x, we fix a relatively compact neighborhood W of x in M. Take any hermitian metric on a neighborhood of $\overline{W}$ and denote by $\|\cdot\|$ the norm defined by this hermitian metric. Put

$$K = \{\xi_y \in \mathbf{T}(M);\ y \in \overline{W},\ \|\xi_y\| = 1\}.$$

Since K is compact in $\mathbf{T}(M) - O$ and F_M is upper semicontinuous on K, F_M attains the maximum A on K. Take any $L > A$. For any $\varepsilon > 0$, put

$$U = \left\{ \xi_y \in \mathbf{T}(M);\ y \in W,\ \|\xi_y\| < \frac{\varepsilon}{L} \right\}.$$

Then U is a neighborhood of O_x *in* $\mathbf{T}(M)$. For any $\xi_y \in U - O$, Lemma (1.2.5) implies

$$\begin{aligned} F_M(\xi_y) = F_M\left(\|\xi_y\| \frac{\xi_y}{\|\xi_y\|} \right) &= \|\xi_y\| F_M\left(\frac{\xi_y}{\|\xi_y\|} \right) \\ &\le \frac{\varepsilon}{L} \cdot A < \varepsilon = F_M(O_x) + \varepsilon. \end{aligned}$$

Therefore F_M is upper semicontinuous at O_x. *Q.E.D.*

(1.2.20) Theorem. *Let* $H: \mathbf{T}(M) \to \mathbf{R}^+$ *be a differential metric satisfying*

$$f^*H \le F_{\Delta(1)} \tag{1.2.21}$$

for all holomorphic mappings $f: \Delta(1) \to M$. *Then* $H \le F_M$. *Namely, the Kobayashi differential metric is the largest one among the differential metrics satisfying* (1.2.21).

Proof. Take any $\xi_x \in \mathbf{T}(M)$, $x \in M$. Let $f: \Delta(1) \to M$ be a holomorphic mapping such that $f(0) = x$ and $f_*(a(\partial/\partial z)_0) = \xi_x$. Since $f^*H \le F_{\Delta(1)}$, Example (1.2.10) implies

$$H(\xi_x) \le F_{\Delta(1)}(a(\partial/\partial z)_0) = |a|.$$

It follows from (1.2.4) that $H(\xi_x) \le F_M(\xi_x)$. *Q.E.D.*

(1.2.22) Proposition. *Let M be a complex manifold and S a complex analytic subset of M with* codim $S \ge 2$. *Then* $F_{M-S} = F_M$ *on* $M - S$.

Proof. Let $f: \Delta(r) \to M$ be an arbitrary holomorphic mapping with $f(0) \notin S$. It is sufficient to show that for any $r' \in (0, r)$ there is a holomorphic mapping $g: \Delta(r') \to M - S$ such that $g'(0) = f'(0)$. Put

$$\tilde{M} = \Delta(r) \times M, \quad \tilde{S} = \Delta(r) \times S.$$

Let $\tilde{f}: z \in \Delta(r) \to (z, f(z)) \in \tilde{M}$ be the graph mapping of f. Then $\tilde{f}$ is an imbedding and by Lemma (1.2.17) there is a holomorphic local imbedding $\tilde{g}: \Delta(r') \times \Delta(1)^m \to \tilde{M}$ such that $\tilde{g}|\Delta(r') \times \{O\} = \tilde{f}$. Then the analytic subset $\tilde{g}^{-1}(\tilde{S})$ of $\Delta(r') \times \Delta(1)^m$ is of codimension ≥ 2 and does not contain the origin. Put

$$\Phi: (z, (w^i)) \in \Delta(r') \times \Delta(1/r'^2)^m \to (z, (z^2 w^i)) \in \Delta(r') \times \Delta(1)^m.$$

Then the critical value of Φ is only the origin and hence

$$\operatorname{codim} \Phi^{-1}(\tilde{g}^{-1}(\tilde{S})) \ge 2.$$

Let $p: \Delta(r') \times \Delta(1/r'^2)^m \to \Delta(1/r'^2)^m$ be the natural projection. Then $p(\Phi^{-1}(\tilde{g}^{-1}(\tilde{S})))$ does not contain a non-empty open subset. Take a point $w_0 \in \Delta(1/r'^2)^m - p(\Phi^{-1}(\tilde{g}^{-1}(\tilde{S})))$ with $w_0 \ne O$ and put

$$g: z \in \Delta(r') \to q(\tilde{g}(z, z^2 w_0)) \in M,$$

where $q: \tilde{M} \to M$ denotes the natural projection. It is clear by the construction that $g(\Delta(r')) \cap S = \emptyset$ and $g'(0) = f'(0)$. *Q.E.D.*

1.3 Kobayashi Pseudo-Distance

Let M be an m-dimensional complex manifold. Any real vector $v_x \in T(M)_x$, $x \in M$ can be uniquely written as $v_x = \xi_x + \bar{\xi}_x$ with $\xi_x \in \mathbf{T}(M)_x$. Set

$$F_M(v_x) = 2F_M(\xi_x).$$

Then $F_M: T(M) \to \mathbf{R}^+$ is upper semicontinuous and satisfies $F_M(av_x) = |a| F_M(v_x)$

for all $v_x \in T(M)_x$ and $a \in \mathbf{R}$.

(1.3.1) *Example.* We have

$$F_{\Delta(1)}(v_x) = \sqrt{g_1(v_x, v_x)} \quad \text{for } v_x \in T(\Delta(1))_x.$$

In fact, take $\xi_x \in \mathbf{T}(\Delta(1))_x$ so that $v_x = \xi_x + \bar{\xi}_x$. Then Example (1.2.10) implies

$$F_{\Delta(1)}(v_x) = 2F_{\Delta(1)}(\xi_x) = \sqrt{2g_1(\xi_x, \bar{\xi}_x)}$$
$$= \sqrt{g_1(v_x, v_x)}.$$

Let $\gamma: [a, b] \to M$ be a piecewise C^∞-curve. We define the length $L_M(\gamma)$ of γ by

$$L_M(\gamma) = \int_a^b F_M(\dot{\gamma}(t))\, dt, \tag{1.3.2}$$

where $\dot{\gamma}(t) = \gamma_*((\partial/\partial t)_t)$. Since $F_M: T(M) \to \mathbf{R}^+$ is upper semicontinuous, $F_M(\dot{\gamma}(t))$ is Lebesgue integrable and $L_M(\gamma)$ is finite. For arbitrary two points $x, y \in M$, set

$$d_M(x, y) = \inf\{L_M(\gamma)\}, \tag{1.3.3}$$

where the infimum is taken for all piecewise C^∞-curves joining x and y. Then it immediately follows that

$$d_M(x, x) = 0, \quad d_M(x, y) = d_M(y, x), \tag{1.3.4}$$
$$d_M(x, z) \le d_M(x, y) + d_M(y, z)$$

for $x, y, z \in M$. In general, a mapping from $M \times M$ into $\mathbf{R}^+$ satisfying (1.3.4) is called a **pseudo-distance,** and called d_M **Kobayashi pseudo-distance** of M, which may identically vanish in general. The continuity of d_M will be proved later. If H is a Finsler metric on M, then, using H instead of F_M, we define a distance d_H on M in the same way as above. It is easy to show that d_H defines the same topology as the original one of M.

The following theorem is called the **distance decreasing principle**, and it is the most fundamental property of the Kobayashi pseudo-distance.

(1.3.5) Theorem. *Let M and N be complex manifolds and $f: M \to N$ a holomorphic mapping. Then we have*

$$d_N(f(x), f(y)) \le d_M(x, y)$$

for all $x, y \in M$. In particular, if f is biholomorphic, then $d_N(f(x), f(y)) = d_M(x, y)$.

Proof. Let $\gamma: [a, b] \to M$ be a piecewise C^∞-curve with $\gamma(a) = x$ and $\gamma(b) = y$.

Then $f \circ \gamma: [a, b] \to N$ is also a piecewise C^∞-curve joining $f(x)$ and $f(y)$. By Theorem (1.2.6) and (1.3.2) we have

$$L_N(f \circ \gamma) \leq L_M(\gamma),$$

so that $d_N(f(x), f(y)) \leq d_M(x, y)$. Moreover, if f is biholomorphic, then

$$\begin{aligned} d_M(x, y) &\geq d_N(f(x), f(y)) \\ &\geq d_M(f^{-1}(f(x)), f^{-1}(f(y))) = d_M(x, y), \end{aligned}$$

so that $d_M(x, y) = d_N(f(x), f(y))$. *Q.E.D.*

(1.3.6) Proposition. *Let $\pi: \tilde{M} \to M$ be an unramified covering of complex manifolds. Let $x, y \in M$ be arbitrary points and take $\tilde{x} \in \tilde{M}$ so that $\pi(\tilde{x}) = x$. Then we have*

$$d_M(x, y) = \inf\{d_{\tilde{M}}(\tilde{x}, \tilde{y}); \tilde{y} \in \tilde{M}, \pi(\tilde{y}) = y\}.$$

Proof. It follows from Theorem (1.3.5) that

$$d_M(x, y) \leq \inf\{d_{\tilde{M}}(\tilde{x}, \tilde{y}); \tilde{y} \in \tilde{M}, \pi(\tilde{y}) = y\}.$$

Let $\gamma: [a, b] \to M$ be a piecewise smooth curve from x to y. Let $\tilde{\gamma}: [a, b] \to \tilde{M}$ be the lifting of γ such that $\tilde{\gamma}(a) = \tilde{x}$. Put $\tilde{y}_0 = \tilde{\gamma}(b)$. Then by Proposition (1.2.8), $L_M(\gamma) = L_{\tilde{M}}(\tilde{\gamma})$. Therefore we have

$$\begin{aligned} \inf\{d_{\tilde{M}}(\tilde{x}, \tilde{y}); \tilde{y} \in \tilde{M}, \pi(\tilde{y}) = y\} &\leq d_{\tilde{M}}(\tilde{x}, \tilde{y}_0) \\ &\leq L_{\tilde{M}}(\tilde{\gamma}) = L_M(\gamma). \end{aligned}$$

Since $d_M(x, y) = \inf_\gamma L_M(\gamma)$, we have

$$d_M(x, y) \geq \inf\{d_{\tilde{M}}(\tilde{x}, \tilde{y}); \tilde{y} \in \tilde{M}, \pi(\tilde{y}) = y\}. \quad Q.E.D.$$

(1.3.7) Proposition. *Let M_i, $i = 1, 2$, be complex manifolds. Then we have*

$$d_{M_1 \times M_2}((x_1, x_2), (y_1, y_2)) = \max\{d_{M_i}(x_i, y_i); i = 1, 2\}$$

for all $x_i, y_i \in M_i$, $i = 1, 2$.

Proof. Let $\pi_i: M_1 \times M_2 \to M_i$, $i = 1, 2$, be the natural projections. Then Theorem (1.3.5) implies that

$$d_{M_1 \times M_2}((x_i), (y_i)) \geq \max\{d_{M_i}(x_i, y_i); i = 1, 2\}.$$

For any $\varepsilon > 0$, there are piecewise C^∞-curves $\gamma_i: [0, 1] \to M_i$ from x_i to y_i, $i = 1, 2$, such that

$$d_{M_i}(x_i, y_i) + \varepsilon > \int_0^1 F_{M_i}(\dot{\gamma}_i(t))\, dt = L_{M_i}(\gamma_i).$$

Take a partition, $0 = t_0 < t_1 < \cdots < t_n = 1$ of $[0, 1]$, so that the restrictions $\gamma_i | [t_{j-1}, t_j]$ of γ_i over $[t_{j-1}, t_j]$, $1 \le j \le n$, are the restrictions of C^∞-curves from neighborhoods of $[t_{j-1}, t_j]$. It follows from Theorem (1.2.16) that for every i and j, the restriction of $F_{M_i}(\dot{\gamma}_i(t))$ over $[t_{j-1}, t_j]$ is the restriction of an upper semicontinuous function defined on a neighborhood of $[t_{j-1}, t_j]$. Thus there is a C^∞-function h_{ij} defined on a neighborhood of $[t_{j-1}, t_j]$ such that $h_{ij}(t) > F_{M_i}(\dot{\gamma}_i(t))$, for $t \in [t_{j-1}, t_j]$ and

$$L_{M_i}(\gamma_i) < \sum_{j=1}^{n} \int_{t_{j-1}}^{t_j} h_{ij}(t)\, dt < d_{M_i}(x_i, y_i) + \varepsilon.$$

Put

$$\tilde{L}_i = \sum_{j=1}^{n} \int_{t_{j-1}}^{t_j} h_{ij}(t)\, dt,$$

$$h_i(t) = \begin{cases} h_{ij}(t) & \text{for } t \in [t_{j-1}, t_j),\ 1 \le j \le n \\ h_{in}(1) & \text{for } t = 1. \end{cases}$$

Then we have

$$\text{(1.3.8)} \qquad \int_0^1 h_i(t)\, dt = \tilde{L}_i < d_{M_i}(x_i, y_i) + \varepsilon,$$

$$h_i(t) > F_{M_i}(\dot{\gamma}_i(t)) \ge 0 \text{ for } t \in [0, 1].$$

Put

$$u_i \colon t \in [0, 1] \to \int_0^t h_i(t)\, dt \in [0, \tilde{L}_i].$$

Then u_i are continuous, piecewise C^∞ and strictly increasing functions, and so are the inverse functions $v_i = u_i^{-1} \colon [0, \tilde{L}_i] \to [0, 1]$. Put

$$\tilde{\gamma}_i \colon t \in [0, 1] \to \gamma_i(v_i(\tilde{L}_i t)),$$

which are piecewise C^∞-curves from x_i to y_i. By the construction we have

$$\begin{aligned} \dot{\tilde{\gamma}}_i(t) &= \dot{\gamma}_i(v_i(\tilde{L}_i t)) v_i'(\tilde{L}_i t) \tilde{L}_i \\ &= \dot{\gamma}_i(v_i(\tilde{L}_i t)) \frac{1}{u_i'(v_i(\tilde{L}_i t))} \tilde{L}_i \\ &= \dot{\gamma}_i(v_i(\tilde{L}_i t)) \frac{1}{h_i(v_i(\tilde{L}_i t))} \tilde{L}_i \end{aligned}$$

for almost all $t \in [0, 1]$. Therefore it follows from (1.3.8) that

$$F_{M_i}(\tilde{\gamma}_i(t)) = \frac{\tilde{L}_i}{h_i(v_i(\tilde{L}_i t))} F_{M_i}(\dot{\gamma}_i(v_i(\tilde{L}_i t))) < \tilde{L}_i. \tag{1.3.9}$$

Put $\tilde{\gamma} = \tilde{\gamma}_1 \times \tilde{\gamma}_2 \colon [0, 1] \to M_1 \times M_2$. Then we have by Proposition (1.2.7), (1.3.9) and (1.3.8)

$$d_{M_1 \times M_2}((x_1, x_2), (y_1, y_2)) \le \int_0^1 F_{M_1 \times M_2}\left(\dot{\tilde{\gamma}}(t)\right) dt$$

$$= \int_0^1 \max_i \left\{ F_{M_i}\left(\dot{\tilde{\gamma}}_i(t)\right) \right\} dt \le \int_0^1 \max_i \{\tilde{L}_i\} \, dt = \max_i \{\tilde{L}_i\}$$

$$\le \max_i \left\{ d_{M_i}(x_i, y_i) \right\} + \varepsilon .$$

Since $\varepsilon > 0$ is arbitrary, we obtain

$$d_{M_1 \times M_2}((x_1, x_2), (y_1, y_2)) \le \max_i \left\{ d_{M_i}(x_i, y_i) \right\} . \quad Q.E.D.$$

(1.3.10) *Example.* $d_{\mathbf{C}^m} \equiv 0$.

(1.3.11) *Example.* We have by Example (1.3.1)

$$d_{\Delta(1)}(x, y) = d_{g_1}(x, y) \quad \text{(cf. (1.1.5))}.$$

Especially, $d_{\Delta(1)}$ is a continuous distance.

(1.3.12) *Example.*

$$d_{\Delta(1)^m}((x^i), (y^i)) = \max\{d_{\Delta(1)}(x^i, y^i); 1 \le i \le m\}$$

for $(x^i), (y^i) \in \Delta(1)^m$. Especially, $d_{\Delta(1)^m}$ is a continuous distance.

(1.3.13) Proposition. *The Kobayashi pseudo-distance* $d_M \colon M \times M \to \mathbf{R}^+$ *is continuous.*

Proof. Let $\{x_n\}_{n=1}^{\infty}$ (resp. $\{y_n\}_{n=1}^{\infty}$) be a sequence of points of M converging to $x \in M$ (resp. $y \in M$). Then it follows from (1.3.4) that

$$|d_M(x_n, y_n) - d_M(x, y)| \le d_M(x_n, x) + d_M(y_n, y).$$

Therefore it is sufficient to show that $d_M(x_n, x) \to 0$ as $n \to \infty$. Let U be a holomorphic local coordinate neighborhood around x which is biholomorphic to $\Delta(1)^m$, where $m = \dim M$. By Example (1.3.12) and Theorem (1.3.5) we see that d_U is continuous, and hence

$$d_M(x_n, x) \le d_U(x_n, x) \to 0 \text{ as } n \to \infty. \quad Q.E.D.$$

(1.3.14) Proposition. *Let M be a complex manifold and S a complex analytic subset of M with* codim $S \geq 2$. *Then $d_{M-S} = d_M$ on $M - S$.*

This immediately follows from Proposition (1.2.22).

1.4 The Original Definition of the Kobayashi Pseudo-Distance

Kobayashi [54] originally defined the Kobayashi pseudo-distance in a different way. Here we present his original definition and show that it coincides with ours.

Let M be a complex manifold and $x, y \in M$ arbitrary points. A **holomorphic chain** $\left[\{f_i\}_{i=0}^n, \{z_i\}_{i=0}^n\right]$ from x to y is the collection of holomorphic mappings $f_i\colon \Delta(1) \to M$ and points $z_i \in \Delta(1)$ such that

$$f_0(0) = x,$$

$$f_i(z_i) = f_{i+1}(0),\ 0 \leq i \leq n-1,$$

$$f_n(z_n) = y.$$

Set

$$d'_M(x, y) = \inf\left\{ \sum_{i=0}^{n} d_{g_1}(0, z_i) \right\}, \tag{1.4.1}$$

where the infimum is taken for all holomorphic chains from x to y. Then it is easy to see that

(1.4.2) d'_M is a pseudo-distance of M (cf. (1.3.4)).

This d'_M is the Kobayashi pseudo-distance defined in Kobayashi [54]. Let $f\colon M \to N$ be a holomorphic mapping into another complex manifold N. Then it easily follows from the definition that

$$d'_M(x, y) \geq d'_N(f(x), f(y)) \text{ for all } x, y \in M. \tag{1.4.3}$$

(1.4.4) Theorem. *We have*

$$d'_M(x, y) = d_M(x, y) \ \textit{for all } x, y \in M.$$

Proof. It follows from Theorem (1.3.5), Example (1.3.11) and (1.4.1) that

$$d'_M(x, y) \geq d_M(x, y) \ \text{ for } x, y \in M.$$

To show the converse, we take an arbitrary $\varepsilon > 0$. Then there is a piecewise C^∞-curve $\gamma\colon [0, 1] \to M$ from x to y such that

$$\int_0^1 F_M(\dot\gamma(t))\, dt < d_M(x, y) + \varepsilon.$$

By Theorem (1.2.16), $F_M(\dot{\gamma}(t))$ is upper semicontinuous at t where $\dot{\gamma}(t)$ is continuous. Hence there is a function $h\colon [0, 1] \to \mathbf{R}^+$ such that for some partition

(1.4.5) $$0 = t_0 < t_1 < \cdots < t_l = 1,$$

the followings hold:

(i) $h(t) > F_M(\dot{\gamma}(t)) \geq 0$;

(ii) $h \mid [t_{j-1}, t_j)$, $1 \leq j \leq l$, are the restrictions of continuous functions defined on neighborhoods of $[t_{j-1}, t_j]$;

(iii) $\int_0^1 F_M(\dot{\gamma}(t))\, dt < \int_0^1 h(t)\, dt < d_M(x, y) + \varepsilon$.

Since $\int_0^1 h(t)\, dt$ can be taken as Riemann integral, there is $\delta > 0$ such that for any partition, $0 = s_0 \leq s_1 \leq \cdots \leq s_k = 1$ with $\max\{s_j - s_{j-1};\ 1 \leq j \leq k\} < \delta$ and for any $p_j \in [0, 1]$, $1 \leq j \leq k$ with $|p_j - s_j| < \delta$

(1.4.6) $$\sum_{j=1}^{k} h(p_j)(s_j - s_{j-1}) < d_M(x, y) + \varepsilon.$$

Take an arbitrary point $p \in [t_{j-1}, t_j]$, $1 \leq j \leq l$. Suppose first that $\dot{\gamma}(p) = O_{\gamma(p)}$. Let $(U, \phi, \Delta(1)^m)$ be a holomorphic local coordinate neighborhood system around $\gamma(p)$ with $\phi(\gamma(p)) = O$, where $m = \dim M$. Then we put

$$F = \phi^{-1}\colon \Delta(1)^m \to U\ (\subset M).$$

Suppose next that $\dot{\gamma}(p) \neq O_{\gamma(p)}$. Then there is a holomorphic mapping $f\colon \Delta(r) \to M$ such that

$$f'(0) + \overline{f'(0)} = \dot{\gamma}(p),$$

$$F_M(\dot{\gamma}(p)) = 2F_M(f'(0)),$$

$$F_M(f'(0)) < \frac{1}{r} < \frac{1}{2} h(p).$$

Taking a little bit smaller r and applying Lemma (1.2.17), we have a holomorphic mapping $F\colon \Delta(r) \times \Delta(1)^{m-1} \to M$ which is iocally biholomorphic around O and satisfies

(1.4.7) $$\frac{1}{r} < \frac{1}{2} h(p),\ F(O) = \gamma(p),$$

$$F_*((\partial/\partial z^1)_O) + \overline{F_*((\partial/\partial z^1)_O)} = \dot{\gamma}(p).$$

In any case, there are a neighborhood I_p of p and a piecewise C^∞-curve

$\alpha: I_p \to \Delta(r) \times \Delta(1)^{m-1}$ such that $\alpha(p) = O$ and $F \circ \alpha = \gamma | I_p$. For $s \in I_p$, $\alpha(s) = O(|s-p|^2)$ or

$$\alpha(s) = (s-p, 0, ..., 0) + O(|s-p|^2).$$

It follows from (1.1.5) that there is an open interval I_p' contained in I_p such that $p \in I_p'$, the length of I_p' is less than δ and

$$d_{\Delta(r) \times \Delta(1)^{m-1}}(\alpha(s), \alpha(s')) \leq (1+\varepsilon)\frac{2}{r}|s-s'|$$

for $s, s' \in I_p'$. By Example (1.3.12) and the definition of d' we easily see that

$$d'_{\Delta(r) \times \Delta(1)^{m-1}} = d_{\Delta(r) \times \Delta(1)^{m-1}}.$$

Therefore, using (1.4.3) and (1.4.7), we obtain

(1.4.8) $$d'_M(\gamma(s), \gamma(s')) = d'_M(F(\alpha(s)), F(\alpha(s')))$$
$$\leq d'_{\Delta(r) \times \Delta(1)^{m-1}}(\alpha(s), \alpha(s')) \leq d_{\Delta(r) \times \Delta(1)^{m-1}}(\alpha(s), \alpha(s'))$$
$$\leq (1+\varepsilon)|s-s'|h(p).$$

Since $[t_{j-1}, t_j]$, $1 \leq j \leq l$, are compact, there is a positive $\eta < \delta$ such that for any $s, s' \in [t_{j-1}, t_j]$ with $|s-s'| < \eta$, there is some $p \in [t_{j-1}, t_j]$ with $s, s' \in I_p'$. Take a partition, $0 = s_0 < s_1 < \cdots < s_k = 1$ of $[0, 1]$ so that it is a refinement of (1.4.5) and $s_j - s_{j-1} < \eta$ for all j. Take $p_j \in [0, 1]$ so that $s_{j-1}, s_j \in I_{p_j}'$. Then we infer from (1.4.2), (1.4.6) and (1.4.8) that

$$d'_M(x, y) = d'_M(\gamma(0), \gamma(1)) \leq \sum_{j=1}^{k} d'_M(\gamma(s_{j-1}), \gamma(s_j))$$
$$\leq \sum_{j=1}^{k}(1+\varepsilon)(s_j - s_{j-1})h(p_j) \leq (1+\varepsilon)(d_M(x, y) + \varepsilon).$$

Letting $\varepsilon \to 0$, we finally get

$$d'_M(x, y) \leq d_M(x, y). \quad Q.E.D.$$

1.5 General Properties of Hyperbolic Manifolds

For a complex manifold M, the Kobayashi pseudo-distance $d_M: M \times M \to \mathbf{R}^+$ may not be a distance, in general. For instance, we know $d_{\mathbf{C}^m} \equiv 0$ by Example (1.3.10).

(1.5.1) *Definition.* A complex manifold M is called a **hyperbolic manifold** if the Kobayashi pseudo-distance d_M is a distance. Moreover, if d_M is a complete distance, then M is called a **complete hyperbolic manifold.**

For a point $x \in M$ and a real positive number r, we set

$$U_M(x\,;r) = \{y \in M\,;\, d_M(x, y) < r\}.$$

The following lemma is a direct consequence of Theorem (1.3.5).

(1.5.2) Lemma. (i) *Let M be a hyperbolic manifold. Then any locally closed complex submanifold of M is also a hyperbolic manifold.*

(ii) *Let M be a complete hyperbolic manifold. Then any closed complex submanifold of M is also a complete hyperbolic manifold.*

We know by Lemma (1.5.2), (i) and Example (1.3.12) that any bounded domain of $\mathbf{C}^m$ is hyperbolic.

For complex manifolds M and N, we denote by Hol(M, N) the set of all holomorphic mappings from M into N. The following proposition due to Royden gives us an interesting criteria for the hyperbolicity of a complex manifold.

(1.5.3) Proposition. *For a complex manifold M, the following five conditions are mutually equivalent.*

(i) *M is hyperbolic.*

(ii) *d_M is a distance and the topology defined by d_M is the same as the original one.*

(iii) *For any complex manifold N, there exists a distance d on M such that the topology defined by d is the same as the original one and* Hol(N, M) *is an equicontinuous family with respect to the distance d; i.e., for any $x \in N$ and any $\varepsilon > 0$, there is a neighborhood U of x such that $d(f(x), f(y)) < \varepsilon$ for all $f \in$* Hol(N, M) *and $y \in U$.*

(iv) Hol($\Delta(1)$, M) *is an even family; i.e., for arbitrary points $z \in \Delta(1)$, and $x \in M$ and for any neighborhood U of x, there are neighborhoods, V of z and W of x such that if $f \in$* Hol($\Delta(1)$, M) *satisfies $f(z) \in W$, then $f(V) \subset U$.*

(v) *Let $H: \mathbf{T}(M) \to \mathbf{R}^+$ be a Finsler metric. Then for any $x \in M$, there are a neighborhood U of x and a constant $C > 0$ such that $F_M(\xi_y) \geq CH(\xi_y)$ for all $\xi_y \in \mathbf{T}(M)_y$ with $y \in U$.*

Proof. (i)⇒(ii). Proposition (1.3.13) implies that open subsets with respect to the topology defined by d_M are also open with respect to the original one. To show the converse, take an arbitrary point $x \in M$ and a relatively compact neighborhood U of x. Put $r = \min\{d_M(y, x);\, y \in \partial U\} > 0$ Then, by the definition of d_M we easily see that $U_M(x\,;r) \subset U$. Hence open subsets with respect to the original topology are open with respect to the one defined by d_M.

(ii)⇒(iii). This immediately follows from the assumption and Theorem (1.3.5).

(iii)⇒(iv). This is a well-known result in the general topology and easily proved.

(iv)⇒(v). Take a holomorphic local coordinate neighborhood system $(U, \phi, \Delta(1)^m)$ around x with $\phi(x) = O$, where $m = \dim M$. Put $\phi = (z^1, \ldots, z^m)$. Since $\mathrm{Hol}(\Delta(1), M)$ is an even family, there exist $\varepsilon > 0$ and $0 < \delta < 1$ such that if $f \in \mathrm{Hol}(\Delta(1), M)$ satisfies $f(0) \in U' = \{y \in U;\ |z^j(y)| < \delta,\ 1 \le j \le m\}$, then $f(\Delta(\varepsilon)) \subset U$. Take an arbitrary vector $\xi_y \in \mathbf{T}(M)_y$ with $y \in U'$. Let $f: \Delta(r) \to M$ be a holomorphic mapping with $f(0) = y$ and $f'(0) = \xi_y$. Put $T: z \in \Delta(1) \to rz \in \Delta(r)$. Then the holomorphic mapping $f \circ T: \Delta(1) \to M$ satisfies $f \circ T(0) = y \in U'$, so that $f \circ T(\Delta(\varepsilon)) \subset U$. Namely, we have $f(\Delta(r\varepsilon)) \subset U$ and hence $F_U(\xi_y) \le 1/(r\varepsilon)$. Therefore we have $F_M(\xi_y) \ge \varepsilon F_U(\xi_y)$. By Theorem (1.2.6) and Example (1.2.15), F_U is a Finsler metric. Since $\overline{U'}$ is compact in U, there is a positive constant A such that $F_U(\xi_y) \ge AH(\xi_y)$ for all $\xi_y \in \mathbf{T}(M)_y$ with $y \in U'$. Thus we have

$$F_M(\xi_y) \ge \varepsilon A H(\xi_y)$$

for all $\xi_y \in \mathbf{T}(M)_y$ with $y \in U'$.

(v)⇒(i). This is clear. *Q.E.D.*

By the above proof, we see that condition (iii) holds if it is true for $N = \Delta(1)$.

(1.5.4) Theorem. *Let M be a hyperbolic manifold. Then M is complete hyperbolic if and only if any bounded subset with respect to d_M is relatively compact.*

Remark. The same statement as above for riemannian or hermitian metrics is well known, and the proof is the same as one given below.

Proof. The "if" part is clear. We show the converse. Assume that M is complete hyperbolic. For any point $x \in M$ there is a sufficiently small $r > 0$ such that $U_M(x; r)$ is relatively compact. Fix any point $x_0 \in M$. It suffices to show that $U_M(x_0; r)$ are relatively compact for all $r > 0$. If not, then we have

$$r_0 = \sup\{r > 0;\ U_M(x_0; r) \text{ is relatively compact}\} < \infty.$$

For any $\varepsilon > 0$, the closure $\overline{U_M(x_0; r_0 - \varepsilon)}$ is compact. Therefore there are finitely many points $y_j \in \overline{U_M(x_0; r_0 - \varepsilon)}$, $1 \le j \le l$ such that

$$\overline{U_M(x_0; r_0 - \varepsilon)} \subset \bigcup_{j=1}^{l} U_M(y_j; \varepsilon).$$

Take any $x \in U_M(x_0; r_0)$. Then there is a piecewise smooth curve $\gamma: [0, 1] \to M$ such that $\gamma(0) = x_0$, $\gamma(1) = x$ and $L_M(\gamma) < r_0$. Suppose that $\gamma(1) \notin U_M(x_0; r_0 - \varepsilon)$. Then there are a value $t_0 \in (0, 1)$ and some y_{j_0} such that $\gamma(t_0) \in \partial U_M(x_0; r_0 - \varepsilon)$ and $\gamma(t_0) \in U_M(y_{j_0}; \varepsilon)$. Since $L_M(\gamma | [0, t_0]) \ge r_0 - \varepsilon$, we have

$$L_M(\gamma | [t_0, 1]) = L_M(\gamma) - L_M(\gamma | [0, t_0])$$
$$< r_0 - (r_0 - \varepsilon) = \varepsilon.$$

Hence we see that $d_M(\gamma(t_0), x) < \varepsilon$. Since $d_M(\gamma(t_0), y_{j_0}) < \varepsilon$, $d_M(x, y_{j_0}) < 2\varepsilon$. Therefore we have

$$\bigcup_{j=1}^{l} U_M(y_j; 2\varepsilon) \supset U_M(x_0; r_0).$$

It follows that there is an integer j_1, $1 \leq j_1 \leq l$ such that $U_M(y_{j_1}; 2\varepsilon)$ is not relatively compact. Now take $\varepsilon = r_0/4$ and put $x_1 = y_{j_1}$. Then $U_M(x_1; r_0/2)$ is not relatively compact. Put

$$r_1 = \sup\{r > 0;\ U_M(x_1; r) \text{ is relatively compact}\}.$$

Then $r_1 \leq r_0/2$, and by the same arguments as above we find a point $x_2 \in U_M(x_1; r_1)$ such that $U_M(x_2; r_1/2)$ is not relatively compact. It follows that $U_M(x_2; r_0/4)$ is not relatively compact. Inductively, we obtain a sequence of points $x_k \in U_M(x_{k-1}; r_0/2^{k-1})$, $k = 1, 2, \ldots$ such that $U_M(x_k; r_0/2^k)$ are not relatively compact. Then the sequence $\{x_k\}_{k=1}^{\infty}$ is a Cauchy sequence, which does not have a limit point. This is a contradiction. *Q.E.D.*

(1.5.5) Proposition. *Let M_1 and M_2 be complex manifolds. Then $M_1 \times M_2$ is hyperbolic (resp. complete hyperbolic) if and only if each of M_1 and M_2 is hyperbolic (resp. complete hyperbolic).*

Proof. The "if part" follows from Proposition (1.3.7). To prove the converse, we assume that $M_1 \times M_2$ is hyperbolic (resp. complete hyperbolic). Take a point $(x_1, x_2) \in M_1 \times M_2$. Then M_1 (resp. M_2) is biholomorphic to $M_1 \times \{x_2\}$ (resp. $\{x_1\} \times M_2$). It follows from Lemma (1.5.2), (ii) and Theorem (1.3.5) that M_1 and M_2 are hyperbolic (resp. complete hyperbolic). *Q.E.D.*

(1.5.6) Proposition. *Let $\pi: \tilde{M} \to M$ be an unramified covering of a complex manifold M. Then M is hyperbolic (resp. complete hyperbolic) if and only if $\tilde{M}$ is hyperbolic (resp. complete hyperbolic).*

Proof. The statement on the hyperbolicity follows from Proposition (1.2.8) and Proposition (1.5.3), (v). Suppose that $\tilde{M}$ is complete hyperbolic. Take arbitrary points $\tilde{x} \in \tilde{M}$ and $x \in M$ so that $\pi(\tilde{x}) = x$. It follows from Proposition (1.3.6) that

$$U_M(x; r) = \pi\left[U_{\tilde{M}}(\tilde{x}; r)\right]$$

for all $r > 0$. By Theorem (1.5.4), $U_{\tilde{M}}(\tilde{x}; r)$ is relatively compact, and so is $U_M(x, r)$. Then it follows again from Theorem (1.5.4) that M is complete hyperbolic. Next suppose that M is complete hyperbolic. Let $\{\tilde{x}_j\}_{j=1}^{\infty}$ be a Cauchy sequence in $\tilde{M}$ with respect to $d_{\tilde{M}}$. Then Theorem (1.3.5) implies that $\{\pi(\tilde{x}_j)\}_{j=1}^{\infty}$ is also a Cauchy sequence in M. Put $x = \lim_{j\to\infty} \pi(\tilde{x}_j)$. Take a sufficiently small $\varepsilon > 0$ so that π gives rise to a biholomorphic mapping between $U_M(x; 2\varepsilon)$ and each connected component of $\pi^{-1}(U_M(x; 2\varepsilon))$. Then there is an integer j_0 such that $\pi(\tilde{x}_j) \in U_M(x; \varepsilon)$ for $j \geq j_0$, and such that $d_{\tilde{M}}(\tilde{x}_j, \tilde{x}_k) < \varepsilon$ for $j, k \geq j_0$. Let $\tilde{U}$ be the connected component of $\pi^{-1}(U_M(x; 2\varepsilon))$ containing $\tilde{x}_{j_0}$. By using Proposition (1.2.8) (or Proposition (1.3.6)) we easily see that

$$U_{\tilde{M}}(\tilde{x}_{j_0}; \varepsilon) \subset \tilde{U}.$$

Therefore $\tilde{x}_j \in \tilde{U}$ for all $j \geq j_0$. Take $\tilde{x} \in \tilde{U}$ with $\pi(\tilde{x}) = x$. Then $\{\tilde{x}_j\}_{j=1}^{\infty}$ converges to $\tilde{x}$. *Q.E.D.*

(1.5.7) *Example.* The disc $\Delta(r)$ is complete hyperbolic. In fact, Example (1.3.11) and (1.1.5) imply that $U_{\Delta(r)}(0; s)$ are relatively compact for all $s > 0$. Then Theorem (1.5.4) implies that $\Delta(r)$ is complete hyperbolic. Furthermore, Propositions (1.5.5) and (1.5.6) imply that $\Delta(r)^m$ and $\Delta^*(r)^m$ are complete hyperbolic (cf. Example (1.2.11)). Moreover, any complex manifold of which universal covering is $\Delta(r)^m$ is complete hyperbolic.

In general, let N and M be paracompact complex manifolds. A family F of $\mathrm{Hol}(N, M)$ is said to be **normal** if any sequence $\{f_j\}_{j=1}^{\infty}$ in F contains a subsequence which converges uniformly on compact subsets to some $f \in \mathrm{Hol}(N, M)$, or if it is compactly divergent; i.e., for any compact subsets $K \subset N$ and $L \subset M$, there is an integer j_0 such that $f_j(K) \cap L = \emptyset$ for all $j > j_0$.

(1.5.8) Theorem. *Let M be a complete hyperbolic manifold and N a paracompact complex manifold. Then* Hol(N, M) *is a normal family.*

Proof. It follows from Theorem (1.3.5) that $\mathrm{Hol}(N, M)$ is equicontinuous with respect to d_M. On the other hand, we see by Theorem (1.5.4) that any bounded subset of M with respect to d_M is relatively compact. Hence Ascoli-Arzelà's theorem implies that $\mathrm{Hol}(N, M)$ is a normal family. *Q.E.D.*

Combining Theorem (1.5.8) with Example (1.5.7), we obtain Montel's theorem.

(1.5.9) Corollary. *Let N be as above. Then* Hol$(N, \Delta(r))$ *is a normal family.*

We proved in general that the Kobayashi differential metric $F_M \colon \mathbf{T}(M) \to \mathbf{R}^+$ is upper semicontinuous (cf. Theorem (1.2.16)), but we do not know if it is continuous. The following lemma due to Royden is interesting in this regard.

(1.5.10) Lemma. *Let M be a paracompact complex manifold. If* Hol$(\Delta(1), M)$ *is a normal family, then $F_M \colon \mathbf{T}(M) \to \mathbf{R}^+$ is continuous.*

Proof. Since F_M is upper semicontinuous, it is continuous at $\xi_x \in \mathbf{T}(M)_x$ with $x \in M$ such that $F_M(\xi_x) = 0$. Take $\xi_x \in \mathbf{T}(M)_x$ such that $F_M(\xi_x) \neq 0$. It is sufficient to show the lower semicontinuity of F_M at ξ_x. Let $\{\xi_j\}_{j=1}^{\infty}$ be a sequence converging to ξ_x such that $\lim_{j\to\infty} F_M(\xi_j) = \varliminf_{\xi \to \xi_x} F_M(\xi)$. Take $r_j > 0$ so that

$$F_M(\xi_j) \leq \frac{1}{r_j} < F_M(\xi_j) + \frac{1}{j}.$$

Then there are holomorphic mappings

$$f_j\colon \Delta(r_j) \to M$$

such that $f_j'(0) = \xi_j$. Put

$$g_j\colon z \in \Delta(1) \to f_j(r_j z) \in M.$$

Then $g_j'(0) = r_j \xi_j$ and $g_j(0) = f_j(0) \to x$ as $j \to \infty$. By the assumption, there is a subsequence of $\{g_j\}$ which converges uniformly on compact subsets to $g \in \mathrm{Hol}(\Delta(1), M)$. Rewrite the subsequence as $\{g_j\}$. Then we have

$$g(0) = \lim_{j\to\infty} g_j(0) = x,$$

$$g'(0) = \lim_{j\to\infty} g_j'(0) = \lim_{j\to\infty} r_j \xi_j$$

$$= \left(\lim_{j\to\infty} F_M(\xi_j) \right)^{-1} \xi_x.$$

Therefore we see by (1.2.4) that

$$F_M(\xi_x) \leq \lim_{j\to\infty} F_M(\xi_j) = \varliminf_{\xi\to\xi_x} F_M(\xi). \quad Q.E.D.$$

Combining Lemma (1.5.10) with Theorem (1.5.8), we immediately have the following theorem.

(1.5.11) Theorem. *Let M be a complete hyperbolic manifold. Then F_M is a Finsler metric. Especially, if M is a compact hyperbolic manifold, then F_M is a Finsler metric.*

(1.5.12) Lemma. *Let M and N be complex manifolds and $H : N \times M \to M$ a holomorphic mapping. Assume that M is hyperbolic and there is a point $t_0 \in N$ such that $H(t_0, \cdot)\colon M \to M$ is biholomorphic. Then we have*

$$H(t, x) = H(t_0, x) \textit{ for all } (t, x) \in N \times M.$$

Proof. We may assume that $N = \Delta(1)$ and $t_0 = 0$. Taking $H^{-1}(0, \cdot) \circ H(t, \cdot)$, we may assume that $H(0, x) = x$ for $x \in M$. Let $x_0 \in M$ be an arbitrary point and $(U, \phi, \Delta(1)^m)$ a holomorphic local coordinate neighborhood system of M around x_0 with $\phi(x_0) = O$. Put $\phi = (z^1, \ldots, z^m)$ and write

$$H(t, z) = (H^1(t, z), \ldots, H^m(t, z))$$

on a small neighborhood of $(0, O) \in \Delta(1) \times \Delta(1)^m$. Then we have

(1.5.13) $$H^j(t, z) = z^j + a^j(z) t^l + O(t^{l+1})$$

as $t \to 0$, where $l \geq 1$. By Proposition (1.5.3), (iv), $\mathrm{Hol}(\Delta(1), M)$ is an even family. Hence there is a positive number $\delta < 1$ such that if $f : \Delta(1) \to M$ is a holomorphic mapping with $f(0) = x_0$, then

$$f(\Delta(\delta)) \subset U' = \left\{ x \in U;\ |z^j(x)| < \frac{1}{2},\ 1 \leq j \leq m \right\}.$$

Put $f^j = z^j \circ f$. By Cauchy's estimate, we have

$$\left| \frac{d^\lambda f^j}{dt^\lambda}(0) \right| \leq A_\lambda,\ \lambda = 1, 2, \ldots, \tag{1.5.14}$$

where A_λ are constants independent of f. Now we inductively define holomorphic mappings ψ_k: $\Delta(1) \to M$ as follows:

$$\psi_0(t) = x_0,$$

$$\psi_k(t) = H(t,\ \psi_{k-1}(t)) \quad \text{for } t \in \Delta(1) \text{ and } k \geq 1.$$

Then $\psi_k(0) = x_0$ for all k and hence $\psi_k(\Delta(\delta)) \subset U'$ for all k. Put

$$\psi_k^j = z^j \circ \psi_k\colon \Delta(\delta) \to \Delta(\frac{1}{2}).$$

Then a simple computation and (1.5.13) imply that

$$\psi_k^j = k a^j(O) t^l + O(t^{l+1}).$$

It follows from (1.5.14) that

$$l!k|a^j(O)| = \left| \frac{d^l \psi_k^j}{dt^l}(0) \right| \leq A_l$$

for $k \geq 1$. Letting $k \to \infty$, we obtain $a^j(O) = 0$. Therefore we deduce that $H^j(t, z) \equiv z^j$, so that $H(t, x) = x$ for all $x \in M$. *Q.E.D.*

(1.5.15) Corollary. *Let M be a complex manifold and G a positive dimensional complex Lie group acting effectively on M. Then M is not hyperbolic.*

(1.5.16) Lemma. *Let U be a domain of $\mathbf{C}^m$ and $\{\alpha_j\}_{j=1}^\infty$ a sequence of non-vanishing holomorphic functions α_j which converges uniformly on compact subsets to a holomorphic function α on U. Then α vanishes either everywhere or nowhere.*

Proof. In the case of $m = 1$, this is known as Hurwitz's theorem in the elementary complex function theory of one variable. In the case of general $m \geq 1$, we first note that in general, the zero-locus of a holomorphic function on U is nowhere dense in U, unless it is identically zero. Hence we easily infer our assertion by taking the restrictions of α_j over complex lines in U. *Q.E.D.*

Note that the above lemma follows also from the fact that $\Delta^*(r)$ with $0 < r < +\infty$ is complete hyperbolic (see Example (1.5.7)). The details of the proof are left to the readers.

(1.5.17) Theorem. *If M is compact hyperbolic, then* $\mathrm{Aut}(M)$ *is finite.*

Proof. A theorem of Bochner-Montgomery [10] says that the group of holomorphic transformations of a compact complex manifold is a complex Lie group. Then Corollary (1.5.15) implies that Aut(M) is discrete. On the other hand, we see by Theorem (1.5.8) that Aut(M) is compact. In fact, let $\{f_j\}_{j=1}^{\infty}$ be any sequence in Aut(M) converging to $f \in \mathrm{Hol}(M, M)$. Let Ω be a positive volume form on M such that $\int_M \Omega = 1$. Then

$$\int_M f_j^* \Omega = 1 \quad \text{for } j \geq 1.$$

Since $\{f_j\}$ converges uniformly on M to f,

$$\int_M f^* \Omega = 1.$$

By using Lemma (1.5.16), we see that the Jacobian of f with respect to holomorphic local coordinates never vanishes. Moreover, for distinct points $x, y \in M$

$$d_M(f(x), f(y)) = \lim_{j \to \infty} d_M(f_j(x), f_j(y)) = d_M(x, y) > 0,$$

so that $f(x) \neq f(y)$. Thus f is injective. Hence we get $f \in \mathrm{Aut}(M)$. Therefore Aut(M) is finite. *Q.E.D.*

Let (P, π, N) be a triple of complex manifolds and a surjective holomorphic mapping $\pi: P \to N$. Let M be a complex manifold. We say that (P, π, N) is a **holomorphic fiber bundle** over N with standard fiber M if for an arbitrary point $x \in N$, there are a neighborhood U of x and a biholomorphic mapping, called a **local trivialization** over U

$$\Phi_U : \pi^{-1}(U) \to U \times M$$

such that $\pi | U = p \circ \Phi_U$, where $p : U \times M \to U$ is the first projection. In case we can take $U = M$, we call (P, π, N) a **trivial bundle.**

(1.5.18) Theorem. *Let* (P, π, N) *be a holomorphic fiber bundle over N with standard fiber M. If M is hyperbolic and N is simply connected, then* (P, π, N) *is a trivial fiber bundle.*

Proof. Lemma (1.5.12) implies that the holomorphic fiber bundle (P, π, N) is defined by a monodromy $\lambda \in \mathrm{Hom}(\pi_1(N), \mathrm{Aut}(M))$, where $\pi_1(N)$ denotes the fundamental group of N. By the assumption, $\pi_1(N) = \{1\}$, so that (P, π, N) is a trivial bundle. *Q.E.D.*

1.6 Holomorphic Mappings into Hyperbolic Manifolds

Let M be $\mathbf{C}-\{0, 1\}$. By the little Picard theorem, we know that any holomorphic mapping $f: \mathbf{C} \to M$ is constant. On the other hand, by the big Picard theorem, any holomorphic mapping $f: \Delta^*(1) \to M$ can be extended to a holomorphic mapping $\tilde{f}: \Delta(1) \to \overline{M}$, where $\overline{M}$ denotes the Riemann sphere or equivalently the 1-dimensional complex projective space $\mathbf{P}^1(\mathbf{C})$. It is known in the theory of holomorphic functions of one complex variable that the universal covering space of M is $\Delta(1)$. Hence Proposition (1.5.6) implies that M is a hyperbolic manifold. In this section, we extend several facts on holomorphic mappings into $\mathbf{C}-\{0, 1\}$ to the case of holomorphic mappings into hyperbolic manifolds.

(1.6.1) Theorem. *Let M be a hyperbolic manifold. Then any holomorphic mapping $f: \mathbf{C}^n \to M$ is constant.*

Proof. For an arbitrary point $z \in \mathbf{C}^n$, we have by Theorem (1.3.5) and Example (1.3.10)

$$d_M(f(z), f(O)) \le d_{\mathbf{C}^n}(z, O) = 0.$$

Since d_M is a distance, $f(z) = f(O)$. *Q.E.D.*

In order to extend the big Picard theorem, we need the following notion due to Kobayashi [54]. Let X be a complex manifold and M a locally closed complex submanifold of X (e.g., M can be a domain of X).

(1.6.2) *Definition.* M is said to be **hyperbolically imbedded** into X if M satisfies the following three conditions:

(i) M is relatively compact in X;

(ii) M is hyperbolic;

(iii) for any point $p \in \partial M = \overline{M} - M$ and any open neighborhood U of p in X, there exists an open neighborhood V of p in X such that $V \subset U$ and

$$\inf\{d_M(x, y);\, x \in M \cap V,\, y \in M - U\} > 0.$$

The above condition (iii) is equivalent to the following one.

(1.6.3) For any distinct points $p_i \in \partial M$, $i = 1, 2$, there are neighborhoods U_i, $i = 1, 2$, in X such that

$$d_M(U_1 \cap M, U_2 \cap M)$$
$$= \inf\{d_M(x_1, x_2);\, x_i \in U_i \cap M,\, i = 1, 2\} > 0.$$

The following lemma is an easy consequence of Proposition (1.3.7) and the definition.

(1.6.4) Lemma. *Let M_1 (resp. M_2) be a locally closed complex submanifold of a* *complex manifold* X_1 *(resp.* X_2*). Then* $M_1 \times M_2$ *is hyperbolically imbedded into* $X_1 \times X_2$ *if and only if so are* M_j *into* X_j *for* $j = 1, 2$.

We give a criterion for M to be hyperbolically imbedded into X.

(1.6.5) Lemma. *Let* M *be relatively compact in* X *and* H *a Finsler metric on* X. *Then* M *is hyperbolically imbedded into* X *if and only if there is a positive constant* C *such that* $F_M(\xi_x) \geq CH(\xi_x)$ *for all* $\xi_x \in \mathbf{T}(M)$.

For the proof we have to prepare the following lemma due to Brody [13], which will play an important role in the next section as well.

(1.6.6) Lemma. *Let* M *be a complex manifold and* $H : \mathbf{T}(M) \to \mathbf{R}^+$ *a continuous differential metric. Let* $f : \Delta(r) \to M$ *be a holomorphic mapping such that* $H(f'(0)) \geq c > 0$. *Then there exists a holomorphic mapping* $g : \Delta(r) \to M$ *satisfying the followings:*

(i) $H(g'(0)) = c/2$;

(ii) $H(g'(z))/\eta_r(z) \leq c/2$ *for all* $z \in \Delta(r)$, *where* $\eta_r(z) = r^2/(r^2 - |z|^2)$;

(iii) $g(\Delta(r)) \subset f(\Delta(r))$.

Proof. For $0 \leq t \leq 1$, define holomorphic mappings $f_t : \Delta(r) \to M$ by $f_t(z) = f(tz)$. Then we have

$$(1.6.7) \qquad \frac{H(f_t'(z))}{\eta_r(z)} = t\frac{r^2 - |z|^2}{r^2 - |tz|^2}\frac{H(f'(tz))}{\eta_r(tz)}.$$

Now we put

$$\mu(t) = \sup_{z \in \Delta(r)} \frac{H(f_t'(z))}{\eta_r(z)}.$$

Then $\mu(t)$ has the following properties for $0 \leq t < 1$ (cf. (1.6.7)):

(i) $0 \leq \mu(t) < \infty$;

(ii) $\mu(t)$ is continuous on $[0, 1)$;

(iii) $\mu(t)$ is an increasing function;

(iv) $\mu(0) = 0$, and $\mu(t) \geq H(f_t'(0))/\eta_r(0) \geq tc$.

Therefore we have that $\lim_{t \to 1} \mu(t) \geq c$. By the intermediate-value theorem, there is a number $0 < t < 1$ with $\mu(t) = c/2$. Since $\lim_{|z| \to r} \eta_r(z) = +\infty$, there exists a point $z_0 \in \Delta(r)$ such that

$$\frac{c}{2} = \mu(t) = \frac{H(f_t'(z_0))}{\eta_r(z_0)}.$$

By (1.1.7), (ii), there is a holomorphic transformation $T \in \mathrm{Aut}(\Delta(r))$ with $T(0) = z_0$.

Set $g = f_t \circ T$. Since $T^* g_r = g_r$ (cf. (1.1.7), (iii)), it follows that

$$\eta_r(T(z)) \left| \frac{dT}{dz}(z) \right| = \eta_r(z).$$

From this we obtain

$$\frac{H(g'(z))}{\eta_r(z)} = \frac{H(f_{t*}(T'(z)))}{\eta_r(z)} = \frac{H\left(\dfrac{dT}{dz}(z) f_t'(T(z)) \right)}{\eta_r(T(z)) \left| \dfrac{dT}{dz}(z) \right|}$$

$$= \frac{H(f_t'(T(z)))}{\eta_r(T(z))} \le \mu(t) = \frac{c}{2}.$$

Clearly it follows that $H(g'(0)) = c/2$ and $g(\Delta(r)) \subset f(\Delta(r))$. *Q.E.D.*

Proof of Lemma (1.6.5). It is clear that if there exists such a constant C, then M is hyperbolically imbedded into X. Conversely, suppose that there is no such C. Then there are tangent vectors $\xi_j \in \mathbf{T}(M)$, $j = 1, 2, \ldots$, such that

$$H(\xi_j) = 1, \quad j = 1, 2, \ldots,$$

$$F_M(\xi_j) \to 0 \quad \text{as } j \to \infty.$$

We may assume that the sequence $\{F_M(\xi_j)\}_{j=1}^{\infty}$ is monotone decreasing. Hence there are a monotone increasing sequence $\{r_j\}_{j=1}^{\infty}$ of positive numbers r_j and holomorphic mappings

$$f_j \colon \Delta(r_j) \to M$$

such that $\lim_{j \to \infty} r_j = +\infty$ and $f_j'(0) = \xi_j$. Hence $H(f_j'(0)) = 1$ for all $j = 1, 2, \ldots$. By making use of Lemma (1.6.6), there are holomorphic mappings $g_j \colon \Delta(r_j) \to M$, $j = 1, 2, \ldots$ such that

$$H(g_j'(0)) = \frac{1}{2},$$

$$H(g_j'(z)) \le \frac{1}{2} \frac{r_j^2}{r_j^2 - |z|^2}.$$

Since $\lim_{j \to \infty} r_j^2 / (r_j^2 - |z|^2) = 1$ and M is relatively compact in X, g_j are equicontinuous on every compact subset of $\mathbf{C}$ for all large j and uniformly bounded as holomorphic mappings into X. By Ascoli-Arzelà's theorem, we may assume that $\{g_j\}$ converges uniformly on compact subsets to a holomorphic mapping $g \colon \mathbf{C} \to X$ such that $H(g'(0)) = 1/2$ and $H(g'(z)) \le 1/2$ for $z \in \mathbf{C}$. Especially, g is not constant. Suppose that there is a point $z_1 \in \mathbf{C}$ with $g(z_1) \in M$. Then there is also another point

$z_2 \in \mathbf{C}$ such that $g(z_2) \in M$ and $g(z_2) \neq g(z_1)$. Then we have

$$d_M(g_j(z_1), g_j(z_2)) \leq d_{\Delta(r_j)}(z_1, z_2) \to 0$$

as $j \to \infty$. Since

$$d_M(g_j(z_1), g_j(z_2)) \to d_M(g(z_1), g(z_2))$$

as $j \to \infty$, we have $d_M(g(z_1), g(z_2)) = 0$. Therefore M is not hyperbolic in this case. Next suppose that $g(\mathbf{C}) \subset \partial M$. Take any two points $z_1, z_2 \in \mathbf{C}$ such that $g(z_1) \neq g(z_2)$. Then for any small neighborhoods U_1 of $g(z_1)$ and U_2 of $g(z_2)$ in X, there is a positive integer j_0 such that

$$g_j(z_i) \in U_i \cap M, \quad i = 1, 2$$

for $j \geq j_0$. Therefore we see that

$$\begin{aligned} d_M(U_1 \cap M, U_2 \cap M) &\leq d_M(g_j(z_1), g_j(z_2)) \\ &\leq d_{\Delta(r_j)}(z_1, z_2) \to 0 \end{aligned}$$

as $j \to \infty$. Hence $d_M(U_1 \cap M, U_2 \cap M) = 0$, so that M is not hyperbolically imbedded into X (cf. (1.6.3)). *Q.E.D.*

For simplicity, we denote by $\{|z| = r\}$ the circle in $\mathbf{C}$ with radius $r > 0$ and center at the origin. By Example (1.2.11) and (1.3.2) we immediately have the following.

(1.6.8) Lemma. *For* $0 < r < 1$, $L_{\Delta^*(1)}(\{|z| = r\}) = \dfrac{2\pi}{|\log r|}$.

For a point $z = (z^1, \ldots, z^m) \in \mathbf{C}^m$, set

$$\|z\| = \sqrt{\sum_{i=1}^{m} |z^i|^2},$$

and

$$B(a; R) = \{z \in \mathbf{C}^m;\ \|z - a\| < R\}$$

for $a \in \mathbf{C}^m$ and $0 < R \leq +\infty$. When $a = O$, we set

$$B(R) = B(O; R).$$

Let S be a complex submanifold of dimension k of an open subset of $\mathbf{C}^m$. We denote by $\mathrm{Vol}(S)$ the real $2k$-dimensional volume of S with respect to the euclidean metric $dz^1 \cdot d\overline{z}^1 + \cdots + dz^m \cdot d\overline{z}^m$; i.e.,

$$\mathrm{Vol}(S) = \frac{\pi^k}{k!} \int_S \alpha^k,$$

where $\alpha = i(2\pi)^{-1}\sum_{j=1}^{m} dz^j \wedge d\bar{z}^j$ and $\alpha^k = \alpha\wedge\cdots\wedge\alpha$ (k-times). We recall some fact from Chapter III, §2, (d) (Corollary (3.2.40)), of which proof can be easily understood without specific preparations of Chapter III in this special case.

(1.6.9) Lemma. *Let S be a complex submanifold of dimension k of $B(R)$ such that $O \in S$. Then we have*

$$\mathrm{Vol}(S \cap B(r)) \geq \frac{\pi^k}{k!} r^{2k}$$

for $0 < r < R$.

For two points $a = (a^1, \ldots, a^n)$ and $b = (b^1, \ldots, b^n)$ of $\Delta^*(1)^n$ we put

$$R\{a, b\} = \{z = (z^1, \ldots, z^n) \in \mathbf{C}^n;\ \min\{|a^i|, |b^i|\} \leq |z^i| \leq \max\{|a^i|, |b^i|\},\ 1 \leq i \leq n\}.$$

(1.6.10) Lemma. *Let M be hyperbolically imbedded into X and $f_j\colon \Delta^*(1)^n \to M$, $j = 1, 2, \ldots$ holomorphic mappings from $\Delta^*(1)^n$ into M. Let $\{a_j\}_{j=1}^{\infty}$ and $\{b_j\}_{j=1}^{\infty}$ be sequences of points of $\Delta^*(1)^n$ converging to the origin O. Suppose that $f_j(a_j) \to P \in X$ as $j \to \infty$. Then we have*

$$f_j(R\{a_j, b_j\}) \to P \quad as\ j \to \infty.$$

Proof. Let $\Delta^*(1)^n \subset \mathbf{C}^n$ be the natural inclusion. Then $\Delta^*(1)^n$ is hyperbolically imbedded into $\mathbf{C}^n$, and so is $\Delta^*(1)^n \times M$ into $\mathbf{C}^n \times X$ by Lemma (1.6.4). Consider the graph mappings

$$G_{f_j}\colon z \in \Delta^*(1)^n \to (z, f_j(z)) \in \Delta^*(1)^n \times M \subset \mathbf{C}^n \times X.$$

Then $\lim_{j\to\infty} G_{f_j}(a_j) = (O, P) \in \mathbf{C}^n \times X$. It is sufficient to prove our lemma for G_{f_j}. Therefore we may assume that f_j are holomorphic imbeddings into M. We may assume that X is paracompact. Let h be a hermitian metric on X and $L_h(\gamma)$ denote the length of a piecewise smooth curve γ in X with respect to h. We use the induction on n. Let $n = 1$. By Lemma (1.6.5), there is a positive constant C_1 such that

$$\sqrt{h} \leq C_1 F_M \text{ on } M. \tag{1.6.11}$$

Since $f_j^* F_M \leq F_{\Delta^*(1)}$, it follows that

$$\sqrt{f_j^* h} \leq C_1 F_{\Delta^*(1)}\,. \tag{1.6.12}$$

Therefore we have by Lemma (1.6.8)

$$L_h(f_j(\{|z| = |a_j|\})) \leq C_1 F_{\Delta^*(1)}(\{|z| = |a_j|\})$$

$$= \frac{2\pi C_1}{|\log |a_j||} \to 0$$

as $j \to \infty$. We see that

(1.6.13) $$f_j(\{|z| = |a_j|\}) \to P \quad (j \to \infty).$$

Suppose that $\{f_j(R\{a_j, b_j\})\}_{j=1}^{\infty}$ does not converge to P. Then we may assume that $\{f_j(b_j)\}_{j=1}^{\infty}$ does not converge to P. Taking a suitable subsequence, we may assume that

$$f_j(b_j) \to Q \in X \quad (j \to \infty), \;\; Q \neq P.$$

In the same way as in (1.6.13), we have

(1.6.14) $$f_j(\{|z| = |b_j|\}) \to Q \quad (j \to \infty).$$

Take a relatively compact neighborhood V of P in X such that V is biholomorphic to the unit ball $B(1)$ of $\mathbf{C}^m$ with $m = \dim X$, P is corresponding to the origin and $Q \notin \bar{V}$. Identify $B(1)$ with V and let $(x^1, \ldots, x^m)$ be the standard complex coordinate system of $\mathbf{C}^m$. Then there is a positive constant C_2 such that

(1.6.15) $$h \geq C_2(dx^1 \cdot \overline{dx}^1 + \cdots + dx^m \cdot \overline{dx}^m)$$

on $V \cap \overline{B(3/4)}$. By (1.6.13) and (1.6.14) there is an integer j_0 such that

(1.6.16) $$f_j(\{|z| = |a_j|\}) \subset V \cap B(1/4),$$

$$f_j(\{|z| = |b_j|\}) \cap \bar{V} = \varnothing$$

for $j \geq j_0$. Therefore there are points $c_j \in R\{a_j, b_j\}$, $j \geq j_0$ such that

$$f_j(c_j) \in \partial(V \cap B(1/2)).$$

It follows from (1.6.16) that $f_j(R\{a_j, b_j\}) \cap B(f_j(c_j); 1/4))$ is a 1-dimensional complex submanifold of $B(f_j(c_j); 1/4)$. Let $\mathrm{Vol}(f_j(R\{a_j, b_j\}) \cap B(f_j(c_j); 1/4))$ (resp. $\mathrm{Vol}_h(f_j(R\{a_j, b_j\}) \cap B(f_j(c_j); 1/4)))$) denote the volume of $f_j(R\{a_j, b_j\}) \cap B(f_j(c_j); 1/4))$ with respect to the metric $\sum_{j=1}^{m} dx^j \cdot \overline{dx}^j$ (resp. h). By Lemma (1.6.9) we have

(1.6.17) $$\mathrm{Vol}(f_j(R\{a_j, b_j\}) \cap B(f_j(c_j); 1/4))) \geq \frac{\pi}{16}.$$

Denote by $\mathrm{Vol}_{\Delta^*(1)}(R\{a_j, b_j\})$ the volume of $R\{a_j, b_j\}$ with respect to the Poincaré metric on $\Delta^*(1)$. Then we have

(1.6.18) $$\mathrm{Vol}_{\Delta^*(1)}(R\{a_j, b_j\}) = 2\pi\left(\frac{1}{\log \min\{|a_j|, |b_j|\}}\right.$$

Continued

$$\left. - \frac{1}{\log \max\{|a_j|, |b_j|\}} \right] \to 0$$

as $j \to \infty$. It follows from (1.6.12) that

(1.6.19) $$\mathrm{Vol}_h(f_j(R\{a_j, b_j\})) \le C_1^2 \mathrm{Vol}_{\Delta^*}(R\{a_j, b_j\}) \to 0 \ \ (j \to \infty).$$

On the other hand, it follows from (1.6.15) and (1.6.17) that

$$\mathrm{Vol}_h(f_j(R\{a_j, b_j\})) \ge \mathrm{Vol}_h(f_j(R\{a_j, b_j\}) \cap B(f_j(c_j); 1/4)))$$
$$\ge C_2 \mathrm{Vol}('f_j(R\{a_j, b_j\}) \cap B(f_j(c_j); 1/4))') \ge C_2 \frac{\pi}{16}.$$

This contradicts (1.6.19).

Assume that our assertion holds for $\Delta^*(1)^{n-1}$. Then we consider the case of $\Delta^*(1)^n$. Suppose that $\{f_j(R\{a_j, b_j\})\}_{j=1}^{\infty}$ does not converge to P. In the same way as above, we may assume that $\{f_j(b_j)\}_{j=1}^{\infty}$ converges to a point $Q \in X$ with $Q \ne P$. For $1 \le k \le n$, we put

$$\Gamma'_{jk} = \{(z^1, \ldots, z^n) \in \Delta^*(1)^n;\ |z^k| = |a_j^k|,\ \min\{|a_j^i|, |b_j^i|\}$$
$$\le |z^i| \le \max\{|a_j^i|, |b_j^i| \min\} \text{ for } i \ne k\},$$
$$\Gamma''_{jk} = \{(z^1, \ldots, z^n) \in \Delta^*(1)^n;\ |z^k| = |b_j^k|,\ \min\{|a_j^i|, |b_j^i|\}$$
$$\le |z^i| \le \max\{|a_j^i|, |b_j^i|\} \text{ for } i \ne k\}.$$

Then $a_j \in \Gamma'_{jk}$ and $b_j \in \Gamma''_{jk}$. Since $f_j(a_j) \to P$ as $j \to \infty$, we have by the induction hypothesis

(1.6.20) $$f_j\Big[\{(z^1, \ldots, z^n) \in \Delta^*(1)^n;\ z^k = a_j^k,\ \min\{|a_j^i|, |b_j^i|\}$$
$$\le |z^i| \le \max\{|a_j^i|, |b_j^i|\} \text{ for } i \ne k\}\Big] \to P$$

as $j \to \infty$. For fixed $z^i \in \Delta^*(1)$, $1 \le i \le n$, $i \ne k$, we have by Lemma (1.6.8) and (1.6.12)

(1.6.21) $$L_h\Big[f_j\{(z^1, \ldots, z^n) \in \Delta^*(1)^n;\ |z^k| = |a_j^k|\}\Big]$$
$$\le \frac{2\pi C_1}{|\log |a_j^k||} \to 0$$

as $j \to \infty$. It follows from (1.6.20) and (1.6.21) that $f_j(\Gamma'_{jk}) \to P$ as $j \to \infty$, so that

(1.6.22) $$f_j\left(\bigcup_{k=1}^{n} \Gamma'_{jk}\right) \to P \ \ (j \to \infty).$$

In the same way as above we have

$$f_j\left(\bigcup_{k=1}^{n}\Gamma''_{jk}\right)\to Q\quad (j\to\infty). \tag{1.6.23}$$

Therefore there is an integer j_0 such that $|a_j^i|\neq|b_j^i|$ for $1\le i\le n$ and $j\ge j_0$. Then $R\{a_j, b_j\}$ are closed domains for $j\ge j_0$, of which boundary consists of $\bigcup_{k=1}^{n}\Gamma'_{jk}$ and $\bigcup_{k=1}^{n}\Gamma''_{jk}$. By making use of (1.6.22), (1.6.23) and Lemma (1.6.9) in the same way as in the case of $n = 1$, we obtain a contradiction. *Q.E.D.*

Let N be an n-dimensional complex manifold. A subset A of N is called an **analytic hypersurface** if for any point $x\in N$ there is an open neighborhood U of x in N and a holomorphic function α on U such that α is not identically zero and

$$A\cap U=\{\alpha=0\}=\{y\in U;\ \alpha(y)=0\}.$$

Necessarily, A is a closed subset. We say that A has **only normal crossings** if for every point $x\in A$, there is a local coordinate neighborhood system $(U, \phi, \Delta(1)^n)$ around x in N such that

$$A\cap U=\{x^1\cdots x^k=0\}\ \text{ with some } k\le n,$$

where $\phi=(x^1, \dots, x^n)$. Now we give the extension and convergence theorem due to Noguchi [86].

(1.6.24) Theorem. *Let M be hyperbolically imbedded into X. Let N be a complex manifold and A a complex hypersurface of N with only normal crossings. Let $f_j\colon N-A\to M$, $j=1, 2, \dots$ be holomorphic mappings which converge uniformly on compact subsets of $N-A$ to a holomorphic mapping $f: N-A\to X$. Then there are unique holomorphic extensions $\bar f_j\colon N\to X$ of f_j and $\bar f\colon N\to X$ of f over N, and $\{\bar f_j\}_{j=1}^{\infty}$ converges uniformly on compact subsets of N to $\bar f$.*

Proof. The uniqueness of the holomorphic extension is clear. Then the problem is local. We may assume that $N=\Delta(1)^n$ and $N-A=\Delta^*(1)^k\times\Delta(1)^{n-k}$ with $1\le k\le n$. Moreover, taking the restrictions $f_j|\Delta^*(1)^n$ of f_j and $f|\Delta^*(1)^n$ of f over $\Delta^*(1)^n$, we may assume that $N-A=\Delta^*(1)^n$. It is sufficient to prove that f_j (resp. f) have holomorphic extensions $\bar f_j$ (resp. $\bar f$) on a neighborhood of the origin and $\{\bar f_j\}$ converges uniformly to $\bar f$ around the origin. Take a sequence $\{a_\mu\}_{\mu=1}^{\infty}$ of points of $\Delta^*(1)^n$ such that $\lim_{\mu\to\infty} a_\mu = O$ and $\lim_{\mu\to\infty} f(a_\mu)=P\in X$. Take a neighborhood V_0 of P in X such that V_0 is biholomorphic to $B(1)\subset\mathbf{C}^m$ with $m=\dim X$ and P corresponds to the origin. Put $V=V_0\cap B(1/2)$. We claim that there are a positive constant $\delta<1$ and an integer j_0 such that

$$f_j(\Delta^*(\delta)^n)\subset V,\quad j\ge j_0. \tag{1.6.25}$$

If (1.6.25) is true, then Riemann's extension theorem implies that f_j (resp. f) have holomorphic extensions $\bar{f}_j$ (resp. $\bar{f}$) over $\Delta(\delta)^n$. In general, let h be a continuous function on $\overline{\Delta(r)}^n$ which is holomorphic on $\Delta(r)^n$. Then we have the following maximum principle:

$$(1.6.26) \qquad \max\{h(z);\ z\in\overline{\Delta(r)}^n\} = \max\{h(z);\ z\in(\partial\Delta(r))^n\}.$$

Since $\{\bar{f}_j|(\partial\Delta(\delta'))^n\}_{j=1}^{\infty}$ converges uniformly to $\bar{f}|(\partial\Delta(\delta'))^n$ for any fixed positive $\delta'<\delta$, the above maximum principle implies the uniform convergence of $\{\bar{f}_j|\Delta(\delta')^n\}_{j=1}^{\infty}$ with limit $\bar{f}|\Delta(\delta')^n$. Suppose that (1.6.25) is not true. Then for any positive integer λ, there is an integer j_λ such that $f_{j_\lambda}(\Delta^*(1/\lambda)^n)\not\subset V$; i.e., there is a point $b_\lambda\in\Delta^*(1/\lambda)^n$ with $f_{j_\lambda}(b_\lambda)\notin V$. On the other hand, there is a subsequence $\{f_{j_{\lambda(\mu)}}\}_{\mu=1}^{\infty}$ of $\{f_{j_\lambda}\}_{\lambda=1}^{\infty}$ such that

$$\lim_{\mu\to\infty} f_{j_{\lambda(\mu)}}(a_\mu) = P.$$

Therefore we get sequences $f_{j_{\lambda(\mu)}}: \Delta^*(1)^n \to M$, $a_\mu\in\Delta^*(1)^n$ and $b_{\lambda(\mu)}\in\Delta^*(1)^n$, $\mu = 1, 2, \ldots$ such that

$$\lim_{\mu\to\infty} a_\mu = \lim_{\mu\to\infty} b_{\lambda(\mu)} = O,$$

$$\lim_{\mu\to\infty} f_{j_{\lambda(\mu)}}(a_\mu) = P \in V,$$

$$f_{j_{\lambda(\mu)}}(b_{\lambda(\mu)}) \notin V.$$

These contradict Lemma (1.6.10). *Q.E.D.*

The following corollary is a special case of Theorem (1.6.24).

(1.6.27) Corollary. *Let the notation be as in Theorem (1.6.24). Then all holomorphic mappings from $N-A$ into M extend to holomorphic mappings from N into X.*

This corollary was first proved by Kwack [60] in the case where M is compact and $\dim N = 1$, by Kobayashi [54] in the case where M is hyperbolically imbedded into X, $\dim N \geq 1$ and A is non-singular, and then by Kiernan [52] in the present form.

Corollary (1.6.27) implies the big Picard Theorem. For we will show in §8 that $\mathbf{C}-\{0, 1\}$ is hyperbolically imbedded into $\mathbf{P}^1(\mathbf{C})$ (see Theorem (1.8.9)). While we can not explain the details, another important example of hyperbolically imbedded manifold is the quotient D/Γ of a symmetric bounded domain D by a fixed-point free arithmetic subgroup Γ of Aut(D). Since D is hyperbolic, so is D/Γ by Proposition (1.5.6). It is known that D/Γ has a compactification $\overline{D/\Gamma}$, so called the Satake compactification. Then Kobayashi-Ochiai [56] and Borel [11] proved that

D/Γ is hyperbolically imbedded into $\overline{D/\Gamma}$.

Now we show another type of extension-convergence theorem for complete hyperbolic manifolds.

(1.6.28) Theorem. *Let M be a complete hyperbolic manifold, N a complex manifold and $E \subset N$ an analytic subset of codimension ≥ 2. Then all holomorphic mappings from $N-E$ into M extend holomorphically over the whole N. Furthermore, if $\{f_\nu\}_{\nu=1}^\infty$ is a sequence of holomorphic mappings from $N-E$ into M and converges uniformly on compact subsets of $N-E$ to $f: N-E \to M$, then the sequence $\{\bar{f}_\nu\}_{\nu=1}^\infty$ converges uniformly on compact subsets of N to $\bar{f}$, where $\bar{f}_\nu$ and $\bar{f}$ stand for the extended holomorphic mappings from N into M.*

Proof. Let $g: N-E \to M$ be an arbitrary holomorphic mapping and $z_0 \in E$ an arbitrary point. Let $\{z_\mu\}_{\mu=1}^\infty$ be a sequence of points of $N-E$, converging to z_0. By Proposition (1.3.14) we have

$$d_M(g(z_\mu), g(z_\nu)) \leq d_{N-E}(z_\mu, z_\nu) = d_N(z_\mu, z_\nu),$$

so that $\{g(z_\mu)\}$ is a Cauchy sequence with respect to the distance d_M. Since M is complete hyperbolic, $\{g(z_\mu)\}$ converges to a point $a_0 \in M$. It is easy to check that a_0 is independent of the choice of the sequence $\{z_\mu\}$ as far as it converges to z_0. Put

$$\bar{g}(z_0) = a_0.$$

It is also easy to see that $\bar{g}$ is continuous and hence holomorphic.

Let $z_0 \in E$ and $a_0 = \bar{f}(z_0)$. We first show that for an arbitrary number $\varepsilon > 0$ there exists a neighborhood V_0 such that $\bar{f}(V_0) \subset U_M(a_0; \varepsilon)$ and $\bar{f}_\nu(V_0) \subset U_M(a_0; \varepsilon)$. Take a point $z_1 \in U_N(z_0; \varepsilon/3) - E$. Then $f(z_1) \in U_M(a_0; \varepsilon/3)$. There is an integer ν_0 such that $f_\nu(z_1) \in U_M(a_0; 2\varepsilon/3)$ for all $\nu \geq \nu_0$. Then we have

$$\bar{f}_\nu(U_N(z_1; \varepsilon/3)) \subset U_M(a_0; \varepsilon).$$

Put

$$V_0 = U_N(z_0; \varepsilon/3) \cap U_N(z_1; \varepsilon/3).$$

Then $z_0 \in V_0$ and

$$\bar{f}(V_0) \subset U_M(a_0; \varepsilon), \quad \bar{f}_\nu(V_0) \subset U_M(a_0; \varepsilon).$$

Let $R(E)$ (resp. $S(E)$) be the set of regular (resp. singular) points of E. Assume that $z_0 \in R(E)$. Take $\varepsilon > 0$ so small that $U_M(a_0; \varepsilon)$ is contained in a holomorphic local coordinate neighborhood of a_0 in M. Let $(V, \phi, \Delta(1)^n)$ ($n = \dim N$) be a holomorphic local coordinate neighborhood system around z_0 such that $V \subset V_0$, $\phi(z_0) = O$ and

$$V \cap E \subset \{z^1 = 0\},$$

where $\phi = (z^1, ..., z^n)$. Take an arbitrary number $0 < r < 1$. Using the maximum principle (1.6.26), we see that $\{f_\nu | \Delta(r)^n\}$ uniformly converges to $\bar{f}|\Delta(r)$. Therefore our assertion holds up to $N - S(E)$. Decompose $S(E) = R(S(E)) \cup S(S(E))$ and repeat the above argument. Thus we inductively see that our assertion holds on N. *Q.E.D.*

1.7 Function Theoretic Criterion of Hyperbolicity

In this section we describe the powerful criterion for hyperbolicity due to Brody [13] and its applications. Lemma (1.6.6) proved in the previous section will play a key role.

(1.7.1) Lemma. *Let M be a complex manifold and $H: \mathbf{T}(M) \to \mathbf{R}^+$ a Finsler metric. Assume that M is not hyperbolic. Then for any $R > 0$, there is a holomorphic mapping $f: \Delta(R) \to M$ such that $H(f'(0)) = 2$.*

Proof. By Proposition (1.5.3), (v) there exists a point $x \in M$ satisfying the following properties: For any $\varepsilon > 0$ and any neighborhood U of x, there is $\xi \in \mathbf{T}(M)_y - \{O_y\}$ with $y \in U$ such that $F_M(\xi) \le \varepsilon H(\xi)$. Put $\varepsilon = 1/(4R)$. Then

$$F_M\left(\frac{2}{H(\xi)}\xi\right) \le \frac{1}{2R} < \frac{1}{R}.$$

Hence there exists a holomorphic mapping $f: \Delta(R) \to M$ such that $f'(0) = (2/H(\xi))\xi$. *Q.E.D.*

(1.7.2) Lemma. *Let N be a complex manifold and M a compact complex submanifold of N. Then one and only one of the following two statements holds.*

(i) *There exists an open neighborhood of M in N which is hyperbolic:*

(ii) *For any Finsler metric $H: \mathbf{T}(M) \to \mathbf{R}^+$, there exists a holomorphic mapping $g: \mathbf{C} \to M$ such that*

$$H(g'(0)) = 1, \quad H(g'(z)) \le 1 \ \textit{for}\ z \in \mathbf{C}.$$

Proof. We see by Lemma (1.5.2), (i) and Theorem (1.6.1) that the statements (i) and (ii) can not hold at the same time. Suppose that no open neighborhood of M in N is hyperbolic. Let H be any Finsler metric on M. Then we may assume that H is defined on a neighborhood U of M in N. Take a sequence $\{U_j\}_{j=1}^{\infty}$ of relatively compact open neighborhoods U_j of M in N such that $U \supset U_j \supset \bar{U}_{j+1}$ and $\bigcap_{j=1}^{\infty} U_j = M$. By Lemma (1.7.1) there are holomorphic mappings $f_j: \Delta(j) \to U_j$ such that $H(f_j'(0)) = 2$, $j = 1, 2, ...$ By Lemma (1.6.6) we have holomorphic mappings $g_j: \Delta(j) \to U_j$ such that $H(g_j'(0)) = 1$ and $H(g_j'(z))/\eta_j(z) \le 1$ for $z \in \Delta(j)$.

We have

$$\frac{1}{\eta_s(z)} = 1 - \frac{|z|^2}{s^2} \leq 1 - \frac{|z|^2}{t^2} = \frac{1}{\eta_t(z)}$$

for $0 < s \leq t$ and $z \in \Delta(s)$. Hence we get

$$H(g_k'(z)) \leq \eta_k(z) \leq \eta_j(z)$$

for $k \geq j$ and $z \in \Delta(j)$. It follows that

$$d_H(g_k(z), g_k(w)) \leq j d_{\Delta(j)}(z, w)$$

for all $z, w \in \Delta(j)$ and $k \geq j$, where d_H stands for the distance on U defined by the Finsler metric H. This implies that for an arbitrarily fixed j, the sequence $\{g_k | \Delta(j)\}_{k=j}^{\infty}$ is equicontinuous. Since $g_k(\Delta(j)) \subset U_j$, $\{g_k | \Delta(j)\}_{k=j}^{\infty}$ is uniformly bounded. By Ascoli-Arzelà's theorem there is a convergent subsequence of $\{g_k | \Delta(j)\}_{k=j}^{\infty}$. By making use of Lebesgue's diagonal argument, we obtain a subsequence $\{g_{j_l}\}_{l=1}^{\infty}$ of $\{g_j\}_{j=1}^{\infty}$ which converges uniformly on compact subsets to a holomorphic mapping $g : \mathbf{C} \to N$. We infer from the above arguments that for all $z \in \mathbf{C}$

$$g(z) = \lim_{l \to \infty} g_{j_l}(z) \in \bigcap_{l=1}^{\infty} U_{j_l} = M,$$

$$H(g'(z)) = \lim_{l \to \infty} H\left(g_{j_l}'(z)\right) \leq \lim_{l \to \infty} \eta_{j_l}(z) = 1,$$

$$H(g'(0)) = \lim_{l \to \infty} H\left(g_{j_l}'(0)\right) = 1. \qquad Q.E.D.$$

The following theorem is a direct consequence of Lemma (1.7.2) and Theorem (1.6.1).

(1.7.3) Theorem. *Let M be a compact complex manifold and $H : \mathbf{T}(M) \to \mathbf{R}^+$ a Finsler metric. Then M is not hyperbolic if and only if there exists a holomorphic mapping $f : \mathbf{C} \to M$ such that $H(f'(0)) = 1$ and $H(f'(z)) \leq 1$ for all $z \in \mathbf{C}$.*

Now we introduce the notion of small deformation of a compact complex manifold M. We write $\underline{M}$ if we disregard the complex structure and consider M merely as a differentiable manifold. Let π: $\Delta(1) \times \underline{M} \to \Delta(1)$ be the first projection. Let W be a complex manifold satisfying the following conditions.

(i) $\underline{W} = \Delta(1) \times \underline{M}$ as differentiable manifolds:

(ii) π: $W \to \Delta(1)$ is holomorphic:

(iii) Set $M(z) = \pi^{-1}(z)$ for $z \in \Delta(1)$. ($M(z)$ are complex submanifolds of W.) Then $M(0)$ is biholomorphic to M.

We call the fiber space $\pi: W \to \Delta(1)$ or the family $\{M(z)\}_{z\in\Delta(1)}$ a **small deformation** of M with $\Delta(1)$ as parameter space. For the sake of simplicity, we write $\bigcup_{z\in\Delta(1)} M(z)$ for the complex manifold W.

(1.7.4) Theorem. *Let M be a compact hyperbolic manifold and $\{M(z)\}_{z\in\Delta(1)}$ a small deformation of M with $\Delta(1)$ as parameter space. Then there exists $0 < \varepsilon < 1$ such that $M(z)$ are hyperbolic for all $z\in\Delta(\varepsilon)$.*

Proof. This follows from Lemma (1.7.2) with $N = \bigcup_{z\in\Delta(1)} M(z)$. *Q.E.D.*

We give another application of Lemma (1.7.2) (or Theorem (1.7.3)) due to Green [40]. A subset Λ of $\mathbf{C}^m$ is called a **lattice** if there is a base $\{v^1, \ldots, v^{2m}\}$ of $\mathbf{C}^m$ considered as a real vector space such that

$$\Lambda = \{n_1 v^1 + \cdots + n_{2m} v^{2m};\ n_j \in \mathbf{Z},\ 1 \le j \le 2m\}.$$

We remark that Λ is an additive subgroup of $\mathbf{C}^m$ and the quotient group $\mathbf{C}^m/\Lambda$ is a connected compact complex manifold with the natural projection $\pi: \mathbf{C}^m \to \mathbf{C}^m/\Lambda$ being holomorphic. Remark also that $\mathbf{C}^m/\Lambda$ is a compact complex Lie group in the natural manner. We call $\mathbf{C}^m/\Lambda$ an m-dimensional **complex torus.** Although we do not go into details, it is known that for different lattices the complex tori are not mutually biholomorphic in general. A complex torus $\mathbf{C}^m/\Lambda$ is said to be **simple** if there is no closed connected complex submanifold A of $\mathbf{C}^m/\Lambda$ such that A is also a subgroup and $A \neq \{O\}$, $\mathbf{C}^m/\Lambda$.

Remark. It is known that if A is a connected complex submanifold of a complex torus $\mathbf{C}^m/\Lambda$ and a subgroup of $\mathbf{C}^m/\Lambda$, then A is a complex torus.

(1.7.5) Proposition. *Let M be a closed connected complex submanifold of a simple complex torus $\mathbf{C}^m/\Lambda$. If $M \neq \mathbf{C}^m/\Lambda$, then M is hyperbolic.*

Proof. Let H be the Finsler metric in $\mathbf{C}^m/\Lambda$ defined by the natural flat metric on $\mathbf{C}^m/\Lambda$ induced from the euclidean metric on $\mathbf{C}^m$. Suppose that M were not hyperbolic. By Theorem (1.7.3), there is a holomorphic mapping $f: \mathbf{C} \to M$ such that $H(f'(z)) \le 1$ and $H(f'(0)) = 1$. Since the natural projection $\pi: \mathbf{C}^m \to \mathbf{C}^m/\Lambda$ is the universal covering mapping, there is a holomorphic mapping $\tilde{f}: \mathbf{C} \to \mathbf{C}^m$ (lifting of f) such that $\pi \circ \tilde{f} = f$. Put $\tilde{f} = (f^1, \ldots, f^m)$. Since $H(f'(z)) \le 1$, we have

$$\sum_{j=1}^{m} \left| \frac{df^j}{dz}(z) \right|^2 \le 1.$$

Since df^j/dz are holomorphic functions on $\mathbf{C}$, Liouville's theorem implies that all df^j/dz are constant functions, $a^j \in \mathbf{C}$. Hence $f^j(z) = a^j z + b^j$ with $b^j \in \mathbf{C}$. Put $a = (a^1, \ldots, a^m)(\neq O)$ and $b = (b^1, \ldots, b^m)$. Define a holomorphic mapping $g: \mathbf{C} \to \mathbf{C}^m$ by $g(z) = \tilde{f}(z) - b = za$. Then $\pi \circ g: \mathbf{C} \to \mathbf{C}^m/\Lambda$ is a non-trivial group homomorphism. Put $G = \pi \circ g(\mathbf{C})$. Then G is an abstract subgroup of $\mathbf{C}^m/\Lambda$.

Now $M - \{\pi(b)\} = \{x - \pi(b);\, x \in M\}$ is a connected closed complex submanifold of $\mathbf{C}^m / \Lambda$. Clearly we have that $G \subset M - \{\pi(b)\}$. In general, we introduce a topology called the **Zariski topology** on any complex manifold by taking analytic subsets as closed subsets (cf. Chapter IV, §1 and Chapter VI, §2). Let $\overline{G}$ be the closure of G with respect to the Zariski topology on $\mathbf{C}^m / \Lambda$; that is, $\overline{G}$ is the smallest analytic subset containing G. Then we know that $\overline{G}$ is an irreducible analytic subset of $M - \pi(b)$. Moreover we show that $\overline{G}$ is a subgroup of $\mathbf{C}^m / \Lambda$. Since the mapping $x \in \mathbf{C}^m / \Lambda \to -x \in \mathbf{C}^m / \Lambda$ is biholomorphic, we obtain $-\overline{G} = \overline{G}$. Take an arbitrary $x \in G$. Then $x + G \subset G \subset \overline{G}$, so that $G \subset \overline{G} - x$. Hence $\overline{G} \subset \overline{G} - x$. It follows that $G + \overline{G} = \overline{G}$. Taking an arbitrary $y \in \overline{G}$ and using the same argument, we see that $\overline{G} + \overline{G} \subset \overline{G}$. Therefore $\overline{G}$ is a subgroup of $\mathbf{C}^m / \Lambda$ and hence non-singular; i.e., $\overline{G}$ is a complex submanifold of $\mathbf{C}^m / \Lambda$. It follows that

$$\{0\} \neq \overline{G} \subset M - \{\pi(b)\} \neq \mathbf{C}^m / \Lambda.$$

This contradicts the assumption of $\mathbf{C}^m / \Lambda$ being simple. *Q.E.D.*

The way to prove Lemma (1.7.2) implies the following.

(1.7.6) Proposition. *Let M be a compact complex manifold and $H : \mathbf{T}(M) \to \mathbf{R}^+$ a Finsler metric. Let $f : \mathbf{C} \to M$ be a non-constant holomorphic mapping with $H(f'(0)) \geq c > 0$. Then there is a holomorphic mapping $g : \mathbf{C} \to M$ such that*

$$H(g'(z)) \leq \frac{c}{2} \text{ for all } z \in \mathbf{C},$$

$$H(g'(0)) = \frac{c}{2}, \quad g(\mathbf{C}) \subset \overline{f(\mathbf{C})}.$$

Proof. Applying Lemma (1.6.6) to the holomorphic mappings $f|\Delta(k)$: $\Delta(k) \to M$, $k = 1, 2, \ldots$, we have holomorphic mappings $g_k : \Delta(k) \to M$ such that $H(g_k'(0)) = c/2$,

$$H(g_k'(z)) \leq \frac{c}{2} \eta_k(z) \text{ for } z \in \Delta(k)$$

and $g_k(\Delta(k)) \subset f(\mathbf{C})$. As in the proof of Lemma (1.6.6), we obtain a subsequence of $\{g_k\}_{k=1}^{\infty}$ which converges uniformly on compact subsets to a holomorphic mapping $g : \mathbf{C} \to M$. Clearly this g satisfies all the requirements. *Q.E.D.*

1.8 Holomorphic Mappings Omitting Hypersurfaces

In this section we shall study holomorphic mappings into compact complex manifolds omitting hypersurfaces, and in particular, those into the complex projective spaces. We begin with recalling an elementary fact from the preceding sections. By Examples (1.2.10), (1.2.11), (1.5.7) and Propositions (1.2.7), (1.5.5) we immediately have the following.

(1.8.1) Lemma. *The open domain* $D = \{(z^1, ..., z^m) \in \Delta(1)^m;\ z^1 \cdots z^k \neq 0\}$ *with* $k \leq m$ *is complete hyperbolic and* F_D *is a Finsler metric.*

(1.8.2) Lemma. *Let M be a complex manifold and X an analytic hypersurface of M. Let* $\{f_j\}_{j=1}^{\infty} \subset \mathrm{Hol}(\Delta(r), M-X)$ *be a sequence converging uniformly on compact subsets to* $f \in \mathrm{Hol}(\Delta(r), M)$. *Then we have either* $f(\Delta(r)) \subset M-X$ *or* $f(\Delta(r)) \subset X$.

Proof. Suppose that there is a point $z \in \Delta(r)$ such that $f(z) \in X$. There are an open neighborhood U of $f(z)$ in M and a holomorphic function α on U such that $X \cap U = \{\alpha = 0\}$. Then there are a sufficiently small number $\varepsilon > 0$ and a large integer j such that

$$\Delta(z;\varepsilon) \subset \Delta(r),\ f(\Delta(z;\varepsilon)) \subset U,$$

$$f_k(\Delta(z;\varepsilon)) \subset U \quad \text{for } k \geq j.$$

Then the sequence $\{\alpha \circ f_k | \Delta(z;\varepsilon)\}_{k=j}^{\infty}$ of non-vanishing holomorphic functions converges uniformly on compact subsets to $\alpha \circ f | \Delta(z;\varepsilon)$ with $\alpha \circ f(z) = 0$. Therefore Lemma (1.5.16) (Hurwitz' theorem) implies that $\alpha \circ f \equiv 0$, so that $f(\Delta(r)) \subset X$. *Q.E.D.*

Let X be an analytic hypersurface of an m-dimensional complex manifold M with only normal crossings. Take a point $x \in X$. Then, by definition, there is a holomorphic local coordinate neighborhood system $(U, \phi, \Delta(1)^m)$ around x such that $\phi(x) = O$ and

$$X \cap U = \{z^1 \cdots z^k = 0\},$$

where $\phi = (z^1, ..., z^m)$ and $1 \leq k \leq m$. The integer k is independent of the particular choice of such a holomorphic local coordinate neighborhood system $(U, \phi, \Delta(1)^m)$. We call k the **multiplicity** of the analytic hypersurface X at x, denoted by $\mathrm{mult}_x(X)$. For $x \in M-X$, we set $\mathrm{mult}_x(X) = 0$. Then we put

$$X^{(k)} = \{x \in M;\ \mathrm{mult}_x(X) = k\}.$$

(1.8.3) Theorem. *Let M be an m-dimensional compact complex manifold and* $H: \mathbf{T}(M) \to \mathbf{R}^+$ *a Finsler metric. Let X be an analytic hypersurface with only normal crossings. Then one and only one of the following cases occurs.*

(i) *$M-X$ is complete hyperbolic and hyperbolically imbedded into M:*

(ii) *There are an integer* $0 \leq k \leq m-1$ *and a non-constant holomorphic mapping* $f: \mathbf{C} \to M$ *such that*

$$f(\mathbf{C}) \subset X^{(k)},\ H(f'(z)) \leq 1,\ H(f'(0)) = 1.$$

Proof. Take a family $\{U_\nu, \phi_\nu, \Delta(2)^m\}_{\nu=1}^l$ of holomorphic local coordinate neighborhood systems of M satisfying the following conditions:

(i) $X \cap U_\nu = \{x \in U_\nu;\ z_\nu^1(x) \cdots z_\nu^{k(\nu)}(x) = 0\}$, where $\phi_\nu = (z_\nu^1, \ldots, z_\nu^m)$;

(ii) $M = \bigcup_{\nu=1}^{l} V_\nu$, where $V_\nu = \{x \in U_\nu;\ |z_\nu^i(x)| < 1,\ 1 \le i \le m\}$.

By Lemma (1.8.1) $V_\nu - X$ is complete hyperbolic and $F_{V_\nu - X}\colon \mathbf{T}(V_\nu - X) \to \mathbf{R}^+$ is a Finsler metric for every ν. More precisely, we have by Proposition (1.2.7) and Examples (1.2.10), (1.2.11)

$$F_{V_\nu - X} = \max\left\{ \frac{|dz_\nu^i|}{|z_\nu^i|\,|\log|z_\nu^i|^2|},\ \frac{|dz_\nu^j|}{1-|z_\nu^j|^2};\ 1 \le i \le k(\nu) < j \le m \right\}. \tag{1.8.4}$$

For $\xi_x \in \mathbf{T}(M-X)_x$ we put

$$G_0(\xi_x) = \inf\{F_{V_\nu - X}(\xi_x);\ x \in U_\nu\}.$$

Then $G_0\colon \mathbf{T}(M-X) \to \mathbf{R}^+$ is a Finsler metric. We claim that G_0 is complete in $M-X$; i.e., the distance defined by G_0 is complete. Let $x_0 \in X$ be an arbitrary point. Then there is a holomorphic local coordinate neighborhood system $(V, \phi, \Delta(1)^m)$ around x_0 such that

$$X \cap V = \{x \in V;\ z^1(x) \cdots z^k(x) = 0\},$$

where $\phi = (z^1, \ldots, z^m)$ and $\phi(x_0) = O$. Take a small relatively compact neighborhood $W \subset V$ of x_0 so that $\overline{W} \subset V_\nu$ for all V_ν containing x_0. Then we see that there is a constant $C > 0$ such that $G_0 \ge C F_{V-X}$ on $W - X$. Note that F_{V-X} is complete in V. Covering X by such V and W, we infer that G_0 is also complete in $M-X$. Choose $\varepsilon > 0$ so that for any $x \in M$, there is some ν such that $U_H(x;\varepsilon) \subset U_\nu$, where $U_H(x;\varepsilon) = \{y \in M;\ d_H(x, y) < \varepsilon\}$. Put

$$G_\varepsilon = \max\left\{H,\ \frac{\varepsilon}{3} G_0\right\} : \mathbf{T}(M-X) \to \mathbf{R}^+.$$

Then G_ε is a complete Finsler metric. If there is a positive constant c_2 such that $F_{M-X} \ge c_2 G_\varepsilon$, then case (i) holds. Suppose that there is no such positive constant c_2. Then for any positive integer $n \in \mathbf{N}$, there is a tangent vector $\xi(n) \in \mathbf{T}(M-X) - O$ such that

$$F_{M-X}(\xi(n)) \le \frac{1}{4n} G_\varepsilon(\xi(n)).$$

Since

$$F_{M-X}\left(\frac{2}{G_\varepsilon(\xi(n))}\xi(n)\right) \leq \frac{1}{2n} < \frac{1}{n},$$

there is a holomorphic mapping $g_n\colon \Delta(n) \to M-X$ satisfying

$$g_n'(0) = \frac{2}{G_\varepsilon(\xi(n))}\xi_n.$$

Since $G_\varepsilon(g_n'(0)) = 2$, Lemma (1.6.6) implies that there is a holomorphic mapping $f_n\colon \Delta(n) \to M-X$ such that $G_\varepsilon(f_n'(0)) = 1$ and $G_\varepsilon(f_n'(z)) \leq \eta_n(z)$ for $z \in \Delta(n)$. Hence we obtain

$$G_\varepsilon(f_k'(z)) \leq \frac{4}{3} \tag{1.8.5}$$

for $k \geq n$ and $z \in \Delta(n/2)$. Since $H \leq G_\varepsilon$, we have that $H(f_k'(z)) \leq 4/3$ for $k \geq n$ and $z \in \Delta(n/2)$. In the same way as in the proof of Lemma (1.7.2) we obtain a subsequence $\{f_{n_j}\}_{j=1}^\infty$ of $\{f_n\}_{n=1}^\infty$ which converges uniformly on compact subsets to a holomorphic mapping $f\colon \mathbf{C} \to M$. We shall prove that $H(f'(0)) = 1$. By (1.8.5) we have

$$H\left(f_{n_j}'(z)\right) \leq G_\varepsilon\left(f_{n_j}'(z)\right) \leq \frac{4}{3}$$

for $z \in \Delta(n_j/2)$. Therefore, for every j with $2\varepsilon/3 < n_j/2$ we have

$$f_{n_j}\left(\Delta\left(\frac{2}{3}\varepsilon\right)\right) \subset U_H\left(f_{n_j}(0); \frac{8}{9}\varepsilon\right).$$

Since $f_{n_j} \to f$ uniformly on $\overline{\Delta(2\varepsilon/3)}$ as $j \to \infty$, there is $j_0 \in \mathbf{N}$ such that

$$f_{n_j}\left(\Delta\left(\frac{2}{3}\varepsilon\right)\right) \subset U_H(f(0); \varepsilon)$$

for $j \geq j_0$. It follows from the choice of ε that there is an integer $1 \leq \nu_0 \leq l$ such that

$$f_{n_j}\left(\Delta\left(\frac{2}{3}\varepsilon\right)\right) \subset U_H(f(0); \varepsilon) \subset U_{\nu_0}$$

for every $j \geq j_0$ Therefore we may consider $f_{n_j}|\Delta(2\varepsilon/3)$ as elements of $\mathrm{Hol}(\Delta(2\varepsilon/3), U_{\nu_0} - X)$. By the definition of $F_{U_{\nu_0}-X}$, we have

$$F_{U_{\nu_0}-X}\left(f_{n_j}'(0)\right) \leq \frac{3}{2\varepsilon},$$

so that

$$\frac{2\varepsilon}{3} F_{U_{\nu 0}-X}\left[f'_{n_j}(0) \right] \leq 1.$$

Hence $(\varepsilon/3)G_0\left[f'_{n_j}(0) \right] \leq 1/2$. Since $G_\varepsilon\left[f'_{n_j}(0) \right] = 1$,

$$H\left[f'_{n_j}(0) \right] = G_\varepsilon\left[f'_{n_j}(0) \right] = 1$$

for all $j \geq j_0$. Finally we get

$$H(f'(0)) = \lim_{j\to\infty} H\left[f'_{n_j}(0) \right] = 1.$$

Especially, f is not a constant mapping. Moreover we have

$$H(f'(z)) = \lim_{j\to\infty} H\left[f'_{n_j}(z) \right]$$

$$\leq \lim_{j\to\infty} G_\varepsilon\left[f'_{n_j}(z) \right] \leq \lim_{j\to\infty} \eta_{n_j}(z) = 1.$$

We put

$$Y^{(k)} = \{x \in M\,;\ \mathrm{mult}_x(X) \geq k\}$$

for $k \geq 0$. Then $Y^{(k)}$ are analytic subset of M and $X^{(k)} = Y^{(k)} - Y^{(k+1)}$. Put

$$k = \sup\{\mathrm{mult}_{f(z)}(X);\ z \in \mathbf{C}\}.$$

Then there is a point $w \in \mathbf{C}$ such that $\mathrm{mult}_{f(w)}(X) = k$. It is sufficient to prove that $f(\mathbf{C}) \subset Y^{(k)}$. This is trivial if $k = 0$. Assume that $k > 0$. Since X has only normal crossings, there is a holomorphic local coordinate neighborhood system $(U, \phi, \Delta(1)^m)$ around $f(w)$ such that $\phi(f(w)) = O$ and

$$X \cap U = \{z^1 \cdots z^k = 0\},$$

where $\phi = (z^1, ..., z^m)$. Taking a sufficiently small number $r > 0$, we have that $f(\overline{\Delta(w\,;r)}) \subset U$. Then $f_{n_j}(\overline{\Delta(w\,;r)}) \subset U$ for all large j. Note that the holomorphic functions $z^i \circ f_{n_j}(z) | \Delta(w\,;r)$, $j = 1, 2, ...,$ have no zero on $\Delta(w\,;r)$ and converge uniformly to $z^i \circ f | \Delta(w\,;r)$ which satisfy $z^i \circ f(w) = 0$ for $1 \leq i \leq k$. We see by Lemma (1.5.16) that $z^i \circ f | \Delta(w\,;r) \equiv 0$ for $1 \leq i \leq k$. Therefore $f(\Delta(w\,;r)) \subset Y^{(k)}$. Since $Y^{(k)}$ is an analytic subset, $f(\mathbf{C}) \subset Y^{(k)}$. Moreover, $k \leq m-1$, for f is not constant. *Q.E.D.*

Now we study holomorphic mappings into the projective spaces. For the time being, we describe the projective spaces. Let E be an m-dimensional complex vector space. We write $P(E)$ for the set of all 1-dimensional linear subspaces (sometimes called lines through the origin O) in E. We define a mapping

$\rho: E - \{O\} \to P(E)$ by

$$\rho(x) = \text{the line passing } O \text{ and } x$$

for $x \in E - \{O\}$. On the other hand, the multiplicative group $\mathbf{C}^* = \mathbf{C} - \{0\}$ acts on $E - \{O\}$ by $x \in E - \{O\} \to ax \in E - \{O\}$ with $a \in \mathbf{C}^*$. Letting $(E - \{O\})/\mathbf{C}^*$ be the quotient space of $E - \{O\}$ by this action of $\mathbf{C}^*$, we have $P(E) = (E - \{O\})/\mathbf{C}^*$. We endow $P(E)$ with the quotient topology of $E - \{O\}/\mathbf{C}^*$. Then $P(E)$ is a compact Hausdorff space. Moreover, we know that $P(E)$ carries the natural complex structure such that $\rho: E - \{O\} \to P(E)$ is holomorphic. In fact, taking a base of E, we identify $E = \mathbf{C}^m$ with the natural coordinate system $(z^1, ..., z^m)$. Put $U_i = \rho(\{z^i \neq 0\})$ $(1 \leq i \leq m)$ and

$$\psi_i: (z_i^0, ..., \hat{z_i^i}, ..., z_i^m) \in \mathbf{C}^m \to \rho(z_i^0, ..., \overset{i\text{-th}}{1}, ..., z^m) \in U_i,$$

where $\hat{\cdot}$ stands for the deletion. Then $P(\mathbf{C}^m) = \bigcup_{i=1}^{m} U_i$, and

$$\psi_i \circ \psi_j^{-1} | \psi_j(U_i \cap U_j): \psi_j(U_i \cap U_j) \to \psi_i(U_i \cap U_j)$$

are biholomorphic mappings. Hence $(U_i, \psi_i, \mathbf{C}^m)$ give rise to holomorphic local coordinate systems of $P(\mathbf{C}^m)$ for all i. We call the complex manifold $P(E)$ the $(m-1)$-dimensional **complex projective space** and the holomorphic mapping $\rho: E - \{O\} \to P(E)$ the **Hopf fibering.** We sometimes write $\mathbf{P}^{m-1}(\mathbf{C})$ instead of $P(\mathbf{C}^m)$. The above coordinate systems $(z_i^0, ..., \hat{z_i^i}, ..., z_i^m)$ are called **affine coordinate systems** on U_i. For a point $x \in \mathbf{P}^{m-1}(\mathbf{C})$, there is $(z^1, ..., z^m) \in \mathbf{C}^m - \{O\}$ such that $\rho((z^1, ..., z^m)) = x$. We write $x = [z^1; \cdots; z^m]$ and call $[z^1; \cdots; z^m]$ the **homogeneous coordinate system** of $\mathbf{P}^{m-1}(\mathbf{C})$. A subset H of $P(E)$ is called a **hyperplane** if there is an $(m-1)$-dimensional linear subspace $\tilde{H}$ of E such that $\rho(\tilde{H} - \{O\}) = H$. If we write E^* for the dual space of E, then there is $\alpha \in E^* - \{O\}$ such that $\tilde{H} = \{\alpha = 0\} = \{x \in E; \alpha(x) = 0\}$. Moreover, $\beta \in E^* - \{O\}$ satisfies $\tilde{H} = \{\beta = 0\}$ if and only if $\alpha = a\beta$ with $a \in \mathbf{C}^*$. Therefore, for each hyperplane H of $P(E)$, there corresponds uniquely a point of $P(E^*)$ and vice versa. Thus we have

$$P(E^*) = \text{the set of all hyperplanes of } P(E).$$

We call $P(E^*)$ the **dual projective space** of $P(E)$. We write $\mathbf{P}^{m-1}(\mathbf{C})^*$ for the dual projective space of $\mathbf{P}^{m-1}(\mathbf{C})$. Let $H_1, ..., H_l$ be hyperplanes in $P(E)$. Let $y_1, ..., y_l$ be the points of $P(E^*)$ corresponding to $H_1, ..., H_l$ respectively. Let $\rho^*: E^* - \{O\} \to P(E^*)$ be the Hopf fibering and take $\alpha_j \in E^* - \{O\}$ so that $\rho^*(\alpha_j) = y_j$. We say the family $\{y_j\}_{j=1}^{l}$ or the points $y_1, ..., y_l$ to be in **general position** if for any choice of indices $1 \leq j_1 < \cdots < j_k \leq l$ with $1 \leq k \leq m$, $\dim <\alpha_{j_1}, ..., \alpha_{j_k}> = k$, where $<\alpha_{j_1}, ..., \alpha_{j_k}>$ denotes the linear subspace of E^* spanned by $\alpha_{j_1}, ..., \alpha_{j_k}$. Clearly, this notion does not depend on the choice of α_1,

..., α_l with $\rho^*(\alpha_j) = y_j$. We say the family $\{H_j\}_{j=1}^l$ or the hyperplanes $H_1, \ldots, H_l$ to be in **general position** if so is $\{y_j\}_{j=1}^l$. It is easy to see that $\{H_j\}_{j=1}^l$ is in general position if and only if

$$\operatorname{codim} H_{j_1} \cap \cdots \cap H_{j_k}$$
$$= \dim P(E) - \dim H_{j_1} \cap \cdots \cap H_{j_k} = k.$$

Therefore $\{H_j\}_{j=1}^l$ is in general position if and only if the analytic hypersurface $\bigcup_{j=1}^{l} H_j$ has only normal crossings.

The following Borel's lemma plays a very important role in the study of holomorphic mappings from $\mathbf{C}$ into the complex projective space. Its proof will be given in Chapter VI, §1 (see Lemma (6.1.20)).

(1.8.6) Lemma. *Let $F_1, \ldots, F_N$ $(N \geq 2)$ be non-vanishing holomorphic functions on $\mathbf{C}$. Suppose that the identity*

$$F_1 + \cdots + F_N = 1$$

holds on $\mathbf{C}$. Then $F_1, \ldots, F_N$ are linearly dependent (over $\mathbf{C}$).

(1.8.7) Lemma. *Let E be an $(m+1)$-dimensional complex vector space and $H_1, \ldots, H_{2m+1}$ hyperplanes in general position of $P(E)$. Then any holomorphic mapping $f : \mathbf{C} \to P(E) - \bigcup_{j=1}^{2m+1} H_j$ is constant.*

Proof. Fix $\{\alpha^1, \ldots, \alpha^{2m+1}\} \subset E^* - \{O\}$ so that $H_j = \rho(\{v \in E - \{O\};\ \alpha^j(v) = 0\})$ for $1 \leq j \leq 2m+1$. By our assumption, $\alpha^1, \ldots, \alpha^{m+1}$ are linearly independent and hence form a base of E^*. By making use of the dual base of $\{\alpha^1, \ldots, \alpha^{m+1}\}$, we identify E with $\mathbf{C}^{m+1}$ and $P(E)$ with $\mathbf{P}^m(\mathbf{C})$. Then

$$H_j = \{[z^1; \cdots; z^{m+1}] \in \mathbf{P}^m(\mathbf{C});\ z^j = 0\} \quad \text{for } 1 \leq j \leq m+1.$$

Define holomorphic functions h^j on $\mathbf{P}^m(\mathbf{C}) - \bigcup_{j=1}^{2m+1} H_j$ by

$$h^j([z^1; \cdots; z^{m+1}]) = \frac{\alpha^j}{z^1} \quad \text{for } 1 \leq j \leq 2m+1.$$

Put $f^j = h^j \circ f$ for $1 \leq j \leq 2m+1$. Then $f^1 \equiv 1$ and all f^j are non-vanishing holomorphic functions on $\mathbf{C}$. Clearly we have

$$f(\zeta) = [f^1(\zeta); f^2(\zeta); \cdots; f^{m+1}(\zeta)] \quad \text{for } \zeta \in \mathbf{C}. \tag{1.8.8}$$

We introduce an equivalence relation in the index set $I = \{1, 2, \ldots, 2m+1\}$ as follows: Two elements $k, l \in I$ are mutually equivalent if f^k / f^l is a constant function. Let $I = \bigcup I_\nu$ be the decomposition into the equivalence classes. We shall show that every I_ν contains at least $m+1$ elements. Suppose that I_ν contained at most m elements. Then $I-I_\nu$ contains at least $m+1$ elements, to say, $i_1, \ldots, i_{m+1}$. Fix one $i_0 \in I_\nu$. Then $\alpha^{i_0}, \alpha^{i_1}, \ldots, \alpha^{i_{m+1}}$ are linearly dependent, but arbitrary $m+1$ of them are linearly independent, since $H_1, \ldots, H_{2m+1}$ are in general position. Therefore, there exist $c_0, \ldots, c_{m+1} \in \mathbf{C}^*$ such that $\sum_{j=0}^{m+1} c_j \alpha^{i_j} = 0$, so that

$$\sum_{j=0}^{m+1} c_j f^{i_j} = 0.$$

Changing the indices $i_1, \ldots, i_{m+1}$ if necessary, we may assume that $f^{i_1}, \ldots, f^{i_l}$ with $1 \le l \le m+1$ are linearly independent and that $f^{i_0}, f^{i_1}, \ldots, f^{i_l}$ are linearly dependent. Moreover, we may assume that l is the smallest one satisfying the above property. Then there exist $a_0, \ldots, a_l \in \mathbf{C}^*$ such that $a_0 f^{i_0} + \cdots + a_l f^{i_l} = 0$. Thus we have

$$-\frac{a_1 f^{i_1}}{a_0 f^{i_0}} - \cdots - \frac{a_l f^{i_l}}{a_0 f^{i_0}} = 1.$$

By the above choice of $f^{i_0}, \ldots, f^{i_l}$, we know that $-a_1 f^{i_1}/(a_0 f^{i_0}), \ldots, -a_l f^{i_l}/(a_0 f^{i_0})$ are linearly independent. From Borel's Lemma (1.8.6) we conclude that $l=1$. Therefore f^{i_1}/f^{i_0} is constant and hence $i_1 \in I_\nu$. This is absurd, because we have chosen $i_1 \notin I_\nu$. What we have proved is that I is the only one equivalence class. Since $f^1 = 1$, all f^j are constant. By (1.8.8) we see that f is a constant mapping. *Q.E.D.*

(1.8.9) Theorem. *Let $P(E)$ be the m-dimensional complex projective space and $H_1, \ldots, H_N$ a finite number of hyperplanes of $P(E)$ which are in general position. Assume that $N \ge 2m+1$. Then $P(E) - \bigcup_{j=1}^{N} H_j$ is complete hyperbolic and hyperbolically imbedded into $P(E)$.*

Proof. Put $X = \bigcup_{j=1}^{N} H_j$. Since X has only normal crossings, we can apply Theorem (1.8.3) to this case. We have only to verify that case (ii) in Theorem (1.8.3) can not happen. Suppose that there were a non-constant holomorphic mapping $f: \mathbf{C} \to X^{(k)}$. By Lemma (1.8.7), $k > 0$. Therefore there are indices $1 \le i_i < i_2 < \cdots < i_k \le m$ such that

$$f(\mathbf{C}) \subset H_{i_1} \cap \cdots \cap H_{i_k} - \left[H_{j_1} \cup \cdots \cup H_{j_{N-k}} \right],$$

where $\{i_1, \ldots, i_k, j_1, \ldots, j_{N-k}\} = \{1, 2, \ldots, N\}$. Since $H_1, \ldots, H_N$ are in general position, $H_1 \cap \cdots \cap H_k - \left[H_{j_1} \cup \cdots \cup H_{j_{N-k}} \right]$ is biholomorphic to $P(\mathbf{C}^{m-k+1})$ minus $N-k$ number of hyperplanes of $P(\mathbf{C}^{m-k+1})$ in general position. Since

$N-k \geq 2m+1-k \geq 2(m-k)+1$, Lemma (1.8.7) again implies that f should be constant. This is a contradiction. *Q.E.D.*

1.9 Geometric Criterion of Complete Hyperbolicity

In the first place, we need to extend Schwarz-Pick-Ahlfors' Lemma (1.1.6) for our purpose of the present section. A hermitian metric $g = 2a(z)\,dz d\bar{z}$ on $\Delta(r)$ is said to be **rotationally symmetric** if $a(z) = a(|z|)$ for all $z \in \Delta(r)$.

(1.9.1) Lemma. *Let $g = 2a(z)\,dz d\bar{z}$ be a rotationally symmetric hermitian metric on $\Delta(r)$.*

(i) *If the Gaussian curvature function K_g of g is non-positive, then the function $a(t)$ is increasing on $[0, 1)$:*

(ii) *If K_g is non-positive and g is complete (i.e., the distance d_g defined by g is complete), then $\lim_{t \to r} a(t) = +\infty$.*

Proof. (i) Using the polar coordinate $z = te^{i\theta}$, we have

$$0 \leq -a(z)K_g(z) = \frac{\partial^2}{\partial z \partial \bar{z}} \log a(z)$$

$$= \frac{1}{4}\left[\frac{\partial^2}{\partial t^2} + \frac{1}{t}\frac{\partial}{\partial t} + \frac{1}{t^2}\frac{\partial^2}{\partial \theta^2}\right] \log a(t)$$

$$= \frac{1}{4}\left[\frac{1}{t}\frac{d}{dt}\left[t\frac{d}{dt}\log a(t)\right]\right].$$

Since the function $t(d\log a(t)/dt)$ is C^∞ on $(0, r)$ and continuous on $[0, r)$ with the value 0 at $t = 0$, the above inequality implies that $t(d\log a(t)/dt) \geq 0$. Hence $a(t)$ is an increasing function.

(ii) Suppose that $a(t)$ were bounded. Take $L > 0$ so that $a(z) \leq L$ for $z \in \Delta(r)$. Then we have

$$d_g(0, z) \leq \sqrt{2L}\,|z| \leq \sqrt{2L}\, r \tag{1.9.2}$$

for $z \in \Delta(r)$. In general, the same method as in the proof of Theorem (1.5.4) yields that a hermitian metric h on a complex manifold is complete if and only if all bounded subsets with respect to the distance d_h are relatively compact (cf. Theorem (1.5.4) and its Remark). Thus (1.9.2) contradicts the completeness of g. *Q.E.D.*

Now we show the following generalization of Lemma (1.1.6) due to Greene-Wu [42].

(1.9.3) Lemma. *Let $k: [0, \infty) \to (0, \infty)$ be a decreasing function satisfying*

$$sk(st) > k(t) \text{ for } t \geq 0 \text{ and } s > 1. \tag{1.9.4}$$

Let $h = 2a(z)\,dzd\bar{z}$ be a rotationally symmetric complete hermitian metric on $\Delta(r)$ such that for $z \in \Delta(r)$

$$-k\left((d_h(0, z))^2\right) \le K_h(z) \le 0. \tag{1.9.5}$$

Let $g = 2b(z)\,dzd\bar{z}$ be a hermitian pseudo-metric on $\Delta(r)$ such that for $z \in \Delta(r)$

$$K_g(z) \le -k\left((d_g(0, z))^2\right). \tag{1.9.6}$$

Then $g \le h$.

Proof. Since $a(z) > 0$, we may put $\mu(z) = b(z)/a(z)$. We shall show that $\mu(z) \le 1$. For $0 < l < 1$, define a holomorphic mapping $T_l\colon \Delta(lr) \to \Delta(r)$ by $T(z) = z/l$. Put $h_l = T_l^* h$. Then h_l is a rotationally symmetric hermitian metric on $\Delta(lr)$. We write $h_l = 2a_l(z)dzd\bar{z}$. Then $a_l(z) = a(z/l)/l^2$, and hence $a_l(z) \to a(z)$ as $l \to 1$. Put $\mu_l(z) = b(z)/a_l(z)$ on $\Delta(lr)$. Then it is sufficient to show that $\mu_l(z) \le 1$ for all $z \in \Delta(lr)$. Since h is complete, h_l is also complete. Moreover, we have by (1.9.5)

$$-k\left((d_{h_l}(0, z))^2\right) \le K_{h_l} \le 0. \tag{1.9.7}$$

By Lemma (1.9.1) we see that $a_l(z) \to +\infty$ as $|z| \to lr$. Thus $\mu_l(z)$ is a non-negative continuous function on $\Delta(lr)$ and attains zero on the boundary of $\Delta(lr)$. Hence there exists a point $w \in \Delta(lr)$ such that μ_l attains its maximum s at w. Assuming $s > 1$, we shall have a contradiction. Since $g \le sh_l$ on $\Delta(lr)$, we have

$$d_g(0, z) \le \sqrt{s}\, d_{h_l}(0, z)$$

for $z \in \Delta(lr)$. It follows from (1.9.4)~(1.9.7) that for $z \in \Delta(lr)$

$$\begin{aligned} K_g(z) &\le -k\left((d_g(0, z))^2\right) = -\frac{1}{s} sk\left(s\frac{1}{s}(d_g(0, z))^2\right) \\ &< -\frac{1}{s} k\left(\frac{1}{s}(d_g(0, z))^2\right) \le -\frac{1}{s} k\left(\left(d_{h_l}(0, z)\right)^2\right) \\ &\le \frac{1}{s} K_{h_l}(z) \le 0. \end{aligned}$$

Hence we have $K_{h_l}(z)/K_g(z) < s$ on $\Delta(lr)$. Since $b \ne 0$ on $\Delta(lr)$, $w \notin \operatorname{Zero}(g)$. Hence $\mu_l(z)$ is C^∞ in a neighborhood of w. Since μ_l is attaining its maximum at w, we have

$$\frac{\partial^2}{\partial z \partial \bar{z}} \log \mu_l(w) = -b(w)K_g(w) + a_l(w)K_{h_l}(w) \le 0.$$

Therefore we have the following contradiction:

$$s = \mu_l(w) = \frac{b(w)}{a_l(w)} \le \frac{K_{h_l}(w)}{K_g(w)} < s. \quad Q.E.D.$$

Let g be a hermitian pseudo-metric on $\Delta(r)$ such that $K_g \le -1$. Put $h = g_r$ and $k \equiv 1$ in Lemma (1.9.3) (cf. (1.1.4)). Then we immediately get $g \le g_r$. This is nothing but Schwarz-Pick-Ahlfors' Lemma (1.1.6).

To make use of Lemma (1.9.3), we need the following lemma, of which proof will be given in the next section.

(1.9.8) Lemma. *If $k : [0, \infty) \to (0, \infty)$ is a C^∞-function satisfying (1.9.4), then there exists a rotationally symmetric complete hermitian metric h on $\Delta(r)$ such that $K_h(z) = -k\left[(d_h(0, z))^2\right]$.*

Let M be a complex manifold with a hermitian metric H. Let $f : \Delta(\varepsilon) \to M$ $(\varepsilon > 0)$ be a holomorphic mapping with $f'(0) \ne O_{f(0)}$. Then f^*H is a hermitian pseudo-metric on $\Delta(\varepsilon)$. Let Π be a complex line through the origin in $\mathbf{T}(M)_x$ with $x \in M$. The **holomorphic sectional curvature** $K_H(\Pi)$ of H with respect to Π is defined by

$$K_H(\Pi) = \sup\{K_{f^*H}(0)\},$$

where the supremum is taken for all holomorphic mappings $f : \Delta(\varepsilon) \to M$ with arbitrary $\varepsilon > 0$ such that $f(0) = x$ and $f'(0) \in \Pi$. Here we note that this definition of the holomorphic sectional curvature is equivalent to the ordinary one defined in terms of curvature tensors of H (cf. Wu [124]). Moreover, we set

$$K_H(x) = \sup\{K_H(\Pi);\ \Pi \in \mathbf{T}(M)_x\}.$$

Now we give a criterion of the complete hyperbolicity due to Greene-Wu [42].

(1.9.9) Theorem. *Let H be a complete hermitian metric on M and $x_0 \in M$ an arbitrarily fixed point. Assume that there are constants $A > 0$ and $B \ge 0$ such that*

$$K_H(x) \le -\frac{A}{1 + B(\rho_H(x))^2}$$

for all $x \in M$, where $\rho_H(x) = d_H(x_0, x)$. Then there exists a positive constant C depending only on A and B such that

$$F_M(\xi_x) \ge \frac{C}{\sqrt{1 + B(\rho_H(x))^2}} \sqrt{H(\xi_x, \overline{\xi}_x)}$$

for all $\xi_x \in \mathbf{T}(M)_x$. Especially, M is complete hyperbolic.

Proof. Take any $\xi_x \in \mathbf{T}(M)_x - \{O\}$. Let $f : \Delta(1) \to M$ be a holomorphic mapping such that $f(0) = x$ and $f_*(\eta_0) = \xi_x$ for some $\eta_0 \in \mathbf{T}(\Delta(1))_0$. Then $h = f^*H$ is a hermitian pseudo-metric. Now take any $z \in \Delta(1)$. Put $y = f(z)$ and $\rho_h(z) = d_h(0, z)$. Then we have

$$\rho_H(y) = d_H(x_0, y) \le d_H(x_0, x) + d_H(x, y)$$
$$\le \rho_H(x) + \rho_h(z).$$

Hence we obtain

$$(\rho_H(y))^2 \le 2\left[(\rho_H(x))^2 + (\rho_h(z))^2\right].$$

Therefore, for $z \in \Delta(1)$

(1.9.10) $$K_h(z) \le K_H(f(z)) \le -\frac{A}{1 + B(\rho_H(y))^2}$$
$$\le -\frac{A}{1 + 2B(\rho_H(x))^2 + 2B(\rho_h(z))^2}.$$

For any $\lambda > 0$, λh is also a hermitian pseudo-metric on $\Delta(1)$ and the following identities hold on $\Delta(1)$:

(1.9.11) $$\lambda K_{\lambda h} = K_h, \quad \rho_{\lambda h}(z) = \sqrt{\lambda}\,\rho_h(z).$$

Now put

$$\lambda = \frac{1}{\left[1 + 2B(\rho_H(x))^2\right]}.$$

Then (1.9.10) and (1.9.11) imply

$$K_{\lambda h}(z) \le -\frac{A}{\lambda\left[1 + 2B(\rho_H(x))^2 + \frac{2}{\lambda}B(\rho_{\lambda h}(z))^2\right]}$$
$$= -\frac{A}{1 + 2B(\rho_{\lambda h}(z))^2}.$$

Applying Lemma (1.9.8) with $k(t) = A/(1 + 2Bt)$, we have a rotationally symmetric hermitian metric $g = 2a(z)\,dzd\bar{z}$ on $\Delta(1)$, depending only on A and B, such that

$$K_g(z) = -\frac{A}{1 + 2B(\rho_g(z))^2}.$$

It follows from Lemma (1.9.3) that $\lambda h \le g$. Hence we infer that

$$\sqrt{H(\xi_x, \bar{\xi}_x)} = \sqrt{h(\eta_0, \bar{\eta}_0)} = \frac{1}{\sqrt{\lambda}} \sqrt{\lambda h(\eta_0, \bar{\eta}_0)}$$

$$\leq \frac{1}{\sqrt{\lambda}} \sqrt{g(\eta_0, \bar{\eta}_0)} = \sqrt{a(0)\left[1 + 2B(\rho_H(x))^2\right]}\, |dz(\eta_0)|$$

$$\leq \sqrt{2a(0)\left[1 + B(\rho_H(x))^2\right]}\, |dz(\eta_0)|.$$

Then we have by (1.2.4)

$$F_M(\xi_x) \geq \frac{1}{\sqrt{2a(0)\left[1 + B(\rho_H(x))^2\right]}} \sqrt{H(\xi_x, \bar{\xi}_x)}.$$

We finally show the completeness of the hyperbolic distance d_M. Let $\phi: [a, b] \to M$ be a piecewise C^∞-curve with $\phi(a) = x_0$ and $\|\dot{\phi}(t)\|_H$ the length of the tangent vector $\dot{\phi}(t)$ with respect to H. Then by the inequality obtained above we have

$$L_M(\phi) = \int_a^b F_M(\dot{\phi}(t))\, dt$$

$$\geq C \int_a^b \frac{\|\dot{\phi}\|_H}{\sqrt{1 + B(\rho_H(\phi(t)))^2}} dt \geq C \int_a^b \frac{\|\dot{\phi}\|_H}{\sqrt{1 + B(L_H(\phi(t)))^2}} dt$$

$$= C \int_0^{L_H(\phi)} \frac{1}{\sqrt{1 + Bs^2}} ds \geq C \log L_H(\phi) + O(1)$$

for large $L_H(\phi)$. Since H is complete, we see that d_M is complete (cf. Theorem (1.5.4) and its Remark). *Q.E.D.*

1.10 Existence of a Rotationally Symmetric Hermitian Metric

The purpose of this section is to prove Lemma (1.9.8) in §9. Let g be a Riemannian metric on the real two-dimensional plane $\mathbf{R}^2$. By introducing the so-called isothermal coordinate on $\mathbf{R}^2$, we consider $\mathbf{R}^2$ as a complex manifold, which is topologically simply connected. By the uniformization theorem in the function theory of one variable we know that this simply connected complex manifold $\mathbf{R}^2$ is biholomorphic to either $\mathbf{C}$ or $\Delta(1)$. We first introduce the result due to Milnor [68], which tells us which case of $\mathbf{C}$ or $\Delta(1)$ occurs according to the behavior of the Gaussian curvature of the Riemannian metric g. We begin with the following basics.

(1.10.1) Lemma. *Let $f(t)$ be a real-valued C^∞-function such that $f^{(k)}(0) = 0$ for $0 \leq k \leq l-1$. Then the function $g(t) = f(t)/t^l$ is C^∞ on $\mathbf{R}$ and $g(0) = f^{(l)}(0)/l!$.*

Proof. We use the induction on l. Let $l = 1$. We have

$$f(t) = \int_0^1 \frac{d}{ds}(f(ts))\,ds = t\int_0^1 f^{(1)}(ts)\,ds.$$

Hence $g(t) = \int_0^1 f^{(1)}(ts)\,ds$. Since $f^{(1)}(ts)$ is a C^∞-function in t, $g(t)$ is also C^∞. Clearly we have $g(0) = f^{(1)}(0)$. Now suppose that our assertion is true up to $(l-1)$. Put $A(t) = f^{(1)}(t)/t^{l-1}$. By the induction hypothesis, $A(t)$ is a C^∞-function with $A(0) = f^l(0)/(l-1)!$. On the other hand, put $B(t) = f(t)/t^{l-1}$. Then, again by the induction hypothesis $B(t)$ is a C^∞-function with $B(0) = f^{(l-1)}(0)/(l-1)! = 0$. Then we have

$$B^{(1)}(t) = A(t) - (l-1)g(t)$$

for $t \neq 0$. Thus $g(t)$ is a C^∞-function on $\mathbf{R}$. Moreover, putting $t = 0$ in the above equality, we get

$$(l-1)g(0) = \frac{1}{(l-1)!}f^{(l)}(0) - g(0),$$

so that $g(0) = f^{(l)}(0)/l!$. *Q.E.D.*

(1.10.2) Lemma. *Let $f(t)$ be an even function (i.e., $f(-t) = f(t)$) of class C^∞ on $\mathbf{R}$. Then the function $f\left(\sqrt{x^2+y^2}\right)$ is C^∞ in $(x, y) \in \mathbf{R}^2$.*

Proof. Set $r = \sqrt{x^2+y^2}$. For $r \neq 0$ we have

(1.10.3)
$$\frac{\partial}{\partial x}f(r) = \frac{f'(r)}{r}x,$$
$$\frac{\partial}{\partial y}f(r) = \frac{f'(r)}{r}y.$$

Since $f'(t) = -f'(-t)$, $f'(0) = 0$. We see by Lemma (1.10.1) that $f'(t)/t$ is C^∞ on $\mathbf{R}$. Therefore $\partial f(r)/\partial x$ and $\partial f(r)/\partial y$ are continuous functions on $\mathbf{R}^2$. Hence $f(r)$ is C^1 on $\mathbf{R}^2$. Since $f'(-t)/(-t) = f'(t)/t$, we know that $f'(r)/r$ is C^1 on $\mathbf{R}^2$. Then it follows from (1.10.3) that $f(r)$ is C^2 on $\mathbf{R}^2$. Repeating this process, we see that $f(r)$ is a C^∞-function on $\mathbf{R}^2$. *Q.E.D.*

Let $k(t)$ be an odd function (i.e., $k(-t) = -k(t)$) of class C^∞ on $\mathbf{R}$ such that $k(t) > 0$ for $t > 0$ and $k'(0) = 1$. Note that $k(0) = 0$, and that the derivative of an odd function is an even function and vice versa. Let (r, θ) be the standard polar coordinate system of $\mathbf{R}^2 - \{O\}$. Let g be the Riemannian metric on $\mathbf{R}^2 - \{O\}$ defined by $g = dr^2 + k(r)^2 d\theta^2$. By the hypothesis on $k(t)$ and Lemma (1.10.1), there exists an even function $b(t)$ of class C^∞ on $\mathbf{R}$ such that

(1.10.4)
$$k(t) = t + t^3 b(t) = t(1 + t^2 b(t)).$$

Let (x, y) be the standard coordinate system of $\mathbf{R}^2$. Then

$$x = r\cos\theta, \quad y = r\sin\theta. \tag{1.10.5}$$

Hence we have

$$dx^2 + dy^2 = dr^2 + r^2 d\theta,$$

$$\begin{aligned} g &= dr^2 + (r + r^3 b(r))^2 d\theta^2 \\ &= dr^2 + r^2 d\theta^2 + r^4(2b(r) + (rb(r))^2)\, d\theta^2 \\ &= dx^2 + dy^2 + (2b(r) + (rb(r))^2)(r^2 dx^2 + r^2 dy^2 - r^2 dr^2) \\ &= dx^2 + dy^2 + (2b(r) + (rb(r))^2)\left[r^2 dx^2 + r^2 dy^2 - \frac{1}{4} d(r^2) \otimes d(r^2)\right]. \end{aligned}$$

By Lemma (1.10.2) we know that $b(r) = b\left(\sqrt{x^2 + y^2}\right)$ is a C^∞-function on $\mathbf{R}^2$, so that g can be extended over $\mathbf{R}^2$ as a Riemannian metric. Thus, from now on, we consider g as a Riemannian metric on $\mathbf{R}^2$. It is easy to see that the C^∞-curve $\gamma\colon t \in [0, r] \to (t, \theta) \in \mathbf{R}^2$ is the shortest geodesic joining the origin and the point (r, θ) with respect to the Riemannian metric g. Therefore we have the following.

(1.10.6) Lemma. *For any $(r, \theta) \in \mathbf{R}^2 - \{O\}$, the C^∞-curve $\gamma\colon t \in [0, r] \to (t, \theta) \in \mathbf{R}^2$ is the shortest geodesic joining the origin and the point (r, θ), and $d_g(O, (r, \theta)) = r$.*

We set

$$h(r) = \int_1^r \frac{dt}{k(t)}, \quad R = \int_1^\infty \frac{dt}{k(t)} \in (0, \infty].$$

Then (1.10.4) implies that

$$h(r) = \int_1^r \left[\frac{1}{t} - \frac{tb(t)}{1 + t^2 b(t)}\right] dt = \log r - \int_1^r \frac{tb(t)}{1 + t^2 b(t)} dt. \tag{1.10.7}$$

Hence we see that $h(0) = -\infty$ and the mapping $h\colon (0, \infty) \to (-\infty, R)$ is a monotone increasing diffeomorphism. We consider the standard complex structure on $\Delta^*(e^R)$ and define a diffeomorphism

$$\Phi\colon (r, \theta) \in \mathbf{R}^2 - \{O\} \to e^{h(r) + i\theta} \in \Delta^*(e^R).$$

Now we shall check the differentiability of Φ at the origin. By (1.10.7) we have

$$e^{h(r)} = r\exp\left[-\int_1^r \frac{tb(t)}{1 + t^2 b(t)} dt\right]. \tag{1.10.8}$$

An easy calculation combined with (1.10.5) yields that

(1.10.9)
$$\Phi(x, y) = \exp\left[\int_0^1 \frac{tb(t)}{1+t^2b(t)}dt\right]$$
$$\times \exp\left[-\int_0^{\sqrt{x^2+y^2}} \frac{tb(t)}{1+t^2b(t)}dt\right](x+iy).$$

Since $\int_0^r tb(t)/(1+t^2b(t))\,dt$ is an even function of class C^∞ in $r \in \mathbf{R}$, Lemma (1.10.2) and (1.10.9) imply that Φ can be extended to a diffeomorphism from $\mathbf{R}^2$ onto $\Delta(e^R)$, denoted also by Φ. By a simple calculation we see that

(1.10.10)
$$\Phi^* dz d\bar{z} = \left(\frac{e^{h(r)}}{k(r)}\right)^2 (dr^2 + k(r)^2 d\theta^2)$$
$$= \left(\frac{e^{h(r)}}{k(r)}\right)^2 g$$

for $r > 0$. Next set $H = (\Phi^{-1})^* g$. Then we can write by (1.10.10)

$$H = a(z)\, dz d\bar{z}, \quad a(z) = a(|z|).$$

If we put $s = e^{h(r)}$ for $r > 0$, then $0 < s < \infty$ and $a(s)$ can be written as

(1.10.11)
$$a(s) = \left(\frac{k \circ (e^h)^{-1}(s)}{s}\right)^2 \text{; i.e., } a(e^{h(r)}) = (k(r)e^{-h(r)})^2.$$

Then function $a(z)$ is C^∞ on $\Delta(e^R)$ and H is rotationally symmetric hermitian metric on $\Delta(e^R)$.

(1.10.12) Lemma. *Let the notation be as above. Then we have*

$$K_H(e^{h(r)+i\theta}) = -\frac{k''(r)}{k(r)}.$$

Proof. Set $z = e^{h(r)+i\theta} = se^{i\theta}$. A straightforward calculation yields that

$$K_H(z) = -\frac{2}{a(|z|)} \frac{\partial^2}{\partial z \partial \bar{z}} \log a(z)$$
$$= -\frac{1}{2a(s)}\left(\frac{\partial^2}{\partial s^2} + \frac{1}{s}\frac{\partial}{\partial s}\right) \log a(s) = -\frac{k''(r)}{k(r)}. \quad Q.E.D.$$

Let $K(t)$ be a C^∞-function on $\mathbf{R}$. Then we know that there uniquely exists a C^∞-function $k(t)$ satisfying

(1.10.13)
$$k''(t) + K(t)k(t) = 0,$$
$$k(0) = 0,\ k'(0) = 1.$$

(1.10.14) Lemma. *Let $K(t)$ be an even function of class C^∞ on $\mathbf{R}$ with $K \leq 0$. Then the function $k(t)$ defined by (1.10.13) is an odd function such that $k(t) > 0$ for $t > 0$.*

Proof. By the uniqueness of $k(t)$ we immediately know that $k(t)$ is an odd function. We deduce from the initial conditions, $k(0) = 0$ and $k'(0) = 1$ that $k(t) > 0$ for sufficiently small $t > 0$. Suppose that $k(t_0) = 0$ for some $t_0 > 0$. Assume that t_0 is the smallest one with this property. Then $k''(t) \geq 0$ on $[0, t_0]$. It follows that $k'(t) \geq 1$ on $[0, t_0]$, so that $k(t) \geq t$ on $[0, t_0]$. This is a contradiction. *Q.E.D.*

Now we give a theorem due to Milnor [68].

(1.10.15) Theorem. *Let $k(t)$ be an odd function of class C^∞ such that $k(t) > 0$ for $t > 0$, $k(0) = 0$ and $k'(t) = 1$. Set*

$$K(t) = -\frac{k''(t)}{k(t)}, \quad R = \int_1^\infty \frac{1}{k(t)} dt \in (0, \infty].$$

(i) *If $K(t) \geq -1/(t^2 \log t)$ for all sufficiently large t, then $R = +\infty$.*

(ii) *If there is $\alpha > 0$ such that $K(t) \leq -(1+\alpha)/(t^2 \log t)$ for all sufficiently large t and $k(t)$ is unbounded, then $R < +\infty$.*

Remark. Consider the Riemannian manifolds $(\mathbf{R}^2, g)$ and $(\Delta(e^R), dzd\bar{z})$, where z is the standard complex coordinate. Then (1.10.10) implies that the diffeomorphism $\Phi: \mathbf{R}^2 \to \Delta(e^R)$ is a conformal mapping. If we regard $(\mathbf{R}^2, g)$ as a Riemann surface by introducing the isothermal coordinates (cf. Chern [19]), Φ gives rise to a biholomorphic mapping from that Riemann surface onto $\Delta(e^R)$. Hence Milnor's Theorem (1.10.15) gives us a criterion to decide whether the Riemann surface $(\mathbf{R}^2, g)$ is biholomorphic to $\mathbf{C}$ or $\Delta(1)$.

Proof. To prove Lemma (1.9.8), we need (ii). So we shall only prove (ii). For the the proof of (i), see Milnor [68].

Let $\varepsilon > 0$ and put

$$k_\varepsilon(t) = t(\log t)^{1+\varepsilon}, \quad K_\varepsilon(t) = -\frac{k_\varepsilon''(t)}{k_\varepsilon(t)}$$

for $t \geq 2$. Then we have

$$K_\varepsilon(t) = -(1+\varepsilon)\left(1 + \frac{\varepsilon}{\log t}\right)\frac{1}{t^2 \log t} \tag{1.10.16}$$

for $t \geq 2$. By a simple calculation we see that

$$\int_2^\infty \frac{dt}{k_\varepsilon(t)} < \infty. \tag{1.10.17}$$

By the assumption and (1.10.16) there is a number $a \geq 2$ such that for sufficiently small ε

$$K(t) \leq K_\varepsilon(t), \quad t \geq a.$$

Since $k(t)$ is unbounded, there is $\tilde{a} > a$ such that $k'(\tilde{a}) > 0$. Take $c > 0$ so that

$$k_\varepsilon(\tilde{a}) < ck(\tilde{a}), \quad k_\varepsilon'(\tilde{a}) < ck'(\tilde{a}). \tag{1.10.18}$$

We claim that $k_\varepsilon(t) < ck(t)$ for $t \geq \tilde{a}$. Suppose that there were $b > \tilde{a}$ such that $k_\varepsilon(b) = ck(b)$. We may assume that b is the smallest one with this property. Then we have

$$k_\varepsilon''(t) = -K_\varepsilon(t)k_\varepsilon(t) \leq -K(t)ck(t) = ck''(t) \tag{1.10.19}$$

for $\tilde{a} \leq t \leq b$. It follows from (1.10.18) and (1.10.19) that $k_\varepsilon(t) < ck(t)$ for $\tilde{a} \leq t \leq b$. This is a contradiction. Therefore we have

$$\int_{\tilde{a}}^{\infty} \frac{dt}{k(t)} \leq c \int_{\tilde{a}}^{\infty} \frac{dt}{k_\varepsilon(t)} < \infty. \quad Q.E.D.$$

Finally we can give the proof of Lemma (1.9.8):

Proof of Lemma (1.9.8). Set $K(t) = -k(t^2)$. Then $K(t)$ is an even function of class C^∞ on $\mathbf{R}$ and $K \leq 0$. Define an odd function $\tilde{k}(t)$ of class C^∞ on $\mathbf{R}$ by

$$\tilde{k}''(t) + K(t)\tilde{k}(t) = 0,$$

$$\tilde{k}(0) = 0, \quad \tilde{k}'(0) = 1.$$

Then Lemma (1.10.14) implies that $\tilde{k}(t) > 0$ for $t > 0$. Therefore there exists a Riemannian metric g on $\mathbf{R}^2$ such that $g = dr^2 + \tilde{k}(t)\,d\theta^2$ on $\mathbf{R}^2 - \{O\}$ with the polar coordinate system (r, θ) of $\mathbf{R}^2 - \{O\}$. Since $K(t) < 0$ for $t > 0$, $\tilde{k}''(t) > 0$ for $t > 0$. Hence $\tilde{k}(t)$ is unbounded. On the other hand, by the assumption on k, we have

$$K(t) = -k(t^2) < -k(1)t^{-2}$$

for $t > 1$. Therefore, for sufficiently large t

$$K(t) < -\frac{2}{t^2 \log t}.$$

Then Theorem (1.10.15), (ii) implies that

$$R = \int_1^\infty \frac{dt}{\tilde{k}(t)} < \infty.$$

Consider the diffeomorphism $\Phi\colon \mathbf{R}^2 \to \Delta(e^R)$ defined by (1.10.9). As observed in (1.10.10), $H = (\Phi^{-1})^* g = a(z)\,dzd\bar{z}$ is a rotationally symmetric hermitian metric on

$\Delta(e^R)$. By definition, $\Phi^*H = dr^2 + \tilde{k}(r)^2 d\theta^2$. Hence Lemma (1.10.6) and (1.10.12) imply that

$$K_H(z) = K(d_H(0, z)) = -k\left[(d_H(0, z))^2\right]. \quad Q.E.D.$$

Notes

The theory of hyperbolic manifolds described in this chapter was first initiated by Ahlfors in the one-dimensional case. The theory of the higher dimensional case was exploited by Kobayashi [54]. The Kobayashi differential metric F_M which is the infinitesimal form of the Kobayashi pseudo-distance d_M and Theorem (1.4.4) are due to Royden [97]. Many of the results obtained prior to the results described in this chapter are summarized in his book [54] and his paper [55]. We strongly recommend the readers to refer to them. Especially, in the last part of the paper [55], many open problems are listed. Theorem (1.8.9) is due to [8], Green and Howard (see [39]), and gives the relation of the theory of hyperbolic manifolds to that of the classical value distribution of holomorphic mappings into the complex projective space. The new proof is based on Lemma (1.8.7) proved by Fujimoto [28] and Green [37], and on Theorem (1.8.3) by Green and Howard [39].

As an application of the theory of hyperbolic manifolds, we can prove the analogue of Mordell's conjecture (now Faltings' theorem [26]) over function fields, which was first proved by Manin [66] and Grauert [35]. Moreover we can show its higher-dimensional analogue (see Noguchi [83]; cf. also Noguchi [81], Riebesehl [96] and Lang [62]).

Theorems (1.6.24) and (1.6.27) hold for singular M and X. Especially, Theorem (1.6.24) has an application to the study of the moduli space of holomorphic mappings into hyperbolically imbedded complex spaces (see Noguchi [86]) and to theorems of Parshin-Arakelov type ([92, 3, 27]).

There is a relation between the hyperbolicity and the existence of bounded plurisubharmonic functions (cf. Chapter III, §3 for plurisubharmonic functions). Note here that we have bounded domains in our mind as the model. For instance, Sibony [103] showed that an open complex manifold M is hyperbolic if M carries a bounded plurisubharmonic exhaustion function. Wu [125] conjectures that M should be complete hyperbolic in this case.

Kobayashi [55] raised the question of the existence of simply connected compact hyperbolic manifolds. Using the criterion Theorem (1.7.3), Brody-Green [14] gave the following example: Let M_ε be the hypersurfaces in $\mathbf{P}^3(\mathbf{C})$ defined by

$$(z^0)^d + (z^1)^d + (z^2)^d + (z^3)^d + (\varepsilon z^0 z^1)^{d/2} + (\varepsilon z^0 z^2)^{d/2} = 0,$$

where d is an even positive integer and $d \geq 50$. If ε is sufficiently small, then M_ε is non-singular and simply connected. If ε is moreover non-zero, then M_ε is hyperbolic. If $\varepsilon = 0$, M_0 contains several $\mathbf{P}^1(\mathbf{C})$, so that M_0 is not hyperbolic. Therefore the family $\{M_\varepsilon\}_{\varepsilon \neq 0}$ gives an example of a deformation family of compact simply connected hyperbolic manifolds of which limit is not hyperbolic.

Besides the Kobayashi pseudo-distance, there is another pseudo-distance named the Carathéodory pseudo-distance which is biholomorphically invariant and satisfies the distance decreasing principle, too (cf. [55]). Lempert [65] proved that the Kobayashi distance coincides with the Carathéodory distance for bounded convex domains of $\mathbf{C}^n$. He defined certain extremal mappings with respect to the Kobayashi metric, and applied it to study complex Monge-Ampère equations.

CHAPTER II

Measure Hyperbolic Manifolds

2.1 Holomorphic Line Bundles and Chern Forms

Let M be an m-dimensional complex manifold. We call a triple $(\mathbf{L}, \pi, M)$ a **holomorphic line bundle** if the following conditions are satisfied:

(2.1.1) (i) $\mathbf{L}$ is an $(m+1)$-dimensional complex manifold.

(ii) $\pi: \mathbf{L} \to M$ is a surjective holomorphic mapping.

(iii) For every $x \in M$, $\pi^{-1}(x)$ is a complex vector space of complex dimension 1.

(iv) For every $x \in M$, there exist an open neighborhood U of x and a biholomorphic mapping $\Phi: \pi^{-1}(U) \to U \times \mathbf{C}$, called a **local trivialization**, such that $p \circ \Phi = \pi$ and $q \circ \Phi|_{\pi^{-1}(y)}: \pi^{-1}(y) \to \mathbf{C}$ are linear isomorphisms for all $y \in U$, where $p: U \times \mathbf{C} \to U$ and $q: U \times \mathbf{C} \to \mathbf{C}$ denote the canonical projections.

If no confusion occurs, we sometimes write $\pi: \mathbf{L} \to M$ or $\mathbf{L}$ instead of $(\mathbf{L}, \pi, M)$. For $x \in M$, $\pi^{-1}(x)$ is sometimes denoted by $\mathbf{L}_x$ and called the **fiber** of $\mathbf{L}$ at x. Let $(\mathbf{L}_j, \pi_j, M)$, $j = 1, 2$, be holomorphic line bundles over M. A holomorphic mapping $\Psi: \mathbf{L}_1 \to \mathbf{L}_2$ is called a **bundle homomorphism** if $\pi_2 \circ \Psi = \pi_1$ and $\Psi | \mathbf{L}_{1x}: \mathbf{L}_{1x} \to \mathbf{L}_{2x}$ is a linear mapping for any $x \in M$ (cf. (2.1.1), (iii) and (iv)). If Ψ is moreover a biholomorphic mapping, then Ψ is called a **bundle isomorphism.** In this case, $\mathbf{L}_1$ and $\mathbf{L}_2$ are said to be **isomorphic.** A holomorphic line bundle is said to be **trivial** if it is isomorphic to the holomorphic line bundle $(M \times \mathbf{C}, p, M)$ with the natural projection $p: M \times \mathbf{C} \to M$, which is denoted by $\mathbf{1}_M$.

Let $(\mathbf{L}, \pi, M)$ be a holomorphic line bundle over M and U an open subset of M. A holomorphic mapping $s: U \to \mathbf{L}$ is called a **holomorphic cross section** (defined over U) if $\pi \circ s(x) = x$ for $x \in U$. We write $\Gamma(U, \mathbf{L})$ for the set of all holomorphic cross sections defined over U. We write O for the mapping which sends each $x \in M$ to the zero vector O_x of $\mathbf{L}_x$. Clearly O belongs to $\Gamma(M, \mathbf{L})$ and is called the **zero section**; we also denote by O its image in $\mathbf{L}$. For $s, t \in \Gamma(U, \mathbf{L})$ and $a, b \in \mathbf{C}$, set

$$as + bt : x \in U \to as(x) + bt(x) \in \mathbf{L}_x \subset \mathbf{L}.$$

Then we have $as + bt \in \Gamma(U, \mathbf{L})$. Thus $\Gamma(U, \mathbf{L})$ is a complex vector space with $O|U$ as the zero element. If $s \in \Gamma(U, \mathbf{L})$ satisfies that $s(x) \neq O_x$ for any $x \in U$, we call s a **holomorphic local frame** over U. Now write $\mathcal{O}(U)$ for the complex vector space of all holomorphic functions on U. For $f \in \mathcal{O}(U)$ and $s \in \Gamma(U, \mathbf{L})$, we define $fs \in \Gamma(U, \mathbf{L})$ by $(fs)(x) = f(x)s(x)$. Then the mapping

$$f \in \mathcal{O}(U) \to fs \in \Gamma(U, \mathbf{L})$$

is linear for every $s \in \Gamma(U, \mathbf{L})$ and a linear isomorphism if s is a holomorphic local frame over U; in this case, the restriction $\mathbf{L}|U = \pi^{-1}(U)$ of $\mathbf{L}$ over U is isomorphic to $\mathbf{1}_M$ by

$$(x, a) \in U \times \mathbf{C} \to as(x) \in \mathbf{L}|U,$$

so that $\mathbf{L}|U$ is trivial. Hence $\mathbf{L}$ is trivial if and only if there exists a holomorphic local frame over M.

Let $\{U_\lambda\}$ be an open covering of M and $s_\lambda \in \Gamma(U_\lambda, \mathbf{L})$ holomorphic local frames over U_λ. We call the pair $(\{U_\lambda\}, \{s_\lambda\})$ a **local trivialization covering** of $\mathbf{L}$. If $U_\lambda \cap U_\mu \neq \varnothing$, there exists uniquely a holomorphic function

$$T_{\lambda\mu} : U_\lambda \cap U_\mu \to \mathbf{C}^*$$

such that

(2.1.2) $$s_\lambda(x) T_{\lambda\mu}(x) = s_\mu(x)$$

for $x \in U_\lambda \cap U_\mu$. Then $\{T_{\lambda\mu}\}$ satisfies the following, so called the **cocycle condition**:

(2.1.3) $$T_{\lambda\mu} T_{\mu\lambda} = 1, \text{ whenever } U_\lambda \cap U_\mu \neq \varnothing,$$

$$T_{\lambda\mu} T_{\mu\nu} T_{\nu\lambda} = 1, \text{ whenever } U_\lambda \cap U_\mu \cap U_\nu \neq \varnothing.$$

Now the pair $(\{U_\lambda\}, \{T_{\lambda\mu}\})$ is called the **system of holomorphic transition functions** subordinated to the local trivialization $(\{U_\lambda\}, \{s_\lambda\})$. Remark that there always exists a local trivialization of $\mathbf{L}$ by (2.1.1), (iv).

The dual holomorphic line bundle $(\mathbf{L}^*, \pi^*, M)$ of $(\mathbf{L}, \pi, M)$ is defined as follows: For any $x \in M$, set

$$\mathbf{L}_x^* = \text{the dual vector space of } \mathbf{L}_x.$$

Set $\mathbf{L}^* = \bigcup_{x\in M} \mathbf{L}_x^*$ (disjoint union). Define the mapping $\pi^*: \mathbf{L}^* \to M$ by the condition $\pi^*(\mathbf{L}_x^*) = x$ for $x\in M$. We see that $\mathbf{L}^*$ naturally becomes an $(m+1)$-dimensional complex manifold. In fact, let $(\{U_\lambda\}, \{s_\lambda\})$ be any local trivialization of $\mathbf{L}$ and $(\{U_\lambda\}, \{T_{\lambda\mu}\})$ the system of holomorphic transition functions subordinated to $(\{U_\lambda\}, \{s_\lambda\})$. For every $x\in U_\lambda$, define $t_\lambda(x)\in\mathbf{L}_x^*$ uniquely by the dual relation $<t_\lambda(x), s_\lambda(x)> = 1$. Then $t_\lambda(x)$ is a basis of $\mathbf{L}_x^*$. Set

$$F_\lambda: at_\lambda(x)\in(\pi^*)^{-1}(U_\lambda)\to(x, a)\in U_\lambda\times\mathbf{C}.$$

Then clearly F_λ is bijective and the following diagram is commutative:

$$\begin{array}{ccc} (\pi^*)^{-1}(U_\lambda) & \overset{F_\lambda}{\rightarrow} & U_\lambda\times\mathbf{C} \\ \downarrow \pi^* & & \downarrow \mathrm{pr} \\ U_\lambda & = & U_\lambda \end{array}$$

where pr: $U_\lambda\times\mathbf{C}\to U_\lambda$ is the natural projection. When $U_\lambda\cap U_\mu\neq\emptyset$, we have $t_\lambda(x) = (T_{\mu\lambda}(x))^{-1}t_\mu(x)$. Therefore the mapping

$$F_\mu\circ F_\lambda^{-1}: (U_\lambda\cap U_\mu)\times\mathbf{C}\to(U_\lambda\cap U_\mu)\times\mathbf{C}$$

is given by $F_\mu\circ F_\lambda^{-1}((x, \xi)) = (x, (T_{\mu\lambda}(x))^{-1}\xi)$. By this observation, $(\mathbf{L}^*, \pi^*, M)$ becomes a holomorphic line bundle in the natural manner so that F_λ are holomorphic bundle isomorphisms. It is easy to see that the above argument is independent of the particular choice of a local trivialization of $\mathbf{L}$. Remark that $(\{U_\lambda\}, \{t_\lambda\})$ is a local trivialization of $(\mathbf{L}^*, \pi^*, M)$ and $(\{U_\lambda\}, \{T_{\lambda\mu}^{-1}\})$ is the system of holomorphic transition functions subordinated to $(\{U_\lambda\}, \{t_\lambda\})$. For an element $\sigma\in\Gamma(M, \mathbf{L}^*)$, define a function σ^* on $\mathbf{L}$ by $\sigma^*(u) = <\sigma(\pi(u)), u>$. Then it is easy to see that σ^* is a holomorphic function. The correspondence

$$\sigma\in\Gamma(M, \mathbf{L}^*)\to\sigma^*\in\{f:\mathbf{L}\to\mathbf{C}; \text{holomorphic and } f(au) = af(u) \text{ for } a\in\mathbf{C}\} \tag{2.1.4}$$

is a linear isomorphism.

In general, let $\{U_\lambda\}$ be an open covering of M and $A_{\lambda\mu}$ non-vanishing holomorphic functions on non-empty $U_\lambda\cap U_\mu$. Assume that $\{A_{\lambda\mu}\}$ satisfies the cocycle condition (2.1.3). When $U_\lambda\cap U_\mu\neq\emptyset$, we define holomorphic mappings

$$F_{\lambda\mu}: (x, \xi)\in(U_\mu\cap U_\lambda)\times\mathbf{C}\to(x, A_{\lambda\mu}(x)\xi)\in(U_\lambda\cap U_\mu)\times\mathbf{C}. \tag{2.1.5}$$

Then we have

$$\begin{aligned} F_{\lambda\mu}\circ F_{\mu\lambda} &= id \quad \text{if } U_\lambda\cap U_\mu\neq\emptyset, \\ F_{\lambda\mu}\circ F_{\mu\nu}\circ F_{\nu\lambda} &= id \quad \text{if } U_\lambda\cap U_\mu\cap U_\nu\neq\emptyset. \end{aligned} \tag{2.1.6}$$

By identifying two points $(x_\lambda, \xi_\lambda) \in U_\lambda \times \mathbf{C}$ and $(x_\mu, \xi_\mu) \in U_\mu \times \mathbf{C}$ if and only if $x_\lambda = x_\mu$ and $\xi_\lambda = A_{\lambda\mu}(x_\mu)\xi_\mu$, we patch $U_\lambda \times \mathbf{C}$ and $U_\mu \times \mathbf{C}$, so that we obtain an $(m+1)$-dimensional complex manifold X. Then there exist an open covering $\{W_\lambda\}$ of X and biholomorphic mapping $\phi_\lambda: W_\lambda \to U_\lambda \times \mathbf{C}$ such that

$$W_\lambda \cap W_\mu \neq \varnothing \iff U_\lambda \cap U_\mu \neq \varnothing$$

and in this case we have $\phi_\lambda(W_\lambda \cap W_\mu) = (U_\lambda \cap U_\mu) \times \mathbf{C}$ and $\phi_\mu \circ \phi_\lambda^{-1} = F_{\mu\lambda}$. We can uniquely define a mapping $\alpha: X \to M$ by the condition

$$\alpha(\phi_\lambda^{-1}(x_\lambda, \xi_\lambda)) = x_\lambda.$$

Since ϕ_λ is holomorphic, α is holomorphic. We have $\alpha^{-1}(U_\lambda) = W_\lambda$ and the commutative diagram:

$$\begin{array}{ccc} \alpha^{-1}(U_\lambda) & \overset{\phi_\lambda}{\rightarrow} & U_\lambda \times \mathbf{C} \\ \downarrow \alpha & & \downarrow \mathrm{pr} \\ U_\lambda & = & U_\lambda \end{array}$$

Therefore $\alpha: X \to M$ becomes a holomorphic line bundle and ϕ_λ are holomorphic bundle isomorphisms. Therefore the pair $(\{U_\lambda\}, \{A_{\lambda\mu}\})$ induces a holomorphic line bundle $\alpha: X \to M$ such that X is trivial over U_λ and $(\{U_\lambda\}, \{A_{\lambda\mu}\})$ gives rise to the subordinated system of holomorphic transition functions.

Let $(\mathbf{L}_j, \pi_j, M)$, $j = 1, 2$, be holomorphic line bundles. We define the holomorphic line bundle $(\mathbf{L}_1 \otimes \mathbf{L}_2, \pi_1 \otimes \pi_2, M)$ as follows. Let $(\{U_\lambda\}, \{s_\lambda\})$, (resp. $(\{U_\lambda\}, \{t_\lambda\})$) be an arbitrary local trivialization covering of $\mathbf{L}_1$ (resp. $\mathbf{L}_2$). Let $(\{U_\lambda\}, \{S_{\lambda\mu}\})$ (resp. $(\{U_\lambda\}, \{T_{\lambda\mu}\})$) be the system of holomorphic transition functions subordinated to $(\{U_\lambda\}, \{s_\lambda\})$ (resp. $(\{U_\lambda\}, \{t_\lambda\})$). Putting $A_{\lambda\mu} = S_{\lambda\mu}T_{\lambda\mu}$ in the above argument, we obtain a holomorphic line bundle $\alpha: X \to M$. Starting from a different local trivialization covering $(\{V_\mu\}, \{f_\mu\})$, (resp. $(\{V_\mu\}, \{g_\mu\})$) of $\mathbf{L}_1$ (resp. $\mathbf{L}_2$), we obtain in the same way a holomorphic line bundle $\alpha': X' \to M$. It is easy to see that two holomorphic line bundles $\alpha: X \to M$ and $\alpha': X' \to M$ are isomorphic. Therefore we set $\mathbf{L}_1 \otimes \mathbf{L}_2 = X$ and $\pi_1 \otimes \pi_2 = \alpha$. We call the holomorphic line bundle $(\mathbf{L}_1 \otimes \mathbf{L}_2, \pi_1 \otimes \pi_2, M)$ the **tensor product** of $\mathbf{L}_1$ and $\mathbf{L}_2$. If we set

$$s_\lambda t_\lambda: x \in U_\lambda \to \phi_\lambda^{-1}(x, 1),$$

then $s_\lambda t_\lambda \in \Gamma(U_\lambda, \mathbf{L}_1 \otimes \mathbf{L}_2)$ and moreover $s_\lambda t_\lambda$ are holomorphic local frames over U_λ. Therefore $(\{U_\lambda\}, \{s_\lambda t_\lambda\})$ is a local trivialization covering of $\mathbf{L}_1 \otimes \mathbf{L}_2$ and $(\{U_\lambda\}, \{S_{\lambda\mu}T_{\lambda\mu}\})$ is the system of holomorphic transition functions subordinated to the local trivialization covering $(\{U_\lambda\}, \{s_\lambda t_\lambda\})$. Let k be a positive integer. For a holomorphic line bundle $(\mathbf{L}, \pi, M)$, we set $\mathbf{L}^k = \mathbf{L} \otimes \cdots \otimes \mathbf{L}$ (k times) and

$\pi^{(k)} = \pi \otimes \cdots \otimes \pi$ (k times). We also set $\mathbf{L}^{-k} = (\mathbf{L}^*)^k$ and $\pi^{(-k)} = (\pi^*)^{(k)}$.

Let $f : N \to M$ be a holomorphic mapping from a complex manifold into M and $\mathbf{L} \to M$ a holomorphic line bundle. We define the **pullback** $f^*\mathbf{L} \to N$ of $\mathbf{L}$ by f as follows. Let $(\{U_\lambda\}, \{s_\lambda\})$ be a local trivialization covering of $\mathbf{L} \to M$ and $\{T_{\lambda\mu}\}$ the subordinated system of holomorphic transition functions. Then $\{f^{-1}U_\lambda\}$ is an open covering of N and $\{f^*T_{\lambda\mu}\}$ satisfies the cocycle condition (2.1.3). Then as in the above arguments, we construct a holomorphic line bundle over N, which is denoted by $f^*\mathbf{L}$. If we start from another local trivialization covering of $\mathbf{L}$, then the resulted holomorphic line bundle is isomorphic to the above one; in this sense, $f^*\mathbf{L}$ is well-defined. By the construction, there is a holomorphic mapping $\alpha: f^*\mathbf{L} \to \mathbf{L}$ such that the diagram

$$\begin{array}{ccc} f^*\mathbf{L} & \xrightarrow{\alpha} & \mathbf{L} \\ \downarrow & & \downarrow \\ N & \xrightarrow[f]{} & M \end{array}$$

commutes and $\alpha|(f^*\mathbf{L})_x : (f^*\mathbf{L})_x \to \mathbf{L}_{f(x)}$ are linear for all $x \in N$.

We call $H = \{H_x\}_{x \in M}$ a **hermitian metric** on a holomorphic line bundle $(\mathbf{L}, \pi, M)$, if the following conditions are satisfied:

(i) H_x is a hermitian inner product on the one-dimensional complex vector space $\mathbf{L}_x$.

(ii) For an open subset $U \subset M$ and a holomorphic cross section $s \in \Gamma(U, \mathbf{L})$, the function $x \in U \to H_x(s(x), s(x)) \in \mathbf{R}$ is a C^∞-function.

For $u, v \in \mathbf{L}_x$, we write $H(u, v)$ instead of $H_x(u, v)$. Let $(\{U_\lambda\}, \{s_\lambda\})$ be any local trivialization covering of $\mathbf{L}$ and $(\{U_\lambda\}, \{T_{\lambda\mu}\})$ the system of holomorphic transition functions subordinated to $(\{U_\lambda\}, \{s_\lambda\})$. Set $H_\lambda(x) = H(s_\lambda(x), s_\lambda(x))$ for $x \in U_\lambda$. Then H_λ is a C^∞-function on U_λ and satisfies the following properties:

(2.1.7) (i) $H_\lambda(x) > 0$ for $x \in U_\lambda$;

(ii) $H_\lambda |T_{\lambda\mu}|^2 = H_\mu$ if $U_\lambda \cap U_\mu \neq \emptyset$.

Conversely, it is easy to see the following.

(2.1.8) Lemma. *Let the notation be as above. Let $\{H_\lambda\}$ be a system of C^∞-functions satisfying (2.1.7). Then there exists uniquely a hermitian metric $H = \{H_x\}$ on $\mathbf{L}$ such that $H_\lambda(x) = H(s_\lambda(x), s_\lambda(x))$ for $x \in U_\lambda$.*

For example, suppose that $\{H_\lambda\}$ is induced from a hermitian metric H on $\mathbf{L}$. For any $k \in \mathbf{Z}$, the system $\{(H_\lambda)^k\}$ satisfies (2.1.7) with respect to the system of holomorphic transition functions $\{(T_{\lambda\mu})^k\}$. Consequently, Lemma (2.1.8) implies that $\{(H_\lambda)^k\}$ determines a hermitian metric $H^{(k)}$ on $\mathbf{L}^k$.

A pair (**L**, H) of a holomorphic line bundle **L** and a hermitian metric H is called a **hermitian line bundle.**

Let (**L**, H) be a hermitian line bundle over M. Let $\{U_\lambda\}$ and $\{H_\lambda\}$ be as above. Let $f: N \to M$ be a holomorphic mapping from a complex manifold N into M. Then we have the pullback $f^*\mathbf{L}$ of **L** by f; moreover, by making use of $(f|f^{-1}(U_\lambda))^* H_\lambda$ on $f^{-1}U_\lambda$, we naturally define a hermitian metric in $f^*\mathbf{L}$, denoted by f^*H.

(2.1.9) Lemma. *Let* (**L**, H) *be a hermitian line bundle over* M. *Then there exists uniquely a real closed 2-form* ω *of type* $(1, 1)$ *on* M *such that for any holomorphic local frame* $s \in \Gamma(U, \mathbf{L})$ *over an open subset* $U \subset M$

$$\omega = -\frac{i}{2\pi}\partial\bar{\partial} \log H(s, s).$$

Proof. Let $(\{U_\lambda\}, \{s_\lambda\})$ be any local trivialization covering of **L** and $(\{U_\lambda\}, \{T_{\lambda\mu}\})$ the system of holomorphic transition functions subordinated to it. Set $H_\lambda = H(s_\lambda, s_\lambda)$ which satisfies (2.1.7). On each U_λ, set

$$\omega_\lambda = -\frac{i}{2\pi}\partial\bar{\partial} \log H_\lambda.$$

If $U_\lambda \cap U_\mu \neq \varnothing$, (2.1.7) implies that $\omega_\lambda | U_\lambda \cap U_\mu = \omega_\mu | U_\lambda \cap U_\mu$. Therefore we can uniquely define a 2-form ω on M by the condition $\omega | U_\lambda = \omega_\lambda$, which is the desired one. *Q.E.D.*

Let H_i, $i = 1, 2$, be hermitian metric on (**L**, π, M). Applying Lemma (2.1.9) to (**L**, H), we have real 2-forms ω_i of type $(1, 1)$ on M. Then there exists a C^∞-function a on M such that

$$\omega_1 - \omega_2 = \partial\bar{\partial} a = d(\bar{\partial} a). \tag{2.1.10}$$

In fact, there exists a C^∞-function $b > 0$ on M such that $H_{1x} = b(x) H_{2x}$ for $x \in M$. Then put $a = -i(2\pi)^{-1} \log b$, which satisfies (2.1.10). As a consequence of the above remark, the de Rham cohomology class $[\omega_i] \in H^2(M, \mathbf{R})$*) is uniquely determined independently from the particular choice of a hermitian metric H_i (and thus depending only on **L**).

(2.1.11) *Definition.* Let (**L**, π, M) be a holomorphic line bundle and H a hermitian metric on **L**. We write $\omega_{(\mathbf{L}, H)}$ for the real closed 2-form of type $(1, 1)$ defined by (**L**, H) as in Lemma (2.1.9), and call $\omega_{(\mathbf{L}, H)}$ the **Chern form** of (**L**, H). Writing $c_1(\mathbf{L})$ for the de Rham cohomology class of $\omega_{(\mathbf{L}, H)}$ in $H^2(M, \mathbf{R})$, we call $c_1(\mathbf{L})$ the **Chern class** of **L**.

---

*) See Note at the end of this section.

(2.1.12) *Remark.* When M is a compact Kähler manifold, it is known that for any real closed 2-form ϕ of type (1, 1) which represents $c_1(\mathbf{L})$, there exists a hermitian metric H on $\mathbf{L}$ such that $\phi = \omega_{(\mathbf{L},\, H)}$ (cf. Weil [120]).

The following Lemma is easily showed.

(2.1.13) Lemma. *Let* $(\mathbf{L}, H)$ *be a hermitian line bundle over* M.

(i) $\omega_{(\mathbf{L}^k,\, H^{(k)})} = k\omega_{(\mathbf{L},\, H)},\ c_1(\mathbf{L}^k) = kc_1(\mathbf{L})\ (k \in \mathbf{Z})$.

(ii) *Let* $f : N \to M$ *be a holomorphic mapping from a complex manifold* N *into* M. *Then*

$$\omega_{(f^*\mathbf{L},\, f^*H)} = f^*\omega_{(\mathbf{L},\, H)}.$$

Now we explain the so-called hyperplane bundle over a complex projective space as an important example of holomorphic line bundle. Let E be an $(m+1)$-dimensional complex vector space and $P(E)$ the projective space associated with E (cf. Chapter I, §8). Reminding that $P(E)$ is the set of all one-dimensional linear subspaces of E, we set

$$\mathbf{L}(E) = \{(V,\ v) \in P(E) \times E;\ v \in V\}.$$

Then $\mathbf{L}(E)$ is an $(m+1)$-dimensional closed complex submanifold of $P(E) \times E$. Define the holomorphic mapping π_E: $\mathbf{L}(E) \to P(E)$ by $\pi_E(V, v) = V$. By definition, $\pi_E^{-1}(V)$ can be naturally identified with V. By this identification, $\pi_E^{-1}(V)$ is a one-dimensional complex vector space. Now let $\{v_0, \ldots, v_m\}$ be a base of E. We identify E with $\mathbf{C}^{m+1}$ through this base $\{v_0, \ldots, v_m\}$. Set $U_\lambda = \{z^\lambda \neq 0\} \subset P(E)$ for $\lambda = 0, 1, \ldots, m$, and define a mapping s_λ: $U_\lambda \to \mathbf{L}(\mathbf{C}^{m+1})$ by

$$s_\lambda([z^0; \cdots; z^m]) = \left[[z^0; \cdots; z^m], \left(\frac{z^0}{z^\lambda}, \ldots, \frac{z^m}{z^\lambda} \right) \right].$$

If we put Ψ_λ: $(x,\ a) \in U_\lambda \times \mathbf{C} \to as_\lambda(x) \in \pi_E^{-1}(U_\lambda)$, then $\Phi_\lambda = \Psi_\lambda^{-1}$: $\pi_E^{-1}(U_\lambda)$ -> $U_\lambda \times \mathbf{C}$ satisfies condition (2.1.1), (iv). Therefore $(\mathbf{L}(E), \pi_E, P(E))$ is a holomorphic line bundle, and s_λ are holomorphic local frames of $\mathbf{L}(E)$ over U_λ. Put

$$T_{\lambda\mu}: [z^0; \cdots; z^m] \in U_\lambda \cap U_\mu \to \frac{z^\lambda}{z^\mu} \in \mathbf{C}^*.$$

Then we have the following:

(2.1.14) $(\{U_\lambda\}, \{s_\lambda\})$ is a local trivialization covering of $\mathbf{L}(E)$, and $(\{U_\lambda\}, \{T_{\lambda\mu}\})$ is a system of holomorphic transition functions subordinated to $(\{U_\lambda\}, \{s_\lambda\})$.

Let h be a hermitian inner product on E. Then we can define a hermitian metric $H_E = \{H_{E,\, x}\}_{x \in \mathbf{L}(E)}$ on $\mathbf{L}(E)$ as follows:

$$H_{E,x}((x, v), (x, w)) = h(v, w)$$

for $(x, v), (x, w) \in \pi_E^{-1}(x)$. Let $\{v_0, \ldots, v_m\}$ be an orthonormal base of E with respect to h; i.e., $h(v_j, v_k) = \delta_{jk}$. Identifying E with $\mathbf{C}^{m+1}$ through this base, we have

$$h((z^0, \ldots, z^m), (w^0, \ldots, w^m)) = \sum_{j=0}^{m} z^j \overline{w}^j.$$

On the other hand, for the Hopf fibering $\rho: E - \{O\} \to P(E)$, we have the following formula:

(2.1.15) $$\rho^* \omega_{(\mathbf{L}(E), H_E)} = -\frac{i}{2\pi} \partial\bar{\partial} \log h(z, z) \quad (z \in E - \{O\}).$$

In fact, we have

$$\rho^* \omega_{(\mathbf{L}(E), H_E)} | \rho^{-1}(U_\lambda) = \rho^* \left(-\frac{i}{2\pi} \partial\bar{\partial} \log H_E(s_\lambda, s_\lambda) \right) \Bigg| \rho^{-1}(U_\lambda)$$

$$= -\frac{i}{2\pi} \partial\bar{\partial} \log H_E(s_\lambda \circ \rho, s_\lambda \circ \rho) | \rho^{-1}(U_\lambda).$$

Since $s_\lambda \circ \rho(z) = (\rho(z), (z^0/z^\lambda, \ldots, z^m/z^\lambda))$ for $z = (z^0, \ldots, z^m) \in \rho^{-1}(U_\lambda)$, we have $H_E(s_\lambda \circ \rho(z), s_\lambda \circ \rho(z)) = h(z, z)/|z^\lambda|^2$. Since $\partial\bar{\partial} \log |z^\lambda|^2 = 0$ on $\rho^{-1}(U_\lambda)$, identity (2.1.15) holds on $\rho^{-1}(U_\lambda)$ and hence on $E - \{O\}$.

Next we shall study $\mathbf{L}(E)^{-1}$. First we determine $\Gamma(P(E), \mathbf{L}(E)^{-1})$. By (2.1.4) we know

$$\Gamma(P(E), \mathbf{L}(E)^{-1}) = \{f : \mathbf{L}(E) \to \mathbf{C};\ \text{holomorphic and}\ f(au) = af(u)\ \text{for}\ a \in \mathbf{C}\}$$

Consider the biholomorphic mapping

$$\alpha: v \in E - \{O\} \to (\rho(v), v) \in \mathbf{L}(E) - O.$$

Here the latter O denotes the zero section of $\mathbf{L}(E)$. Take an element $f \in \Gamma(P(E), \mathbf{L}(E)^{-1})$. Then $f \circ \alpha: E - \{O\} \to \mathbf{C}$ is a holomorphic function. Since the zero section O is compact, f is bounded on a neighborhood of O. Therefore $f \circ \alpha$ is bounded on a neighborhood of the origin, and so $f \circ \alpha$ can be extended to a holomorphic function on E. Since $f \circ \alpha(av) = a(f \circ \alpha(v))$ for $a \in \mathbf{C}$, we see that $f \circ \alpha \in E^*$. Therefore we have

(2.1.16) $$\Gamma(P(E), \mathbf{L}(E)^{-1}) = E^*$$

as complex vector spaces and $\dim \Gamma(P(E), \mathbf{L}(E)^{-1}) = m + 1$. Moreover if we set $\operatorname{Zero}(\sigma) = \{x \in P(E); \sigma(x) = 0\}$ for $\sigma \in \Gamma(P(E), \mathbf{L}(E)^{-1})$ with $\sigma \neq O$, then we have

Zero $(\sigma) = \{\rho(v);\ v \in E - \{O\},\ \sigma(v) = 0\}$. Namely Zero (σ) is a hyperplane. We call $(\mathbf{L}(E)^{-1}, \pi^{(-1)}, P(E))$ the **hyperplane bundle** over $P(E)$. The hyperplane bundle $\mathbf{L}(E)^{-1}$ carries the hermitian metric $H_E^{(-1)}$, such that

$$\omega_{(\mathbf{L}(E)^{-1},\ H_E^{(-1)})} = -\omega_{(\mathbf{L}(E),\ H_E)}.$$

Let $(z_\lambda^0, \ldots, \hat{z}_\lambda^\lambda, \ldots, z_\lambda^m)$ be affine coordinate systems on U_λ. Then we have

$$\text{(2.1.17)} \qquad \omega_{(\mathbf{L}(E)^{-1},\ H_E^{(-1)})} = \frac{i}{2\pi}\partial\bar{\partial}\log\left[1 + \sum_{\mu\neq\lambda} |z_\lambda^\mu|^2\right] \quad \text{on } U_\lambda.$$

Putting

$$\omega_{(\mathbf{L}(E)^{-1},\ H_E^{(-1)})} = \frac{i}{2\pi}\sum_{\substack{\mu\neq\lambda \\ \nu\neq\lambda}} A_{\mu\bar{\nu}}(x)\, dz_\lambda^\mu \wedge d\bar{z}_\lambda^\nu \quad (x \in U_\lambda),$$

we see by a direct computation that

(2.1.18) the hermitian matrix $(A_{\mu\bar{\nu}}(x))$ is positive definite at all $x \in U_\lambda$.

We call $\omega_{(\mathbf{L}(E)^{-1},\ H_E^{(-1)})}$ the **Fubini-Study Kähler form** on $P(E)$.

By now our notation becomes so messy that we adopt the following convention. If there is no danger of confusion, we write $\mathbf{L}_0$ for $\mathbf{L}(\mathbf{C}^{m+1})$ and $\mathbf{H}_0$ for $\mathbf{L}_0^{-1}$. Let h be the usual hermite inner product on $\mathbf{C}^{m+1}$. Set

$$\omega_0 = -\omega_{(\mathbf{L}_0,\ H_{\mathbf{C}^{m+1}})},$$

which is the Fubini-Study Kähler form on $\mathbf{P}^m(\mathbf{C})$.

Now let us go back to a general holomorphic line bundle $(\mathbf{L}, \pi, M)$. For a section $s \in \Gamma(M, \mathbf{L}) - \{O\}$, set

$$\text{Zero}\,(s) = \{x \in M;\ s(x) = 0\}.$$

If Zero $(s) \neq \varnothing$, then we know by (2.1.1), (iv) that Zero (s) is an analytic hypersurface of M. Let $E \subset \Gamma(M, \mathbf{L})$ be a finite dimensional complex vector subspace with $\dim E \geq 2$. Set

$$B(E) = \bigcap_{s \in E} \text{Zero}\,(s).$$

Then $B(E) \neq M$ and $B(E)$ is a (possibly empty) analytic subset of M. We call $B(E)$ the set of **base points** of E. For a point $x \in M$, we set

$$E_x = \{s \in E;\ s(x) = 0\}.$$

Then $\dim E_x \geq \dim E - 1$ and we have

$$x \notin B(E) \iff \dim E_x = \dim E - 1.$$

Since E_x is a $(\dim E - 1)$-dimensional linear subspace of E for all $x \in M - B(E)$, the point $\{E_x\}$ of $P(E^*)$ is defined by

$$\{E_x\} = \{\sigma \in E^*;\ \sigma(E_x) = 0\}.$$

Define a mapping

$$\Phi_E\colon X \in M - B(E) \to \{E_x\} \in P(E^*).$$

We shall show that Φ_E is holomorphic. In fact, fix arbitrarily a base $\{t_0, \dots, t_m\}$, where $\dim E = m+1$. Let $(\{U_\lambda\}, \{s_\lambda\})$ be a local trivialization covering of $\mathbf{L}$ and $(\{U_\lambda\}, \{T_{\lambda\mu}\})$ the system of holomorphic transition functions subordinated to it. Define holomorphic functions $\phi_{j\lambda}$ on U_λ with $j = 0, 1, \dots, m$ by $t_j | U_\lambda = \phi_{j\lambda} s_\lambda$. Then we have

$$\phi_{j\lambda} = T_{\lambda\mu}\phi_{j\mu} \text{ on } U_\lambda \cap U_\mu \neq \emptyset. \tag{2.1.19}$$

For $x \in M - B(E)$, take any U_λ so that $x \in U_\lambda$ and set

$$\Phi(x) = [\phi_{0\lambda}(x); \cdots; \phi_{m\lambda}(x)] \in P(\mathbf{C}^{m+1}).$$

Here remark that $x \notin B(E)$ implies $(\phi_{0\lambda}(x), \dots, \phi_{m\lambda}(x)) \neq O$. We know by (2.1.19) that $\Phi(x)$ is well defined independently from the particular choice of U_λ with $x \in U_\lambda$, and hence we have the holomorphic mapping

$$\Phi\colon M - B(E) \to P(\mathbf{C}^{m+1}).$$

We identify E with $\mathbf{C}^{m+1}$ through the base $\{t_0, \dots, t_m\}$, and E^* with $\mathbf{C}^{m+1}$. Then for $z = (z^0, \dots, z^m) \in E$ and $w = (w^0, \dots, w^m) \in E^*$, we have $w(z) = \sum_{j=0}^{m} w^j z^j$. Hence we have the following

$$\begin{array}{ccc} M - B(E) & \xrightarrow{\Phi_E} & P(E^*) \\ \| & & \| \\ M - B(E) & \xrightarrow{\Phi} & P(\mathbf{C}^{m+1}). \end{array}$$

Therefore we know that Φ_E is holomorphic.

Now we study the case where M is compact. Then $\Gamma(M, \mathbf{L})$ is of finite dimension (cf. Wells [121]). We call $\mathbf{L}$ a **very ample** line bundle if $\dim \Gamma(M, \mathbf{L}) \geq 2$, $B(\Gamma(M, \mathbf{L})) = \emptyset$ and the holomorphic mapping

$$\Phi_{\Gamma(M, \mathbf{L})}\colon M \to P(\Gamma(M, \mathbf{L})^*)$$

is an imbedding. We call $\mathbf{L}$ an **ample** line bundle if there exists $k \in \mathbf{N}$ such that $\mathbf{L}^k$ is very ample. The manifold M is called a complex **projective algebraic** manifold if M carries an ample line bundle. Let ω be a real 2-form of type $(1, 1)$ on M. Let

$(z^1, \ldots, z^m)$ be any holomorphic local coordinate system and write

$$\omega = i \sum a_{j\bar{k}} dz^j \wedge d\bar{z}^k .$$

We call ω **positive** (resp. **non-negative**) if the hermitian matrix $(a_{j\bar{k}})$ is positive definite (resp. positive semidefinite) at every point. We write $\omega > 0$ (resp. $\omega \geq 0$) if ω is positive (resp. non-negative). We say that ω is **negative** (resp. **non-positive**) if $-\omega > 0$ (resp. $-\omega \geq 0$). A holomorphic line bundle $\mathbf{L}$ over M is called **positive** (resp. **negative**), denoted by $\mathbf{L} > 0$ (resp. $\mathbf{L} < 0$) if there exists a hermitian metric H on $\mathbf{L}$ such that the Chern form $\omega_{(\mathbf{L}, H)}$ is positive (resp. negative). It is noted that

$$\mathbf{L} > 0 \iff \mathbf{L}^{-1} < 0.$$

The following is fundamental.

(2.1.20) Theorem. *Let E be an $(m+1)$-dimensional complex vector space and h a hermitian inner product on E. Let $\mathbf{L}(E)^{-1} \to P(E)$ be the hyperplane bundle and $H_E^{(-1)}$ be the hermitian metric naturally defined by h. Then we have the following:*

(i) $\Gamma(P(E), \mathbf{L}(E)^{-1}) = E^*$ *and* $B(E^*) = \varnothing$;

(ii) $\Phi_{E^*}: P(E) \to P((E^*)^*) = P(E)$ *is the identity mapping;*

(iii) $\omega_{(\mathbf{L}(E)^{-1}, H_E^{(-1)})} = -\omega_{(\mathbf{L}(E), H_E)} > 0$ *and*

$$\int_{P(E)} \overset{m}{\wedge}\, \omega_{(\mathbf{L}(E)^{-1}, H_E^{(-1)})} = 1.$$

Proof. It is easy to see (i) and (ii) by (2.1.16) and the definitions. The positivity of $\omega_{(\mathbf{L}(E)^{-1}, H_E^{(-1)})}$ is shown by (2.1.18). We deduce from (2.1.17) that

$$\overset{m}{\wedge}\, \omega_{(\mathbf{L}(E)^{-1}, H_E^{(-1)})} = \frac{m!}{\left(1 + \sum_{\mu \neq \lambda} |z_\lambda^\mu|^2\right)^{m+1}} \bigwedge_{\mu \neq \lambda} \left(\frac{i}{2\pi} dz_\lambda^\mu \wedge d\bar{z}_\lambda^\mu \right). \tag{2.1.21}$$

Hence it suffices to show that

$$\int_{\mathbf{C}^m} \frac{m!}{\left(1 + \sum_{\mu \neq \lambda} |z_\lambda^\mu|^2\right)^{m+1}} \bigwedge_{\mu=1}^{m} \left(\frac{i}{2\pi} dz^\mu \wedge d\bar{z}^\mu \right) = 1.$$

By making use of polar coordinates, we easily see this. *Q.E.D.*

In regard to the positivity and the ampleness of holomorphic line bundles, the following theorem of Kodaira is famous and very important.

(2.1.22) Theorem. *Let $\mathbf{L}$ be a holomorphic line bundle over a compact complex* *manifold* M. *Then* $\mathbf{L}$ *is ample if and only if* $\mathbf{L} > 0$.

For the proof of this theorem, see Wells [121]. The necessity which is needed for the later purpose is proved as follows. Take $k \in \mathbf{N}$ so that $\mathbf{L}^k$ is very ample. Set $E = \Gamma(M, \mathbf{L}^k)$. Then $\Phi_E: M \to P(E^*)$ is a holomorphic imbedding. Fix a hermitian metric h on E^*. Set $\omega = \omega_{(\mathbf{L}(E^*)^{-1}, H_{E^*}^{(-1)})}$. Then it follows from Theorem (2.1.20), (iii) that $\omega > 0$. By the definitions of ω and Φ_E, $\Phi_E^*\omega > 0$ and $\Phi_E^*\omega$ is the Chern form of $\mathbf{L}^k$ with respect to the hermitian metric $\Phi_E^* H_{E^*}^{(-1)}$ on $\mathbf{L}^k$. Therefore we see that $\mathbf{L}^k > 0$. It is easy to see that $\mathbf{L}^k > 0 \Longleftrightarrow \mathbf{L} > 0$. Hence we have $\mathbf{L} > 0$.

Next we define the so-called canonical line bundle $\mathbf{K}(M) \to M$. Let $\mathbf{T}(M) = \bigcup_{x \in M} \mathbf{T}(M)_x$ be the holomorphic tangent bundle of M. Here $\mathbf{T}(M)_x$ is the holomorphic tangent space of M at x and $\dim \mathbf{T}(M) = m$ ($m = \dim M$). Let $\mathbf{T}(M)_x^*$ be the dual space of $\mathbf{T}(M)_x$. Set $\mathbf{K}(M)_x = \overset{m}{\wedge}\, \mathbf{T}(M)_x^*$. Then $\mathbf{K}(M)_x$ is a one-dimensional complex vector space. As a set, $\mathbf{K}(M)_x$ consists of all the skew symmetric multilinear functionals of degree m on $\mathbf{T}(M)_x$. Set $\mathbf{K}(M) = \bigcup_{x \in M} \mathbf{K}(M)_x$ (disjoint union) and define a mapping $\pi: \mathbf{K}(M) \to M$ by the condition $\pi(\mathbf{K}(M)_x) = x$ for any $x \in M$. Let $\{(U_\lambda, \phi_\lambda, \Delta(1)^m)\}$ be an open covering of M by holomorphic local coordinate neighborhood systems. Set $\phi_\lambda = (z_\lambda^1, \ldots, z_\lambda^m)$ and $s_\lambda = dz_\lambda^1 \wedge \cdots \wedge dz_\lambda^m$. If $U_\lambda \cap U_\mu \neq \varnothing$, we set

$$K_{\lambda\mu} = \det\left(\frac{\partial z_\mu^k}{\partial z_\lambda^j}\right)_{j,k}.$$

Then $K_{\lambda\mu}$ is nowhere zero and $\{K_{\lambda\mu}\}$ satisfies the cocycle condition (2.1.3). Moreover we have

$$s_\lambda K_{\lambda\mu} = s_\mu$$

if $U_\lambda \cap U_\mu \neq \varnothing$. Define a mapping $F_\lambda: a s_\lambda(x) \in \pi^{-1}(U_\lambda) \to (x, a) \in U_\lambda \times \mathbf{C}$. Then F_λ is bijective, and if $U_\lambda \cap U_\mu \neq \varnothing$, we have

$$F_\lambda \circ F_\mu^{-1}(x, \xi) = (x, K_{\lambda\mu}(x)\xi).$$

Therefore we can uniquely define an $(m+1)$-dimensional complex structure on $\mathbf{K}(M)$ so that F_λ become biholomorphic mappings. Then $(\mathbf{K}(M), \pi, M)$ forms a holomorphic line bundle and F_λ are bundle isomorphisms such that the following diagram commutes:

$$\begin{array}{ccc} \pi^{-1}(U_\lambda) & \xrightarrow{F_\lambda} & U_\lambda \times \mathbf{C} \\ \downarrow \pi & & \downarrow \mathrm{pr} \\ U_\lambda & = & U_\lambda \end{array}$$

Clearly we have $s_\lambda \in \Gamma(U_\lambda, \mathbf{K}(M))$. Then $(\{U_\lambda\}, \{s_\lambda\})$ is a local trivialization covering of $\mathbf{K}(M)$ and $(\{U_\lambda\}, \{K_{\lambda\mu}\})$ is a system of holomorphic transition functions subordinated to it. It is easy to see that

$$\Gamma(M, \mathbf{K}(M)) = \{\text{holomorphic } m\text{-forms on } M\}.$$

The holomorphic line bundle $\mathbf{K}(M)$ is called the **canonical bundle** over M.

Let H be a hermitian metric on $\mathbf{K}(M)$. Put $H_\lambda = H(s_\lambda, s_\lambda)$. Define real $2m$-forms Ω_λ on U_λ by

$$\Omega_\lambda = \frac{1}{H_\lambda}\left(\frac{i}{2} dz_\lambda^1 \wedge d\overline{z}_\lambda^1\right) \wedge \cdots \wedge \left(\frac{i}{2} dz_\lambda^m \wedge d\overline{z}_\lambda^m\right).$$

If $U_\lambda \cap U_\mu \neq \emptyset$, then $H_\lambda |K_{\lambda\mu}|^2 = H_\mu$ by (2.1.7) and so $\Omega_\lambda = \Omega_\mu$ on $U_\lambda \cap U_\mu$. Therefore there exists uniquely a volume element Ω_H on M such that $\Omega_H | U_\lambda = \Omega_\lambda$. Conversely, if Ω is a volume element on M, then there exists uniquely a hermitian metric H on $\mathbf{K}(M)$ such that $\Omega_H = \Omega$. For a volume element Ω on M, set

$$\Omega | U_\lambda = a_\lambda \left(\frac{i}{2} dz_\lambda^1 \wedge d\overline{z}_\lambda^1\right) \wedge \cdots \wedge \left(\frac{i}{2} dz_\lambda^m \wedge d\overline{z}_\lambda^m\right)$$

on U_λ. Then we have that $a_\lambda = |K_{\lambda\mu}|^2 a_\mu$ on $U_\lambda \cap U_\mu \neq \emptyset$. Hence if we set $\omega_\lambda = -i\partial\overline{\partial} \log a_\lambda$ on U_λ, we have $\omega_\lambda = \omega_\mu$ on $U_\lambda \cap U_\mu$. Therefore there exists a real closed 2-form ω of type $(1, 1)$ such that $\omega | U_\lambda = \omega_\lambda$. From now on we write $\operatorname{Ric} \Omega$ for this ω and call it the **Ricci form** of Ω. As explained above, Ω defines a hermitian metric H on $\mathbf{K}(M)$. If we write $\omega_{(\mathbf{K}(M), H)}$ for the Chern form of the hermitian line bundle $(\mathbf{K}(M), H)$, then we have

$$-\frac{1}{2\pi} \operatorname{Ric} \Omega = \omega_{(\mathbf{K}(M), H)}. \tag{2.1.23}$$

In what follows, we compute $\mathbf{K}(M)$ and $\operatorname{Ric} \Omega$ for several typical examples.

(2.1.24) *Example.* Let $M = \mathbf{C}^m$ with the standard coordinate system $(z^1, \ldots, z^m)$. Then $dz^1 \wedge \cdots \wedge dz^m$ is a holomorphic local frame over $\mathbf{C}^m$, so that $\mathbf{K}_{\mathbf{C}^m} = \mathbf{1}_{\mathbf{C}^m}$. Take a volume form

$$\Omega = \frac{i}{2} dz^1 \wedge d\overline{z}^1 \wedge \cdots \wedge \frac{i}{2} dz^m \wedge d\overline{z}^m.$$

Then $\operatorname{Ric} \Omega \equiv 0$.

(2.1.25) *Example.* Let $M = \mathbf{C}^m/\Lambda$ be a complex torus and $p: \mathbf{C}^m \to \mathbf{C}^m/\Lambda$ the universal covering. Then the above $dz^1 \wedge \cdots \wedge dz^m$ and Ω are translation-invariant and so induce respectively a holomorphic local frame of $\mathbf{K}(\mathbf{C}^m/\Lambda)$ over $\mathbf{C}^m/\Lambda$ and a volume form Ω_0. Thus $\mathbf{K}(\mathbf{C}^m/\Lambda) = \mathbf{1}_{\mathbf{C}^m/\Lambda}$ and $\operatorname{Ric} \Omega_0 \equiv 0$.

(2.1.26) *Example.* Let $M = \mathbf{P}^m(\mathbf{C})$ with homogeneous coordinate system $[z^0; \cdots; z^m]$. Let $U_i = \{z^i \neq 0\}$ $(0 \leq i \leq m)$ and (z_i^k) $(0 \leq k \neq i \leq m)$ affine coordinate systems on U_i. Put

$$A_i = (-1)^i dz_i^0 \wedge \cdots \wedge \widehat{dz_i^i} \wedge \cdots \wedge dz_i^m.$$

Then A_i are holomorphic local frames of $\mathbf{K}(\mathbf{P}^m(\mathbf{C}))$ over U_i and a direct computation implies that

$$A_i = (z_i^j)^{m+1} A_j = \left(\frac{z^j}{z^i} \right)^{m+1} A_j \quad \text{on } U_i \cap U_j.$$

Hence

$$\mathbf{K}(\mathbf{P}^m(\mathbf{C})) = \mathbf{H}_0^{-(m+1)}, \quad c_1(\mathbf{K}(\mathbf{P}^m(\mathbf{C}))) = -(m+1)c_1(\mathbf{H}_0).$$

We have a volume form $\omega_0^m = \overset{m}{\wedge}\, \omega_0$ on $\mathbf{P}^m(\mathbf{C})$ and by (2.1.20) and (2.1.17)

$$\frac{1}{2\pi} \mathrm{Ric}\, \omega_0^m = (m+1)\omega_0.$$

(2.1.27) *Example.* Let $F(z^0, \ldots, z^m)$ be a homogeneous polynomial of degree d without multiple factor. We can uniquely define a closed subset V_d of $P(\mathbf{C}^{m+1})$ by

$$\rho^{-1}(V_d) = \{(z^0, \ldots, z^m) \in \mathbf{C}^{m+1} - \{O\}; F(z^0, \ldots, z^m) = 0\}.$$

Then V_d is called an analytic hypersurface of **degree** d. Assume that V_d has no singularity, or equivalently that V_d is a complex submanifold. Let $\mathbf{H}_0 | V_d$ denote the restriction of $\mathbf{H}_0$ over V_d. Then we have

$$\mathbf{K}(V_d) = (\mathbf{H}_0 | V_d)^{k-m-1}, \quad c_1(\mathbf{K}(V_d)) = (k-m-1)\{\iota_{V_d}^* \omega_0\},$$

where $\iota_{V_d}: V_d \to P(\mathbf{C}^{m+1})$ denotes the natural inclusion mapping and $[\iota_{V_d}^* \omega_0]$ denotes the cohomology class of $\iota_{V_d}^* \omega_0$ in $H^2(V_d, \mathbf{R})$. For the proof, see Appendix I. Hence, if $d > m+1$, then $\mathbf{K}(V_d)$ is very ample and $c_1(\mathbf{K}(V_d)) > 0$.

Note

Here we explain the de Rham cohomology group. Let M be an m-dimensional differentiable manifold. We write $A^p(M, \mathbf{R})$ for the real vector space of real p-forms of class C^∞ on M. Take any $\alpha \in A^p(M, \mathbf{R})$. Let $(x^1, \ldots, x^m)$ be a local coordinate system of M defined on an open subset $U \subset M$. Then $\alpha | U$ can be written uniquely as

$$\alpha | U = \sum_{1 \leq i_1 < \cdots < i_p \leq m} a_{i_1 \cdots i_p} dx^{i_1} \wedge \cdots \wedge dx^{i_p}.$$

Then, as is well known, we can define uniquely the so-called exterior derivative $d: A^p(M, \mathbf{R}) \to A^{p+1}(M, \mathbf{R})$ by

$$d\alpha | U = \sum_{1 \le i_1 < \cdots < i_p \le m} \sum_{k=1}^{m} \frac{\partial a_{i_1 \cdots i_p}}{\partial x^k} dx^k \wedge dx^{i_1} \wedge \cdots \wedge dx^{i_p}.$$

Then by this definition, we have $d(d\alpha) = 0$ and the following sequence:

$$0 \to \mathbf{R} \to A^0(M, \mathbf{R}) \xrightarrow{d} A^1(M, \mathbf{R}) \xrightarrow{d} \cdots \xrightarrow{d} A^m(M, \mathbf{R}) \to 0.$$

If we set

$$Z^p(M, \mathbf{R}) = \{\alpha \in A^p(M, \mathbf{R}); d\alpha = 0\},$$

$$B^p(M, \mathbf{R}) = dA^{p-1}(M, \mathbf{R}),$$

then $B^p(M, \mathbf{R})$ is a linear subspace of $Z^p(M, \mathbf{R})$. We set

$$H^p(M, \mathbf{R}) = Z^p(M, \mathbf{R}) / B^p(M, \mathbf{R}),$$

where the right hand side denotes the quotient vector space. We call $H^p(M, \mathbf{R})$ the p-th real **de Rham cohomology group.**

Now let M be an m-dimensional complex manifold (so that M is, in particular, a $2m$-dimensional differentiable manifold). So we have the p-th real de Rham cohomology group $H^p(M, \mathbf{R})$. Moreover, consider the complex vector space $A^p(M, \mathbf{C})$ of complex valued p-forms on M and naturally extend the exterior derivative $d: A^p(M, \mathbf{R}) \to A^{p+1}(M, \mathbf{R})$ to the complex linear mapping $d: A^p(M, \mathbf{C}) \to A^{p+1}(M, \mathbf{C})$. Then in the same way as above, we obtain the p-th complex de Rham cohomology group $H^p(M, \mathbf{C})$.

2.2 Pseudo-Volume Elements and Ricci Curvature Functions

Let M be an m-dimensional complex manifold and Ω a real $2m$-form on M. Let x be an arbitrary point of M and $(U, (z^1, \ldots, z^m))$ a holomorphic local coordinate system. Then we can write $\Omega | U$ as

(2.2.1) $$\Omega | U = a_U \left(\frac{i}{2} dz^1 \wedge d\bar{z}^1 \right) \wedge \cdots \wedge \left(\frac{i}{2} dz^m \wedge d\bar{z}^m \right),$$

where a_U is a real-valued function on U. We write $\Omega(x) > 0$ (resp. $\Omega(x) \ge 0$) if $a_U(x) > 0$ (resp. $a_U(x) \ge 0$). It is easily checked that this definition is independent of the particular choice of a holomorphic local coordinate system $(U, (z^1, \ldots, z^m))$. We write $\Omega > 0$ (resp. $\Omega \ge 0$) if $\Omega(x) > 0$ (resp. $\Omega \ge 0$) for all $x \in M$. For two real $2m$-forms Ω_1 and Ω_2 on M, we write $\Omega_1(x) > \Omega_2(x)$ (resp. $\Omega_1(x) \ge \Omega_2(x)$) if $\Omega_1(x) - \Omega_2(x) > 0$ (resp. $\Omega_1(x) - \Omega_2(x) \ge 0$). If $\Omega_1(x) > \Omega_2(x)$ (resp.

$\Omega_1(x) \geq \Omega_2(x)$) for all $x \in M$, then we write $\Omega_1 > \Omega_2$ (resp. $\Omega_1 \geq \Omega_2$). For any real $2m$-form Ω on M, set

$$\text{Zero}\,(\Omega) = \{x \in M;\ \Omega(x) = 0\}.$$

We call Ω a **pseudo-volume form,** if $\Omega \geq 0$. If Ω is continuous on M and C^∞ on $M - \text{Zero}\,(\Omega)$, we call Ω a **pseudo-volume element.** Moreover, if $\Omega > 0$, then we call Ω a **volume element.**

Let Ω be a pseudo-volume element on M. We have a real closed 2-form Ric Ω of type (1, 1) on $M - \text{Zero}\,(\Omega)$ by

$$\text{Ric}\,\Omega = -i\partial\bar{\partial} \log a_U | U - \text{Zero}\,(\Omega),$$

where a_U is as in (2.2.1) (see §1). We deal only with the case where Ric $\Omega \leq 0$. We define the **Ricci curvature function** K_Ω: $M \to [-\infty, \infty)$ of such Ω as follows:

$$K_\Omega(x) = \begin{cases} -\infty & \text{if } x \in \text{Zero}\,(\Omega), \\ -\dfrac{(-\text{Ric}\,\Omega)^m}{m!\,\Omega} & \text{if } x \in M - \text{Zero}\,(\Omega). \end{cases}$$

(2.2.2) *Example.* Let $r_j > 0$, $j = 1, \ldots, m$. We define the **Poincaré volume element** $\Omega(r_1, \ldots, r_m)$ on $\prod_{j=1}^m \Delta(r_j)$ by

$$\Omega(r_1, \ldots, r_m) = \bigwedge_{j=1}^{m} \frac{4r_j^2}{\left(r_j^2 - |z^j|^2\right)^2} \frac{i}{2} dz^j \wedge d\bar{z}^j.$$

We shall list the fundamental properties of $\Omega(r_1, \ldots, r_m)$. For $r_j' > 0$, $1 \leq j \leq m$, consider a biholomorphic mapping T: $(z^j) \in \prod \Delta(r_j) \to (r_j' z^j / r_j) \in \prod \Delta(r_j')$. Then we have

$$T^*\Omega(r_1', \ldots, r_m') = \Omega(r_1, \ldots, r_m). \tag{2.2.3}$$

Now we compute the Ricci form of $\Omega(r_1, \ldots, r_m)$:

$$\begin{aligned} \text{Ric}\,\Omega(r_1, \ldots, r_m) &= -i\partial\bar{\partial} \log \prod_{j=1}^{m} \frac{4r_j^2}{\left(r_j^2 - |z^j|^2\right)^2} \\ &= -\sum_{j=1}^{m} \frac{4r_j^2}{\left(r_j^2 - |z^j|^2\right)^2} \frac{i}{2} dz^j \wedge d\bar{z}^j. \end{aligned}$$

Hence we have

$$\text{Ric}\,\Omega(r_1, \ldots, r_m) < 0, \tag{2.2.4}$$

$$(-\mathrm{Ric}\,\Omega(r_1, \ldots, r_m))^m = m!\Omega(r_1, \ldots, r_m),$$

$$K_{\Omega(r_1, \ldots, r_m)} \equiv -1.$$

Let $T_j \in \mathrm{Aut}(\Delta(r_j))$, $1 \le j \le m$. Then $T = (T_1, \ldots, T_m) \in \mathrm{Aut}(\prod \Delta(r_j))$ and (1.1.7), (iii) implies the following:

(2.2.5) $$T^*\Omega(r_1, \ldots, r_m) = \Omega(r_1, \ldots, r_m).$$

(2.2.6) *Example.* For $r_j > 0$, $1 \le j \le m$, we define the **Poincaré volume element** $\Omega^*(r_1, \ldots, r_m)$ on

$$\Delta^*(r_1) \times \prod_{j=2}^{m} \Delta(r_j) = \{(z^1, \ldots, z^m) \in \mathbf{C}^m;$$

$$0 < |z^1| < r_1,\ |z^j| < r_j,\ 2 \le j \le m\}$$

as follows:

$$\Omega^*(r_1, \ldots, r_m) = \frac{4}{|z^1|^2(\log|z^1/r_1|^2)^2} \frac{i}{2} dz^1 \wedge d\bar{z}^1$$

$$\wedge \left(\bigwedge_{j=2}^{m} \frac{4r_j^2}{\left(r_j^2 - |z^j|^2\right)^2} \frac{i}{2} dz^j \wedge d\bar{z}^j \right).$$

For $r_j' > 0$, $1 \le j \le m$, we put

$$T : (z^j) \in \Delta^*(r_1) \times \prod_{j=2}^{m} \Delta(r_j) \to \left(\frac{r_j'}{r_j} z^j \right) \in \Delta^*(r_1') \times \prod_{j=2}^{m} \Delta(r_j').$$

Then we can easily see the following:

(2.2.7) $$T^*\Omega^*(r_1', \ldots, r_m') = \Omega^*(r_1, \ldots, r_m),$$

$$\mathrm{Ric}\,\Omega^*(r_1, \ldots, r_m) < 0,$$

$$(-\mathrm{Ric}\,\Omega^*(r_1, \ldots, r_m))^m = m!\Omega^*(r_1, \ldots, r_m),$$

$$K_{\Omega^*(r_1, \ldots, r_m)} \equiv -1.$$

The following lemma is of fundamental importance in this chapter.

(2.2.8) Lemma. *Let Ω be a pseudo-volume element on $\prod \Delta(r_j)$ $(0 < r_j < \infty)$ such that* $-\mathrm{Ric}\,\Omega \ge 0$ *on* $\prod \Delta(r_j) - \mathrm{Zero}\,(\Omega)$ *and there exists a constant $A > 0$ with* $K_\Omega \le -A$ *on* $\prod \Delta(r_j)$. *Then we have*

$$\Omega \le \frac{1}{A} \Omega(r_1, \ldots, r_m).$$

Proof. Since $A\Omega$ is also a pseudo-volume element and $K_{A\Omega} = A^{-1}K_{\Omega}$, we may assume that $A = 1$. Define a function $\mu: \prod\Delta(r_j) \to \mathbf{R}$ by $\Omega(z) = \mu(z)\,\Omega(r_1, \dots, r_m)(z)$. It is sufficient to show that $\mu \le 1$. For any $0 < t < 1$, define a function $\mu_t: \prod\Delta(tr_j) \to \mathbf{R}$ by $\Omega(z) = \mu_t(z)\Omega(tr_1, \dots, tr_m)$. Then we have

$$\mu_t(z) = \frac{\mu(z)}{t^{2m}} \prod_{j=1}^{m} \frac{\left((tr_j)^2 - |z^j|^2\right)^2}{\left(r_j^2 - |z^j|^2\right)^2}.$$

Therefore $\mu_t(z) \to \mu(z)$ uniformly on compact subsets of $\prod\Delta(r_j)$ as $t \to 1$. Hence it suffices to show that $\mu_t \le 1$ for any fixed $0 < t < 1$. The function μ_t has the following properties:

(2.2.9) (i) μ_t is a non-negative continuous function;

(ii) μ_t is a C^∞-function on $\prod\Delta(tr_j) - \mathrm{Zero}\,(\Omega)$;

(iii) $\mu_t(z) \to 0$ uniformly as $z \to \partial\prod\Delta(tr_j)$.

If $\mu_t \equiv 0$, nothing is to be proved. Therefore we assume that $\mu_t \not\equiv 0$. By (2.2.9) there exists $w \in \prod\Delta(r_j) - \mathrm{Zero}\,(\Omega)$ such that μ_t attains its maximum at w. Then we have

$$0 \le -\partial\bar{\partial} \log \mu_t(w) \le \mathrm{Ric}\,\Omega(w) - \mathrm{Ric}\,\Omega(tr_1, \dots, tr_m)(w),$$

so that

$$0 \le -\mathrm{Ric}\,\Omega(w) \le -\mathrm{Ric}\,\Omega(tr_1, \dots, tr_m)(w).$$

Hence we have

$$0 \le (-\mathrm{Ric}\,\Omega)^m(w) \le (-\mathrm{Ric}\,\Omega(tr_1, \dots, tr_m))^m(w),$$

$$0 \le \frac{(-\mathrm{Ric}\,\Omega)^m(w)}{m!\,\Omega^m(w)} \le \frac{(-\mathrm{Ric}\,\Omega(tr_1, \dots, tr_m))^m(w)}{\mu_t(w) m!\,\Omega(tr_1, \dots, tr_m)(w)},$$

$$1 \le -K_\Omega(w) \le \frac{1}{\mu_t(w)}\left[-K_{\Omega(tr_1, \dots, tr_m)}(w)\right]$$

$$= \frac{1}{\mu_t(w)} \quad \text{(by (2.2.4))}.$$

Thus we see that $\mu_t(w) \le 1$. Since $\mu_t(w)$ is the maximum value of μ_t, $\mu_t \le 1$. *Q.E.D.*

(2.2.10) Corollary. *Let Ω be a pseudo-volume element on $\Delta^*(1) \times \Delta(1)^{m-1}$ such that $-\mathrm{Ric}\,\Omega \ge 0$ on $\Delta^*(1) \times \Delta(1)^{m-1} - \mathrm{Zero}\,(\Omega)$ and there exists a constant $A > 0$ with $K_\Omega \le -A$. Then $\Omega \le A^{-1}\Omega^*(1, \dots, 1)$.*

Proof. Put $\pi: z \in \Delta(1) \to \exp(2\pi(z+1)/(z-1)) \in \Delta^*(1)$, which is an unramified covering mapping (cf. Example (1.2.11)). Then we have by (1.2.14)

$$\pi^*\left(\frac{4}{|z^1|^2(\log|z^1|^2)^2}\frac{i}{2}dz^1\wedge d\bar{z}^1\right)=\frac{4}{(1-|z^1|^2)^2}\frac{i}{2}dz^1\wedge d\bar{z}^1.$$

Put $\alpha=\pi\times\mathrm{id}\colon \Delta(1)\times\Delta(1)^{m-1}\to\Delta^*(1)\times\Delta(1)^{m-1}$, which is an unramified covering mapping. Then we have that $\alpha^*\Omega^*(1,\ldots,1)=\Omega(1,\ldots,1)$. Applying Lemma (2.2.8) to $\alpha^*\Omega$, we deduce that $A\alpha^*\Omega\le\Omega(1,\ldots,1)$. Hence $A\Omega\le\Omega^*(1,\ldots,1)$. *Q.E.D.*

2.3 Hyperbolic Pseudo-Volume Form

Let M be an m-dimensional complex manifold. A holomorphic mapping $f\colon\Delta(1)^m\to M$ is said to be **non-degenerate** at $z\in\Delta(1)^m$, if $f_*\colon \mathbf{T}(\Delta(1)^m)_z\to\mathbf{T}(M)_{f(z)}$ is a linear isomorphism. Suppose f is non-degenerate at the origin. Then the inverse function theorem implies that there is a positive number $r<1$ and a neighborhood U of $f(O)$ such that $f|\Delta(r)^m\colon\Delta(r)^m\to U$ is a biholomorphic mapping. For the sake of simplicity, we write Ω_0 for the Poincaré volume element $\Omega(1,\ldots,1)$ defined in Example (2.2.2). Now $((f|\Delta(r)^m)^{-1})^*\Omega_0$ is a volume element on U. Set $x=f(O)$ and

$$\Psi_{M,f}(x)=((f|\Delta(r)^m)^{-1})^*((\Omega_0(O))\in\bigwedge^{2m}T(M)_x^*.$$

For an arbitrary point $x\in M$, set

$$\Psi_M(x)=\inf\{\Psi_{M,f}(x)\},$$

where the infimum is taken for all holomorphic mappings $f\colon\Delta(1)^m\to M$ with $f(O)=x$ which are non-degenerate at the origin. Then Ψ_M is a pseudo-volume form on M. We call Ψ_M the **hyperbolic pseudo-volume form** of M. As a basic fact, we shall prove the following.

(2.3.1) Theorem. *Ψ_M is an upper semicontinuous $2m$-form on M.*

Proof. Take any $x\in M$. Let $(z^1,\ldots,z^m)$ be a holomorphic local coordinate system defined on a neighborhood U of x. Setting

$$\Psi_M|U=a\bigwedge_{j=1}^m\left(\frac{i}{2}dz^j\wedge d\bar{z}^j\right).$$

We prove that the function a is upper semicontinuous at x. Take any $\varepsilon>0$. By definition there exists a holomorphic mapping $f\colon\Delta(1)^m\to M$, such that $f(O)=x$, f is non-degenerate at O and

$$\Psi_{M,f}(x)<\Psi_M+\varepsilon\left(\bigwedge_{j=1}^m\frac{i}{2}dz^j\wedge d\bar{z}^j\right)_x.$$

Since f is non-degenerate at the origin, there exist $0 < r < 1$ and a neighborhood $W \subset U$ such that $f|\Delta(r)^m\colon \Delta(r)^m \to W$ is biholomorphic. Now set

$$((f|\Delta(r)^m)^{-1})^*\Omega_0 = b \bigwedge_{j=1}^{m} \frac{i}{2} dz^j \wedge d\bar{z}^j.$$

Since $b(x) \bigwedge_{j=1}^{m} (i/2)(dz^j \wedge d\bar{z}^j)_x = \Psi_{M,f}(x)$, we have $b(x) < a(x) + \varepsilon$. Since b is a C^∞-function on W, we may assume, by taking sufficiently small W, that

$$b(y) < a(x) + \varepsilon$$

for all $y \in W$. For any $y \in W$, take $z \in \Delta(r)^m$ so that $f(z) = y$. By (1.1.7), (ii), there is a biholomorphic mapping $T\colon \Delta(1)^m \to \Delta(1)^m$ with $T(O) = z$. Set $g = f \circ T\colon \Delta(1)^m \to M$. Then $g(O) = y$ and g is non-degenerate at the origin. Reminding $T^*\Omega_0 = \Omega_0$ by (2.2.5), we have

$$\Psi_{M,g}(y) = ((g|g^{-1}(W))^{-1})^*\Omega_0(O)$$

$$= ((f|\Delta(r)^m)^{-1})^*\Omega_0(z) = b(y) \bigwedge_{j=1}^{m} \frac{i}{2}(dz^j \wedge d\bar{z}^j)_y.$$

Hence $a(y) < b(y) < a(x) + \varepsilon$. Therefore a is upper semicontinuous at x. *Q.E.D.*

The hyperbolic pseudo-volume forms satisfy the **decreasing principle** as the Kobayashi pseudo-distances do:

(2.3.2) Theorem. *Let M and N be m-dimensional complex manifolds and $F\colon M \to N$ a holomorphic mapping. Then we have $F^*\Psi_N \le \Psi_M$. If F is, moreover, a biholomorphic mapping, then $F^*\Psi_N = \Psi_M$.*

Proof. Take any $x \in M$. When $F_{*x}\colon \mathbf{T}(M)_x \to \mathbf{T}(N)_{F(x)}$ is not an isomorphism, then $(F^*\Psi_N)(x) = 0$. So we have $F^*\Psi_N(x) \le F_M(x)$ in this case. Next consider the case when $F_{*x}\colon \mathbf{T}(M)_x \to \mathbf{T}(N)_{F(x)}$ is isomorphic. Let $f\colon \Delta(1)^m \to M$ be a holomorphic mapping with $f(O) = x$ which is non-degenerate at the origin. Then the holomorphic mapping $F \circ f\colon \Delta(1)^m \to N$ is non-degenerate at the origin and $F \circ f(O) = F(x)$. Since $F^*\Psi_{N, F\circ f}(F(x)) = \Psi_{M,f}(x)$, $F^*\Psi_N(x) \le \Psi_M(x)$. If F is biholomorphic, we clearly have $F^*\Psi_N = \Psi_M$. *Q.E.D.*

(2.3.3) *Example.* For $r_j > 0$, let $\Omega(r_1, \ldots, r_m)$ be the Poincaré volume element defined in (2.2.2). Then we have

$$\Psi_{\prod_{j=1}^m \Delta(r_j)} = \Omega(r_1, \ldots, r_m).$$

Therefore we have $\Psi_{\prod \Delta(r_j)} = \Psi_{\Delta(r_1)} \wedge \cdots \wedge \Psi_{\Delta(r_m)}$. To show (2.3.3), we may assume by (2.2.5) and Theorem (2.3.2) that all $r_j = 1$. It follows from (1.1.7), (ii) that for any $z \in \Delta(1)^m$ there is a biholomorphic mapping $T \in \mathrm{Aut}(\Delta(1)^m)$ with

$T(O)=z$. For the sake of simplicity, we write Ω_0 for $\Omega(1, ..., 1)$. Since $T^*\Omega_0 = \Omega_0$, $\Psi_{\Delta(1)^m}(z) \leq (T^{-1})^*\Omega_0(O) = \Omega_0(z)$. On the other hand, let $f: \Delta(1)^m \to \Delta^m$ be a holomorphic mapping with $f(O) = z$ which is non-degenerate at the origin. Put $\Lambda = f^*\Omega_0$. Then Λ is a pseudo-volume element on $\Delta(1)^m$ satisfying the following properties:

(i) Zero $(\Lambda) = \{x \in \Delta(1)^m; f_{*x}$ is degenerate$\}$;

(ii) $-\mathrm{Ric}\,\Lambda = -f^*\mathrm{Ric}\,\Omega_0 > 0$ on $\Delta(1)^m - \mathrm{Zero}\,(\Lambda)$;

(iii) $K_\Lambda = -(-\mathrm{Ric}\,\Lambda)^m/(m!\Lambda^m) = -1$.

Applying Lemma (2.2.8) to Λ, we have $\Lambda \leq \Omega_0$. Therefore $\Psi_{\Delta(1)^m, f}(z) \geq \Omega_0(z)$, so that $\Psi_{\Delta(1)^m}(z) \geq \Omega_0(z)$. Hence we see that $\Psi_{\Delta(1)^m} = \Omega_0$.

(2.3.4) *Example.* In the same way as above, we have by Corollary (2.2.10)

$$\Psi_{\Delta^*(r_1) \times \prod_{j=1}^{m} \Delta(r_j)} = \Omega^*(r_1, ..., r_m).$$

(2.3.5) Proposition. *Let M be an m-dimensional complex manifold and Λ a pseudo-volume form on M. Assume that for any holomorphic mapping f: $\Delta(1)^m \to M, f^*\Lambda \leq \Psi_{\Delta(1)^m}$ $(= \Omega_0)$. Then we have $\Lambda \leq \Psi_M$.*

Proof. Let $x \in M$ be an arbitrary point of M. Let $f: \Delta(1)^m \to M$ be a holomorphic mapping with $f(O) = x$ which is non-degenerate at the origin. Since $f^*\Lambda \leq \Omega_0$, we have $\Lambda(x) \leq \Psi_{M,f}(x)$. Hence $\Lambda(x) \leq \Psi_M(x)$. *Q.E.D.*

2.4 Measure Hyperbolic Manifolds

In this section, all complex manifolds are assumed to satisfy the second countability axiom. Let M be an m-dimensional complex manifold. A subset B of M is said to be Lebesgue measurable if for any holomorphic local coordinate neighborhood system $(U, \phi, \Delta(1)^m)$, $\phi(U \cap B)$ is a Lebesgue measurable subset of $\Delta(1)^m$. Let $(V, \psi, \Delta(1)^m)$ be a holomorphic local coordinate neighborhood system such that $U \cap V \neq \varnothing$. Then $\phi(U \cap V \cap B)$ is a Lebesgue measurable subset if and only if so is $\psi(U \cap V \cap B)$. Moreover, if the Lebesgue measure of $\phi(U \cap B)$ is 0 for any $(U, \phi, \Delta(1)^m)$, we say that B is of Lebesgue measure 0, and this is well defined. We know by Theorem (2.3.1) that the integral

$$\mu_M(B) = \int_B \Psi_M$$

is always defined for any Lebesgue measurable subset B of M, and this μ_M defines a measure on the family of Lebesgue measurable subsets of M. We call the measure μ_M the **hyperbolic measure** of M. The most fundamental property of the hyperbolic measure μ_M is given by the following:

(2.4.1) Theorem. *Let M and N be m-dimensional complex manifolds and $F : M \to N$ a holomorphic mapping. Let B be a Lebesgue measurable subset of M. Then $F(B)$ is a Lebesgue measurable subset of N and*

$$\mu_N(F(B)) \leq \mu_M(B).$$

Proof. Let $C \subset N$ be the set of critical values of F. Then C is of Lebesgue measure 0 by Sard's theorem. If the differential F_* is degenerate at all points of M, then $F(M) = C$, and hence $F(B)$ is a Lebesgue measurable subset of Lebesgue measure 0. If F_* is non-degenerate at some point of M, then $F|M - F^{-1}(C)$: $M - F^{-1}(C) \to N - C$ is locally biholomorphic. Therefore $F(B - F^{-1}(C))$ is Lebesgue measurable. Since $F(B) = F(B - F^{-1}(C)) \cup (F(B) \cap C)$, $F(B)$ is Lebesgue measurable. Then, Theorem (2.3.2) yields the required inequality. *Q.E.D.*

Let $B \subset M$ be a Lebesgue measurable subset. Let $f_j \colon \Delta(1)^m \to M$, $j = 1, 2, \ldots$, be holomorphic mappings and $E_j \subset \Delta(1)^m$ Lebesgue measurable subsets such that $B \subset \bigcup_{j=1}^{\infty} f_j(E_j)$. Let Ω_0 be the Poincaré volume element of $\Delta(1)^m$. Set

$$\tilde{\mu}_M(B) = \inf \left\{ \sum_{j=1}^{\infty} \int_{E_j} \Omega_0 \right\},$$

where the infimum is taken over all such $\{(f_j, E_j)\}_{j=1}^{\infty}$ as above (cf. Eisenman [25] and Chapter IX of Kobayashi [54]). Then it is easy to see that $\tilde{\mu}_M$ defines a measure on the family of Lebesgue measurable subsets of M.

(2.4.2) Theorem. *For any Lebesgue measurable subset B of M, we have* $\mu_M(B) = \tilde{\mu}_M(B)$.

Proof. Take any volume element Ω on M. We write λ for the measure on M defined by Ω. First we shall show that $\tilde{\mu}_M$ is absolutely continuous with respect to λ. Namely, $\tilde{\mu}_M(B) = 0$ for any Lebesgue measurable subset B with $\lambda(B) = 0$. Take any family of holomorphic local coordinate neighborhood systems $\{(U_\alpha, \phi_\alpha, \Delta(1)^m)\}$ $(\alpha = 1, 2, \ldots)$ such that $M = \bigcup_\alpha U'_\alpha$, where $U'_\alpha = \{x \in U_\alpha;\ \phi_\alpha(x) \in \Delta(1/2)^m\}$. Put $B_\alpha = B \cap U'_\alpha$. Then B_α are Lebesgue measurable subsets with $\lambda(B_\alpha) = 0$. Put $E_\alpha = \phi_\alpha(B_\alpha)$. Then E_α are Lebesgue measurable subsets of $\Delta(1/2)^m$. Since two volume elements $(\phi_\alpha^{-1})^*\Omega$ and Ω_0 are equivalent*) on $\Delta(1/2)^m$, $\lambda(B_\alpha) = 0$ implies that $\int_{E_\alpha} \Omega_0 = 0$. Hence we have

$$\tilde{\mu}_M(B) \leq \sum_\alpha \tilde{\mu}_M(B_\alpha) \leq \sum_\alpha \int_{E_\alpha} \Omega_0 = 0.$$

---

*) Two volume elements Ω_1 and Ω_2 are said to be equivalent if there are positive constants A_1 and A_2 such that $A_1\Omega_1 \leq \Omega_2 \leq A_2\Omega_1$.

Then it follows from the well-known theorem of Radon-Nikodym that there exists a Lebesgue measurable function a on M such that $\tilde{\mu}_M(B') = \int_M a\Omega$ for any Lebesgue measurable subset B' of M. Let $f: \Delta(1)^m \to M$ be a holomorphic mapping which is non-degenerate at $z \in \Delta(1)^m$. Set $f^*(a\Omega) = b\Omega_0$. Then we have

$$b(z) = \lim_{\varepsilon \to 0} \left[\int_{B(z;\,\varepsilon)} f^*(a\Omega) \Big/ \int_{B(z;\,\varepsilon)} \Omega_0 \right]$$

$$= \lim_{\varepsilon \to 0} \left[\tilde{\mu}_M(f(B(z;\,\varepsilon))) \Big/ \int_{B(z;\,\varepsilon)} \Omega_0 \right]$$

$$\leq \lim_{\varepsilon \to 0} \left[\int_{B(z;\,\varepsilon)} \Omega_0 \Big/ \int_{B(z;\,\varepsilon)} \Omega_0 \right] = 1.$$

Therefore, $f^*(a\Omega) \leq \Omega_0$. By Proposition (2.3.5), we have $a\Omega \leq \Psi_M$, so that $\tilde{\mu}_M(B) \leq \mu_M(B)$. To show the converse, we take an arbitrary Lebesgue measurable subset B with $\tilde{\mu}_M(B) < +\infty$. Take any $\varepsilon > 0$. By definition, there exist holomorphic mappings $f_j: \Delta(1)^m \to M$, and Lebesgue measurable subsets $E_j \subset \Delta(1)^m$, $j = 1, 2, \ldots$, such that $B \subset \bigcup_{j=1}^{\infty} f_j(E_j)$ and

$$\sum_{j=1}^{\infty} \int_{E_j} \Omega_0 < \tilde{\mu}_M(B) + \varepsilon.$$

By Theorem (2.3.2) and Example (2.3.3), we have $f_j^* \Psi_M \leq \Psi_{\Delta(1)^m} = \Omega_0$. Then it follows that

$$\sum_{j=1}^{\infty} \int_{E_j} \Omega_0 \geq \sum_{j=1}^{\infty} \Psi_M \geq \int_B \Psi_M = \mu_M(B).$$

Therefore, $\mu_M(B) \leq \tilde{\mu}_M(B) + \varepsilon$. So we have $\mu_M(B) \leq \tilde{\mu}_M(B)$. *Q.E.D.*

Let M be an m-dimensional complex manifold. We call M a **measure hyperbolic manifold** if $\mu_M(B) > 0$ for any non-empty open subset $B \subset M$.

(2.4.3) Proposition. *Let M be a compact measure hyperbolic manifold and $f: M \to M$ a holomorphic mapping. Then either f is generically injective or the rank of f_*: $\mathbf{T}(M) \to \mathbf{T}(M)$ is everywhere less than m.*

Remark. Here $f: M \to M$ is said to be generically injective if there is an analytic subset $A \underset{\neq}{\subset} M$ such that $f|M - f^{-1}(A): M - f^{-1}(A) \to M - A$ is injective.

Proof. It follows from Theorem (2.3.2) that

$$\mu_M(M) \geq \int f^*\Psi_M.$$

Suppose that the rank of f_* is m at some point. Then

$$\mu_M(M) \geq \int f^*\Psi_M > 0.$$

Let $C = \{x \in M\,;\ \text{rank of } f_{*x} < m\}$ ($\neq M$). Then C is an analytic subset of M and so is $A = f(C)$ (cf. Theorem (4.3.3)). Put

$$M' = M - f^{-1}(A), \quad M'' = M - A.$$

Then $f|M' \colon M' \to M''$ is an unramified finite covering, of which covering number is denoted by k. We have

$$\mu_M(M) = \mu_M(M') \geq \int_{M'} f^*\Psi_M$$

$$= k\int_{M''}\Psi_M = k\mu_M(M'') = k\mu_M(M) > 0,$$

so that $k = 1$. *Q.E.D.*

2.5 Differential Geometric Criterion of Measure Hyperbolicity

Here we shall give an interesting criterion of the measure hyperbolicity.

(2.5.1) Theorem. *Let M be an m-dimensional complex manifold and Ω a volume element on M such that* $\mathrm{Ric}\,\Omega \leq 0$. *If there is a positive constant A such that $K_\Omega \leq -A$, then $\Psi_M \geq A\Omega$. In particular, M is measure hyperbolic.*

Proof. Let x be an arbitrary point of M. Take any holomorphic mapping $f : \Delta(1)^m \to M$ with $f(O) = x$, which is non-degenerate at the origin. Then $f^*\Omega$ is a pseudo-volume element on $\Delta(1)^m$ and

$$\mathrm{Zero}\,(f^*\Omega) = \{z \in \Delta(1)^m;\ f \text{ is degenerate at } z\}.$$

On $\Delta(1)^m$, we have

$$-\mathrm{Ric}\,f^*\Omega = f^*(-\mathrm{Ric}\,\Omega) \geq 0.$$

If $z \in \mathrm{Zero}\,(f^*\Omega)$, then $K_{f^*\Omega}(z) = -\infty$; otherwise, we have

$$K_{f^*\Omega}(z) = -\frac{(-\mathrm{Ric}\,f^*\Omega)^m(z)}{m!(f^*\Omega)(z)}$$

$$= -\frac{f^*(-\mathrm{Ric}\,\Omega)^m(z)}{m!(f^*\Omega)(z)} = K_\Omega(f(z)).$$

Consequently, we have $K_{f^*\Omega} \leq -A$. By Lemma (2.2.8) we see that $f^*\Omega \leq (1/A)\Omega_0$, where Ω_0 is the Poincaré volume element $\Omega(1, ..., 1)$ on $\Delta(1)^m$. Hence $A\Omega(x) \leq \Psi_{M,f}(x)$, so that $A\Omega(x) \leq \Psi_M(x)$. *Q.E.D.*

(2.5.2) Corollary. *Let M be an m-dimensional compact complex manifold and* Ω *a volume element on M. If* $-\mathrm{Ric}\,\Omega > 0$ *on M, then there exists a positive constant A such that* $\Psi_M \geq A\Omega$*; consequently, M is measure hyperbolic.*

Proof. There exists a C^∞-function a such that $(-\mathrm{Ric}\,\Omega)^m = a\Omega$. Since $-\mathrm{Ric}\,\Omega > 0$, it follows that $a > 0$. Take a positive constant A so that $m!A \leq a$ on M. Then $K_\Omega \leq -A$. Then our assertion follows from Theorem (2.5.1). *Q.E.D.*

The essential part of the following Theorem will be proved in Chapter V. For the details, refer to Lemma (5.5.1) in Chapter V, §5.

(2.5.3) Theorem. *Let M be an m-dimensional compact complex manifold. Let D be a finite union of non-singular hypersurfaces of M which has only normal crossings. If the holomorphic line bundle* $[D] \otimes \mathbf{K}(M)$ *is positive, then* $M - D$ *is measure hyperbolic.*

Proof. By Lemma (5.4.1), there exists a volume element Ω on $M - D$ with $K_\Omega \leq -1$. Hence Theorem (2.5.1) implies that $M - D$ is measure hyperbolic. *Q.E.D.*

2.6 Meromorphic Mappings into a Measure Hyperbolic Manifold

In this section, M denotes an m-dimensional compact complex manifold. If $\mathbf{K}(M)$ is ample, then we have $\mathbf{K}(M) > 0$ by Theorem (2.1.8), and Corollary (2.5.2) (also cf. (2.1.9)) implies that M is measure hyperbolic. In this section we study meromorphic mappings into M such that $\mathbf{K}(M)$ is ample. The notion of meromorphic mappings is described in details in Chapter IV. If the reader does not know meromorphic mappings, he should use Chapter IV as a dictionary for properties of meromorphic mappings. First of all, we study the problem of extension of those meromorphic mappings.

(2.6.1) Lemma. *Suppose that* $\mathbf{K}(M)$ *is ample. Let* $f: \Delta^*(1) \times \Delta(1)^{m-1} \to M$ *be a non-degenerate meromorphic mapping into M; i.e., the differential* f_{*x} *has the maximal rank at some point* $x \in M$ *where f is holomorphic. Then f can be extended to a meromorphic mapping from* $\Delta(1) \times \Delta(1)^{m-1}$ *into M.*

Proof. By the assumption there exists a positive integer k such that $\mathbf{K}(M)$ is very ample. Namely

$$\Phi = \Phi_{\Gamma(M,\, \mathbf{K}(M)^k)}: M \to P(\Gamma(M,\, \mathbf{K}(M)^k)^*)$$

is a biholomorphic mapping. Let $\{s_0, ..., s_N\}$ is a basis of $\Gamma(M, \mathbf{K}(M)^k)$ and we identify $P(\Gamma(M, \mathbf{K}(M)^k)^*)$ with $\mathbf{P}^N(\mathbf{C})$ with respect to this basis. Let $(w^1, ..., w^m)$

be a holomorphic local coordinate system of M defined on an open subset U. Then we can write each $s_j|U$ as

$$s_j|U = a_j(dw^1 \wedge \cdots \wedge dw^m)^k,$$

where a_j is a holomorphic function on U. Then for any $w \in U$, we have

$$\text{(2.6.2)} \qquad \begin{aligned} \Phi(w) &= [s_0(w); \cdots; s_N(w)] \\ &= [a_0(w); \cdots; a_N(w)] \in \mathbf{P}^N(\mathbf{C}). \end{aligned}$$

We can define a volume element Ω on M by

$$\Omega|U = \left(\sum_{j=0}^{N} |a_j|^2 \right)^{\frac{1}{k}} \bigwedge_{l=1}^{m} \left(\frac{i}{2} dw^l \wedge d\overline{w}^l \right).$$

For the sake of simplicity we write Ω as

$$\Omega = \left(\sum_{j=0}^{N} s_j \wedge \overline{s}_j \right)^{\frac{1}{k}}.$$

Now we know $\Phi^* \omega_0 > 0$, and (2.6.2) implies

$$\Phi^* \omega_0 | U = \frac{i}{2\pi} \partial \overline{\partial} \log \left(\sum_{j=0}^{N} |a_j|^2 \right).$$

Therefore $-\text{Ric}\,\Omega = (2\pi/k)\Phi^* \omega_0$ and so we see that $-\text{Ric}\,\Omega > 0$. Since M is compact, there is a constant $A > 0$ such that

$$\text{(2.6.4)} \qquad \text{Ric}\,\Omega < 0, \quad K_\Omega \leq -A.$$

By a general property of meromorphic mappings, (ii) of §5, Chapter IV, we have

$$f^* s_j \in \Gamma(\Delta^*(1) \times \Delta(1)^{m-1}, \mathbf{K}(\Delta^*(1) \times \Delta(1)^{m-1})^k).$$

Since f is non-degenerated at some point, we know that $f^* s_j \neq 0$. Let $(z^1, ..., z^m)$ be the natural coordinate of $\Delta^*(1) \times \Delta(1)^{m-1}$. Then we can write $f^* s_j$ as

$$f^* s_j = b_j(dz^1 \wedge \cdots \wedge dz^m)^k,$$

where b_j is a holomorphic function on $\Delta^*(1) \times \Delta(1)^{m-1}$. Then the meromorphic mapping $\Phi \circ f : \Delta^* \times \Delta(1)^{m-1} \to \Phi(M) \subset \mathbf{P}^N(\mathbf{C})$ is given by

$$\Phi \circ f(z) = [b_0(z); \cdots; b_N(z)].$$

Therefore it is sufficient to show that every $b_j(z)$ can be extended to a meromorphic function on $\Delta(1) \times \Delta(1)^{m-1}$ (cf. Theorem (4.5.3)). Now set

$$\Psi = \left[\sum_{j=0}^{N} f^* s_j \wedge \overline{f^* s_j} \right]^{\frac{1}{k}} = \left[\sum_{j=0}^{N} |b_j|^2 \right]^{\frac{1}{k}} \bigwedge_{l=1}^{m} \left[\frac{i}{2} dz^l \wedge d\bar{z}^l \right].$$

Then Ψ is a pseudo-volume element on $\Delta^*(1) \times \Delta(1)^{m-1}$. By (2.6.4) we have the following:

(2.6.5) $$\text{Ric}\,\Psi \leq 0 \text{ on } \Delta^* \times \Delta(1)^{m-1} - \text{Zero}\,(\Psi),$$

$$K_\Psi \leq -A.$$

The Poincaré volume element $\Omega^*(1, ..., 1)$ on $\Delta^*(1) \times \Delta(1)^{m-1}$ is given by

$$\Omega^*(1, ..., 1) = \frac{4}{|z^1|^2 (\log |z^1|^2)^2} \prod_{l=2}^{m} \frac{4}{(1 - |z^l|^2)^2} \bigwedge_{l=1}^{m} \left[\frac{i}{2} dz^l \wedge d\bar{z}^l \right].$$

From Corollary (2.2.10) and (2.6.5), we obtain $\Psi \leq A\Omega^*(1, ..., 1)$. Therefore we have

(2.6.6) $$|b_j|^2 \leq \sum_{j=0}^{N} |b_j|^2 \leq A^k \frac{4^k}{|z^1|^{2k} (\log |z^1|^2)^{2k}} \prod_{l=2}^{m} \frac{4^k}{(1 - |z^l|^2)^{2k}}$$

for all $0 \leq j \leq N$. We consider the Laurent expansion of $b_j(z)$ on $\Delta^*(1) \times \Delta(1)^{m-1}$:

(2.6.7) $$b_j(z) = \sum_{\nu=-\infty}^{\infty} b_{j\nu}(z')(z^1)^\nu,$$

where $b_{j\nu}(z)$ are holomorphic functions in $z' = (z^2, ..., z^m) \in \Delta(1)^{m-1}$ given by

$$b_{j\nu}(z') = \frac{1}{2\pi i} \int_{\{|z^1| = r\}} \frac{b_j(z^1, z')}{(z^1)^{\nu+1}} dz^1 \quad (0 < r < 1).$$

Fix $z' \in \Delta(1)^{m-1}$ and set $z^1 = re^{i\theta}$ $(0 < r < 1)$. Integrating both the sides of (2.6.6) with respect to $\theta \in [0, 2\pi]$, we have

(2.6.8) $$\frac{1}{2\pi} \int_0^{2\pi} |b_j(re^{i\theta}, z')|^2 d\theta = \sum_{\nu=-\infty}^{\infty} |b_{j\nu}(z')|^2 r^{2\nu}$$

$$\leq (4^m A)^k r^{-2k} (\log r^{-2})^{-2k} \prod_{l=2}^{m} (1 - |z^l|^2)^{-2k}.$$

Comparing both the sides of inequality (2.6.8) as $r \to 0$, we see that $|b_{j\nu}(z')| = 0$ for all $\nu \leq -k$. Therefore $(z^1)^{k-1} b_j(z)$ can be extended to a holomorphic function on $\Delta(1) \times \Delta(1)^{m-1}$ and so $b_j(z)$ to a meromorphic function on $\Delta(1) \times \Delta(1)^{m-1}$. *Q.E.D.*

(2.6.9) Theorem. *Let M be an m-dimensional compact complex manifold such that* $\mathbf{K}(M)$ *is ample. Let N be an m-dimensional complex manifold and* $D \subset N$ *a* *proper analytic subset of N. If a meromorphic mapping $f: N-D \to M$ is non-degenerated at some point of $N-D$, then f extends to a meromorphic mapping from N into M.*

Proof. We write D' for the union of all the $(m-1)$-dimensional irreducible components of D and D'' for the union of the rest of all irreducible components of D. We write $S(D')$ for the set of singular points of D' and set $E = D'' \cup S(D')$ and $N' = N - E$. Then $N' \cap D = N' \cap D'$ is an $(m-1)$-dimensional closed complex submanifold of N' and the meromorphic mapping $f|(N'-D'): N'-D' \to M$ is non-degenerated at a certain point. Therefore it follows from Lemma (2.6.1) that $f|(N'-D')$ can be extended to a meromorphic mapping from N' into M, denoted by the same f. Now let

$$\Phi = [s_0; \cdots; s_N]: M \to \mathbf{P}^N(\mathbf{C}),$$
$$s_j \in \Gamma(M, \mathbf{K}(M)^k), \; 0 \le j \le N$$

be the holomorphic imbedding given in (2.6.2). Since $f: N-E \to M$ is a meromorphic mapping, we have $f^* s_j \in \Gamma(N-E, \mathbf{K}(N)^k)$ by (iv) of §5, Chapter IV. Since $\dim E \le m-2$, Corollary (3.3.43) implies that $f^* s_j$ extend to $t_j \in \Gamma(N, \mathbf{K}(N)^k)$. Clearly, we have $\Phi \circ f = [t_0; \cdots; t_N]: N \to \Phi(M) \subset \mathbf{P}^N(\mathbf{C})$. Since Φ is a holomorphic imbedding, we see that f extends to a meromorphic mapping from N into M. *Q.E.D.*

Let N be an m-dimensional complex manifold and set

$$\mathrm{Mer}^*(N, M) = \{f: N \to M;\ \text{meromorphic mappings,}$$
$$\text{non-degenerate at some points}\}.$$

Let X be an m-dimensional compact complex manifold and $D \subset X$ a proper analytic subset. Set $N = X - D$.

(2.6.10) Theorem. *Let M be an m-dimensional compact complex manifold such that $\mathbf{K}(M)$ is ample. Then* $\mathrm{Mer}^*(N, M)$ *is a finite set.*

Proof. Theorem (2.6.9) implies that $\mathrm{Mer}^*(N, M) = \mathrm{Mer}^*(X, M)$. Hence we may assume that N is compact. Let

$$\Phi = [s_0; \cdots; s_N]: M \to P(\Gamma(M, \mathbf{K}(M)^k)^*) \cong \mathbf{P}^N(\mathbf{C}),$$
$$s_j \in \Gamma(M, \mathbf{K}(M)^k), \; 0 \le j \le N$$

be the holomorphic imbedding constructed in (2.6.2). Set $V = \Gamma(M, \mathbf{K}(M)^k)$. Let Ω be the volume element on M constructed in (2.6.3). We remark that (2.6.4) holds. Set $W = \Gamma(N, \mathbf{K}(N)^k)$. Then any $f \in \mathrm{Mer}^*(N, M)$ defines a linear mapping

$$f^*: \xi \in V \to f^*\xi \in W.$$

Since f is non-degenerated at a certain point, f^* is injective. Set $t_j = f^* s_j \in W$ for $0 \le j \le N$. We have

$$f^*\Omega = \left(\sum_{j=0}^{N} t_j \wedge \bar{t}_j \right)^{\frac{1}{k}},$$

and $f^*\Omega$ is a pseudo-volume element on N. By (2.6.4) we have

$$\text{(2.6.11)} \qquad \operatorname{Ric} f^*\Omega \le 0, \quad K_{f^*\Omega} \le -A$$

on $N - \operatorname{Zero}(f^*\Omega)$. Let $\{(U_\alpha, \phi_\alpha, \Delta(1)^m)\}$ be a finite open covering of N by holomorphic local coordinate neighborhood systems. Let Ω_0 be the Poincaré volume element on $\Delta(1)^m$, and $\{c_\alpha\}$ be a partition of unity subordinated to $\{U_\alpha\}$. Then we have a volume element Ψ on N defined by

$$\Psi = \frac{1}{A} \sum c_\alpha \phi_\alpha^* \Omega_0.$$

We obtain from (2.6.11) and Lemma (2.2.8)

$$A f^*\Omega \le \phi_\alpha^* \Omega_0$$

on every U_α. Therefore we have

$$\text{(2.6.12)} \qquad f^*\Omega = \sum c_\alpha f^*\Omega \le \frac{1}{A} \sum c_\alpha \phi_\alpha^* \Omega_0 = \Phi$$

for all $f \in \operatorname{Mer}^*(N, M)$. For $\eta \in W$, set

$$\| \eta \|_N = \int_N (\eta \wedge \bar{\eta})^{\frac{1}{k}}.$$

For $\xi \in V$, set

$$\| \xi \|_M = \int_M (\xi \wedge \bar{\xi})^{\frac{1}{k}}.$$

Clearly $\| \xi \|_M$ is continuous in $\xi \in V$. Next we shall show that $B = \{\xi \in V; \| \xi \|_M \le 1\}$ is compact in V. Note that $O \in B$ and any point of B can be joined with O by a line segment in B. Any element $\xi \in V$ can be uniquely written as

$$\xi = \sum_{j=0}^{N} a^j s_j$$

with $(a^j) \in \mathbf{C}^{N+1}$. Set

$$S = \left\{ \xi = \sum_{j=0}^{N} a^j s_j \in V; \ \sum_{j=0}^{N} |a^j|^2 = 1 \right\}.$$

Then S is a compact subset of V. Set

$$C_1 = \min\{\|\xi\|_M;\ \xi \in S\} > 0.$$

Take any $\xi = \sum_{j=0}^{N} a^j s_j \in B - \{O\}$. Then $(\sum |a^j|^2)^{-1/2}\xi \in S$. Therefore we have

$$\begin{aligned} C_1 &\le \|(\sum |a^j|^2)^{-1/2}\xi\|_M = (\sum |a^j|^2)^{-\frac{1}{k}} \|\xi\|_M \\ &\le \left(\sum |a^j|^2\right)^{-\frac{1}{k}}. \end{aligned}$$

Setting $C_2 = C_1^{-k}$, we have

$$B \subset \left\{ \sum_{j=0}^{N} a^j s_j \in V;\ \sum |a^j|^2 \le C_2 \right\}. \tag{2.6.13}$$

Hence B is compact. By (2.6.13) we have

$$\begin{aligned} (\xi \wedge \bar{\xi})^{\frac{1}{k}} &= \left[(\sum a^j s_j) \wedge \overline{(\sum a^j s_j)} \right]^{\frac{1}{k}} \\ &\le \left[(\sum |a^j|^2)(\sum s_j \wedge \bar{s}_j) \right]^{\frac{1}{k}} \le C_2^{\frac{1}{k}}\, \Omega \end{aligned} \tag{2.6.14}$$

for any $\xi = \sum_{j=0}^{N} a^j s_j \in B$. Since $f: N \to M$ is surjective for $f \in \mathrm{Mer}^*(N, M)$, we have

$$\| f^*\xi \|_N \ge \|\xi\|_M \tag{2.6.15}$$

for $\xi \in B$. On the other hand, we deduce from (2.6.14) and (2.6.12) that

$$\| f^*\xi \| \le C_2^{\frac{1}{k}} \int_N f^*\Omega \le C_2^{\frac{1}{k}} \int_N \Psi = C_3$$

for $\xi \in B$. Therefore we have $\| f^*\xi \|_N \le C_3 \|\xi\|_M$. Combining this fact with (2.6.15), we obtain

$$\|\xi\|_M \le \| f^*\xi \|_N \le C_3 \|\xi\|. \tag{2.6.16}$$

Since $\Phi: M \to P(V^*)$ is a holomorphic imbedding, the mapping

$$\tau: f \in \mathrm{Mer}^*(N, M) \to f^* \in \mathrm{Hom}(V, W)$$

is injective. Now we shall show that $\tau(\mathrm{Mer}^*(N, M))$ is a compact subset of $\mathrm{Hom}(V, W)$. Set

$$K = \{\phi \in \mathrm{Hom}(V, W);\ \|\xi\|_M \le \|\phi(\xi)\|_N$$

$$\leq C_2 \|\xi\|_M \text{ for all } \xi \in V\}.$$

Clearly this is a compact subset. We see by (2.6.16) that $\tau(\mathrm{Mer}^*(N, M)) \subset K$. Therefore, if $\{f_\nu\}_{\nu=1}^{\infty}$ is a sequence in $\mathrm{Mer}^*(N, M)$, then $\{\tau(f_\nu)\}_{\nu=1}^{\infty}$ is a sequence in K. Hence there exists a subsequence $\{\tau(f_{\nu_\mu})\}$ and $\phi \in K$ such that $\tau(f_{\nu_\mu}) \to \phi$ as $\mu \to \infty$. Since ϕ is in K, ϕ is injective and its dual $\phi^*: W^* \to V^*$ is surjective. Therefore it induces a surjective meromorphic mapping $[\phi^*]: P(W^*) \to P(V^*)$. Consider the following diagram:

$$(2.6.17)\qquad \begin{array}{ccc} P(W^*) & \overset{[\phi^*]}{\to} & P(V^*) \\ \uparrow \Phi_{\Gamma(N,\, \mathbf{K}(N)^k)} & & \uparrow \Phi \\ N & & M \end{array}$$

Now take $x \in N$ and $\xi \in V$ so that $\phi(\xi(x) \neq 0$. Then $\Phi_{\Gamma(N,\, \mathbf{K}(N)^k)}$ is holomorphic at x. Take $\alpha \in W^* - \{O\}$ so that $\rho(\alpha) = \Phi_{\Gamma(N,\, \mathbf{K}(N)^k)}(x)$, where $\rho: W^* - \{O\} \to P(V^*)$ is the Hopf fibering. Since $\alpha(\phi(\xi)) \neq 0$, we have $\phi^*(\alpha) \neq 0$. Hence the meromorphic mapping $[\phi^*]: P(W^*) \to P(V^*)$ is holomorphic at $\Phi_{\Gamma(N,\, \mathbf{K}(N)^k)}(x)$, and we have the composition of meromorphic mappings

$$[\phi^*] \circ \Phi_{\Gamma(N,\, \mathbf{K}(N)^k)}: N \to P(V^*).$$

Since $[\tau(f^*_{\nu_\mu})] \circ \Phi_{\Gamma(N,\, \mathbf{K}(N)^k)} = \Phi \circ f_{\nu_\mu}$ and $\lim_{\mu\to\infty} \tau(f_{\nu_\mu}) = \phi$, we get

$$[\phi^*] \circ \Phi_{\Gamma(N,\, \mathbf{K}(N)^k)}(N) \subset \Phi(M).$$

Since Φ is a holomorphic imbedding, we have the following meromorphic mapping

$$f = (\Phi^{-1} | \Phi(M)) \circ [\phi^*] \circ \Phi_{\Gamma(N,\, \mathbf{K}(N)^k)}: N \to M.$$

By definition, $f^* = \phi$, so that f is non-degenerate at some point at which f is holomorphic. Thus $f \in \mathrm{Mer}^*(N, M)$ and $\lim_{\mu\to\infty} \tau(f_{\nu_\mu}) = \tau(f)$. Therefore $\tau(\mathrm{Mer}^*(N, M))$ is compact. We denote by $(\tau(\mathrm{Mer}^*(N, M)))^*$ the image of the linear isomorphism $\phi \in \mathrm{Hom}(V, W) \to \phi^* \in \mathrm{Hom}(W^*, V^*)$. Clearly, $(\tau(\mathrm{Mer}^*(N, M)))^*$ is compact in the linear space $H = \mathrm{Hom}(W^*, V^*)$. Let

$$\rho_0: H - \{O\} \to P(H)$$

be the Hopf fibering and set

$$Z = \rho_0((\tau(\mathrm{Mer}^*(N, M)))^*) \subset P(H).$$

Then Z is a compact subset of $P(H)$ and its points are corresponding one-to-one to those of $\mathrm{Mer}^*(N, M)$. Every element $\gamma \in P(H)$ naturally induces a meromorphic mapping $\tilde{\gamma}: P(W^*) \to P(V^*)$. Set $N' = \Phi_{\Gamma(N,\, \mathbf{K}(N)^k)}(N)$ and

$$Y = \{\gamma \in P(H);\ \tilde{\gamma}(N') \subset \Phi(M)\}$$

(cf. (2.6.17)). Then $\dim N' = m$ and Y is an analytic (actually algebraic) subset of $P(H)$. Furthermore, those $\gamma \in Y$ such that $\tilde{\gamma}$ are holomorphic at some points of N' and the restrictions $\tilde{\gamma}|N'$ are non-degenerate at certain points where $\tilde{\gamma}|N'$ are holomorphic, form an open subset Z' of Y. It is clear that $Z = Z'$. Since Z is compact, Z coincides with a union of connected components of Y and hence is an analytic subset of $P(H)$. We claim that $\dim Z = 0$. Then Z is a finite set, and we see the finiteness of $\mathrm{Mer}^*(N, M)$. Assume that $\dim Z > 0$. Let Z_1 be an irreducible component of Z with $\dim Z_1 > 0$. Take a hyperplane E of $P(H)$ so that $E \not\supset Z_1$ and $E \cap Z_1 \neq \varnothing$. Then $\dim Z_1 \cap E < \dim Z_1$. Repeating this process, we have a 1-dimensional irreducible analytic subset X of Y. Then we have a compact Riemann surface X_0 and a surjective holomorphic mapping $\pi: X_0 \to X$. For every element $z \in X_0$, we have a unique meromorphic mapping $f_z \in \mathrm{Mer}^*(N, M)$, since $\pi(z) \in X \subset \rho_0((\tau(\mathrm{Mer}^*(N, M)))^*)$. By the above construction the mapping

$$(z, x) \in X_0 \times N \to f_z(x) \in M$$

is a meromorphic mapping. Therefore we have a non-constant holomorphic mapping

$$\delta: z \in X_0 \to f_z^* \in \mathrm{Hom}(V, W) \cong \mathbf{C}^{(\dim V) \times (\dim W)}.$$

On the other hand, sine X_0 is a compact Riemann surface, δ must be constant. This is a contradiction. *Q.E.D.*

Notes

D. A. Eisenman [25] defined and discussed various intrinsic measures on complex manifolds (cf. also Graham-Wu [33]). In Chapter IX of Kobayashi [54] there is also a description on more fundamental properties of measure hyperbolic manifolds than those given in this book. We advise the readers to read Chapter IX of his book.

The extension Theorem (2.6.9) was first proved by Griffiths [43] for M with very ample $\mathbf{K}(M)$. It is now known that this theorem holds for M of general type (Kobayashi-Ochiai [57]). Here M is said to be of general type if for a large l the dimension of the image of the meromorphic mapping $\Phi_{\Gamma(M, \mathbf{K}(M)^l)}$: $M \to P((\Gamma(M, \mathbf{K}(M)^l))^*$ is equal to $\dim M$. It remains as an open problem whether the similar extension theorem holds for measure hyperbolic M. By Kobayashi-Ochiai [57] we know that if M is of general type, then M is measure hyperbolic. It is also an open problem to show the converse. For an algebraic surface M, the converse was recently proved by Mori-Mukai [71] (cf. also [41]).

In the above we dealt with the case where a meromorphic mapping $f: \Delta^*(1) \times \Delta(1)^{m-1} \to M$ ($m = \dim m$) has rank m. Let $g: \Delta^*(1) \times \Delta(1)^{k-1} \to M$

$(k \leq m)$ be a meromorphic mapping with rank k. If $\overset{k}{\wedge}\mathbf{T}(M)$ is negative (in the sense of Grauert), then f extends to a meromorphic mapping from $\Delta(1) \times \Delta(1)^{k-1}$ into M (Carlson [15], and cf. also Noguchi [78]).

The finiteness theorem, Theorem (2.6.9), also holds for M of general type (cf. Kobayashi-Ochiai [57]). This result is motivated by the following conjecture due to Lang [61]:

Assume that M and N are projective algebraic complex manifolds and that M is hyperbolic. Then, are there only finitely many surjective meromorphic mappings from N onto M?

It has been conjectured that hyperbolic manifolds are of general type, but not yet proved. (By the above result of Mori-Mukai [71] this is true for $\dim M \leq 2$.) The above conjecture of Lang was proved for Kähler M with $c_1(M) \leq 0$ in Noguchi [83], and lately for general compact hyperbolic Moisezon M by Horst [49]. It arose in connection with the problem of the higher dimensional analogue of Mordell's conjecture (now Faltings' Theorem) over function fields. This subject was dealt with by Noguchi [81, 83] and Riebesehl [96]. For the references of this subject, cf. those of the above papers.

Now we consider meromorphic mappings $f: N \to M$ with $\operatorname{rank} f \leq \dim M$. Set

$$F_k(N, M) = \{ f : N \to M, \text{ meromorphic; rank } f \geq k \}$$

for $k \leq \dim M$. Noguchi [78] proved that if $\overset{k}{\wedge}\mathbf{T}(M)$ is negative, then $F_k(N, M)$ is compact, but moreover, Noguchi-Sunada [87] proved that $F_k(N, M)$ is finite in this case. Urata [118] and Kalka-Shiffman-Wong [50] obtained similar results in the case of $k = 1$ and related interesting ones.

Graham-Wu [34] proved that if Ψ_M of a complex manifold M coincides with the density of the Carathéodory volume form of M at one point, then M is biholomorphic to the ball in $\mathbf{C}^m$ (cf. also, [33]).

CHAPTER III

Currents and Plurisubharmonic Functions

3.1 Currents

In this section we describe the notion of currents due to de Rham and their fundamentals.

(a) Notation. For the simplicity we introduce the following notation. Set $\mathbf{Z}^+ = \{m \in \mathbf{Z};\ m \geq 0\}$ and

$$|\alpha| = \alpha_1 + \cdots + \alpha_m$$

for $\alpha = (\alpha_1, \ldots, \alpha_m) \in (\mathbf{Z}^+)^m$. Let $(x^1, \ldots, x^m)$ be the standard coordinate system of $\mathbf{R}^m$ and set

$$D^\alpha = \left(\frac{\partial}{\partial x^1}\right)^{\alpha_1} \cdots \left(\frac{\partial}{\partial x^m}\right)^{\alpha_m}.$$

For a natural number k with $1 \leq k \leq m$ we set

$$\{m\,;k\} = \{(j_1, \ldots, j_k) \in \mathbf{N}^k;\ 1 \leq j_1 < \cdots < j_k \leq m\}.$$

Moreover, for $J = (j_1, \ldots, j_k) \in \{m\,;k\}$ we set

$$dx^J = dx^{j_1} \wedge \cdots \wedge dx^{j_k}$$

and define $\hat{J} = (i_1, \ldots, i_{m-k}) \in \{m\,; m-k\}$ by

$$\{j_1, \ldots, j_k, i_1, \ldots, i_{m-k}\} = \{1, 2, \ldots, m\}.$$

For a convenience we put $\{m\,;0\} = \{0\}$ and $dx^0 = 1$.

(b) Function spaces. Let U be an open subset of $\mathbf{R}^m$ and $C(U)$ (resp. $E(U)$) denote the complex vector space of all complex-valued continuous (resp. C^∞-)

functions on U. Then $E(U)$ is a vector subspace of $C(U)$. We define the support $\operatorname{supp}\phi$ of $\phi \in C(U)$ by

$$\operatorname{supp}\phi = \overline{\{x \in U;\ \phi(x) \neq 0\}}.$$

We set

$$K(U) = \{\phi \in C(U);\ \operatorname{supp}\phi \text{ is compact}\},$$

$$D(U) = K(U) \cap E(U).$$

Then $D(U)$ is a vector subspace of $E(U)$ and $K(U)$. Moreover, for a subset $A \subset U$ we set

$$K_A(U) = \{\phi \in C(U);\ \operatorname{supp}\phi \subset A\},$$

$$D_A(U) = K_A(U) \cap D(U).$$

A Lebesgue measurable function $\phi\colon U \to \mathbf{C}$ is said to be **locally integrable** if for any compact subset $A \subset U$

$$\int_{x \in A} |\phi(x)|\, dx < \infty,$$

where $dx = dx^1 \cdots dx^m$ denotes the Lebesgue measure on $\mathbf{R}^m$. We denote by $L_{\text{loc}}(U)$ the vector space of all locally integrable functions on U. Then the function spaces $C(U)$, $K(U)$, $D(U)$, etc. are subspaces of $L_{\text{loc}}(U)$.

(c) Convolution. For $x = (x^1, \dots, x^m) \in \mathbf{R}^m$ we define, as usual, the norm $\|x\|$ by

$$\|x\| = \sqrt{\sum_{j=1}^{m} |x^j|^2}.$$

We take and fix a function $\chi \in D(\mathbf{R}^m)$ satisfying the following conditions:

(i) $0 \leq \chi \leq 1$, $\operatorname{supp}\chi \subset \{x \in \mathbf{R}^m;\ \|x\| < 1\}$.

(ii) $\chi(x) = \chi(x')$ for $x, x' \in \mathbf{R}^m$ with $\|x\| = \|x'\|$.

(iii) $\displaystyle\int_{\mathbf{R}^m} \chi dx = 1$.

For $\varepsilon > 0$ we set $\chi_\varepsilon(x) = \chi(x/\varepsilon)/\varepsilon^m$ and

$$U_\varepsilon = \{x \in U;\ \inf_{y \in \partial U} \|x - y\| > \varepsilon\}.$$

For $\phi \in L_{\text{loc}}(U)$ we define the **convolution** $\phi * \chi_\varepsilon$ by

$$\phi * \chi_\varepsilon(x) = \int_U \phi(y)\chi_\varepsilon(y - x)dy \tag{3.1.1}$$

$$= \int_{\mathbf{R}^m} \phi(x+y)\chi_\varepsilon(y)dy \quad (x \in U_\varepsilon).$$

Then $\phi * \chi_\varepsilon \in E(U_\varepsilon)$ and $\phi * \chi_\varepsilon$ converges to ϕ as $\varepsilon \to 0$ in the sense of $L_{\text{loc}}(U)$; i.e.,

(3.1.2)
$$\lim_{\varepsilon \to 0} \int_A |\phi * \chi_\varepsilon(x) - \phi(x)| dx = 0$$

for an arbitrary compact subset $A \subset U$. Moreover, if $\phi \in C(U)$, then $\phi * \chi_\varepsilon$ converges uniformly on compact subsets to ϕ as $\varepsilon \to 0$. For $\phi \in E(U)$ and $\alpha \in (\mathbf{Z}^+)^m$ we have

(3.1.3)
$$D^\alpha(\phi * \chi_\varepsilon) = (D^\alpha \phi) * \chi_\varepsilon \quad \text{on } U_\varepsilon.$$

(d) Distribution. We define the seminorms

$$\| \phi \|_0 = \sup\{ |\phi(x)|; x \in U \}$$

for $\phi \in K(U)$ and

$$\| \phi \|_l = \sup\{ |D^\alpha \phi(x)|; x \in U, \ \alpha \in (\mathbf{Z}^+)^m, \ |\alpha| \le l \},$$
$$l = 0, 1, 2, \ldots$$

for $\phi \in D(U)$. We say that a sequence $\{\phi_j\}_{j=1}^\infty \subset K(U)$ converges to $\phi \in K(U)$ if the following two conditions hold:

(i) There is a compact subset $A \subset U$ such that $\operatorname{supp} \phi_j \subset A$ for all j.

(ii) $\lim_{j \to \infty} \| \phi_j - \phi \|_0 = 0$.

Similarly, a sequence $\{\phi_j\}_{j=1}^\infty \subset D(U)$ is said to converge to $\phi \in D(U)$ if the following two conditions hold:

(i) There is a compact subset $A \subset U$ such that $\operatorname{supp} \phi_j \subset A$ for all j.

(ii) For every $l \in \mathbf{Z}^+$, $\lim_{j \to \infty} \| \phi_j - \phi \|_l = 0$.

We henceforth endow $K(U)$ and $D(U)$ with the above topologies and consider them as topological vector spaces, if nothing is specially mentioned.

(3.1.4) *Definition.* A continuous linear functional $T: D(U) \to \mathbf{C}$ is called a **distribution** on U; i.e., for an arbitrary compact subset $A \subset U$ there are a constant $C_A \ge 0$ and $l_A \in \mathbf{Z}^+$ such that

$$|T(\phi)| \le C_A \| \phi \|_{l_A}$$

for all $\phi \in D_A(U)$.

If there is $l \in \mathbf{Z}^+$ such that $l_A \le l$ for any above A, then T is said to have the **order less than or equal to** l. If T has the order less than or equal to l and not

$l-1$, then l is called the **order** of T.

We write $\mathcal{D}(U)'$ for the complex vector space of all distributions on U. For $T \in \mathcal{D}(U)'$ we define a partial derivative $\partial T / \partial x^j \in \mathcal{D}(U)'$ by

$$\frac{\partial T}{\partial x^j}(\phi) = -T\left(\frac{\partial \phi}{\partial x^j}\right) \quad (\phi \in \mathcal{D}(U)).$$

For D^{α} ($\alpha \in (\mathbf{Z}^+)^m$) we set

$$D^{\alpha}T(\phi) = (-1)^{|\alpha|}T(D^{\alpha}\phi) \quad (\phi \in \mathcal{D}(U)).$$

For $f \in \mathcal{E}(U)$ we define $fT \in \mathcal{D}(U)'$ by

$$fT(\phi) = T(f\phi) \quad (\phi \in \mathcal{D}(U)).$$

Next we consider a continuous linear functional $T: \mathcal{K}(U) \to \mathbf{C}$. That is, T satisfies that for an arbitrary compact subset $A \subset U$ there is a constant $C_A \geq 0$ with

$$|T(\phi)| \leq C_A \|\phi\|_0 \quad (\phi \in \mathcal{K}_A(U)).$$

We denote by $\mathcal{K}(U)'$ the complex vector space of all those T. Since $\mathcal{K}(U) \supset \mathcal{D}(U)$ as vector spaces and the inclusion $\mathcal{D}(U) \to \mathcal{K}(U)$ is continuous, the restriction $T|\mathcal{D}(U): \mathcal{D}(U) \to \mathbf{C}$ of $T \in \mathcal{K}(U)'$ is an element of $\mathcal{D}(U)'$ of order 0. By this correspondence we have the natural identification

$$\mathcal{K}(U)' = \{T \in \mathcal{D}(U)';\, T \text{ is of order } 0\} \subset \mathcal{D}(U)'.$$

For $f \in \mathcal{L}_{\mathrm{loc}}(U)$ we define the distribution $[f] \in \mathcal{K}(U)'$ by

$$[f](\phi) = \int f\phi\, dx \quad (\phi \in \mathcal{K}(U)).$$

If $[f] = [g]$ for $f, g \in \mathcal{L}_{\mathrm{loc}}(U)$, then $f(x) = g(x)$ for almost all x with respect to the Lebesgue measure. Therefore the mapping $f \in \mathcal{L}_{\mathrm{loc}}(U) \to [f] \in \mathcal{K}(U)'$ is injective. Hence we have the following identifications:

$$\text{(3.1.5)} \qquad \mathcal{D}(U) \subset \left\{\begin{matrix} \mathcal{K}(U) \\ \\ \mathcal{E}(U) \end{matrix}\right\} \subset \mathcal{C}(U) \subset \mathcal{L}_{\mathrm{loc}}(U) \subset \mathcal{K}(U)' \subset \mathcal{D}(U)'.$$

For $g \in \mathcal{E}(U)$ we have that $g[f] = [gf]$, and if $f \in \mathcal{E}(U)$, then by repeating the partial integrations we see that

$$D^{\alpha}[f] = [D^{\alpha}f].$$

Now we consider the topology of $\mathcal{D}(U)'$. We say that a sequence

$\{T_j\}_{j=1}^{\infty} \subset D(U)'$ converges to $T \in D(U)'$ if $\lim_{j\to\infty} T_j(\phi) = T(\phi)$ for every $\phi \in D(U)$. In this case we write $\lim_{j\to\infty} T_j = T$ and clearly have

$$\lim_{j\to\infty} D^{\alpha} T_j = D^{\alpha} T \tag{3.1.6}$$

for $\alpha \in (\mathbf{Z}^+)^m$. The following theorem is a direct consequence of the so-called Banach-Steinhaus theorem and a fundamental fact for the convergence of distributions, while we will not use it in this book.

(3.1.7) Theorem. *Let $\{T_j\}_{j=1}^{\infty} \subset D(U)'$ be a sequence of distributions. If $\lim_{j\to\infty} T_j(\phi)$ have limits for all $\phi \in D(U)$, then there exists the limit $\lim_{j\to\infty} T_j \in D(U)'$ of $\{T_j\}_{j=1}^{\infty}$.*

Let $V \subset U$ be an open subset. Then we consider an element $\phi \in D(V)$ as an element of $D(U)$ by extending ϕ to be 0 on $U - V$. By this identification we get $D(V) \subset D(U)$. For $T \in D(U)'$ we define the restriction $T|V \in D(V)'$ by

$$T|V(\phi) = T(\phi) \quad (\phi \in D(V) \subset D(U)).$$

(3.1.8) Lemma. *It is a necessary and sufficient condition for a sequence $\{T_j\} \subset D(U)'$ to converge to $T \in D(U)'$ that for every point $x \in U$ there is a neighborhood $V \subset U$ of x such that $\{T_j|V\}_{j=1}^{\infty} \in D(V)'$ converges to $T|V \in D(V)'$.*

Proof. The necessity is clear. We show the sufficiency. By the assumption there is a locally finite open covering $\{V_\nu\}_{\nu=1}^{\infty}$ of U such that $\{T_j|V_\nu\}_{j=1}^{\infty}$ converge to $T|V_\nu$ for all ν. Take a partition $\{c_\nu\}_{\nu=1}^{\infty}$ of unity subordinated to $\{V_\nu\}_{\nu=1}^{\infty}$. Then $\phi = \sum_\nu c_\nu \phi$ for $\phi \in D(U)$, where the summation is actually a finite one and $c_\nu \phi \in D(V_\alpha)$. By definition we have

$$\begin{aligned} T(\phi) &= \sum_\nu T(c_\nu \phi) = \sum_\nu T|V_\nu(c_\nu \phi) \\ &= \sum_\nu \lim_{j\to\infty} T_j|V_\nu(c_\nu\phi) = \lim_{j\to\infty} \sum_\nu T_j(c_\nu \phi) \\ &= \lim_{j\to\infty} T_j(\phi). \end{aligned}$$

Hence it follows that $T = \lim_{j\to\infty} T_j$. *Q.E.D.*

By the similar method to the above proof we have the following:

(3.1.9) Proposition. *Let T, $S \in D(U)'$. Then $T = S$ if and only if every point of U has a neighborhood $V \subset U$ such that $T|V = S|V \in D(U)'$.*

(3.1.10) Proposition. *A distribution $T \in D(U)'$ has order $\leq l$ if and only if every point of U has a neighborhood $V \subset U$ such that $T|V$ has order $\leq l$.*

We define the **support** of $T \in \mathcal{D}(U)'$ as follows. Let W be a subset of U consisting those $x \in U$ such that $T|V = 0$ for a neighborhood $V \subset U$ of x. We set

$$\operatorname{supp} T = U - W.$$

Then, of course, $\operatorname{supp} T$ is a closed subset of U.

Let $T \in \mathcal{D}(U)'$ and define the **smoothing** $T_\varepsilon \in \mathcal{D}(U_\varepsilon)'$ ($\varepsilon > 0$) by

$$(3.1.11) \qquad T_\varepsilon(\phi) = T(\phi * \chi_\varepsilon) = [T_y(\chi_\varepsilon(x-y))]_x(\phi(x))$$

for $\phi \in \mathcal{D}(U_\varepsilon)$. Here $T_y(\chi_\varepsilon(x-y))$ stands for the value of T at the function $\chi_\varepsilon(x-y)$ in variable y with fixed x. Then it is a C^∞-function in x and defines the distribution $[T_y(\chi_\varepsilon(x-y))]$, which takes the value $[T_y(\chi_\varepsilon(x-y))](\phi(x))$ at the function $\phi \in \mathcal{D}(U_\varepsilon)$ in variable x. We also write $T * \chi_\varepsilon(x)$ for $T_y(\chi_\varepsilon(x-y))$. By making use of (3.1.5), (3.1.11) is the same as

$$T_\varepsilon = T * \chi_\varepsilon \in \mathcal{E}(U_\varepsilon).$$

For an arbitrary $\phi \in \mathcal{D}(U)$ we have $\lim_{\varepsilon \to 0} \phi * \chi_\varepsilon = \phi$, so that

$$\lim_{\varepsilon \to 0} T_\varepsilon(\phi) = T(\phi).$$

Therefore, $\lim_{\varepsilon \to 0} T_\varepsilon | V = T|V$ for any relatively compact subset V of U. For an arbitrary D^α ($\alpha \in (\mathbf{Z}^+)^m$) we have by (3.1.3)

$$(3.1.12) \qquad D^\alpha T_\varepsilon = (D^\alpha T)_\varepsilon \quad (\varepsilon > 0)$$

and for $f \in L_{\mathrm{loc}}(U)$

$$[f]_\varepsilon = [f * \chi_\varepsilon] \quad (\varepsilon > 0).$$

(e) $\mathcal{K}(U)'$ and Radon measures. Let $T \in \mathcal{K}(U)'$. Then, by definition, for each compact subset $A \subset U$ there is a constant $C_A \geq 0$ such that

$$|T(\phi)| \leq C_A \|\phi\|_0 \quad (\phi \in \mathcal{K}_A(U)).$$

We set

$$\|T\|(\phi) = \sup\{|T(\eta)|;\ \eta \in \mathcal{D}(U),\ |\eta| \leq \phi\}$$

for $\phi \in \mathcal{K}(U)$ with $\phi \geq 0$. For a real-valued function $\phi \in \mathcal{K}(U)$, we set

$$\|T\|(\phi) = \|T\|(\phi^+) - \|T\|(\phi^-),$$

where $\phi^+(x) = \max\{\phi(x), 0\}$ and $\phi^-(x) = \max\{-\phi(x), 0\}$, and extend it for complex-valued $\phi \in \mathcal{K}(U)$ by linearity. Put $A = \operatorname{supp} \phi$. If $\eta \in \mathcal{D}(U)$ and $|\eta| \leq \phi$, then $\eta \in \mathcal{D}_A(U)$, so that

$$|T(\eta)| \le C_A \|\eta\|_0 \le C_A \|\phi\|_0.$$

Hence we have

$$\|T\|(\phi) \le C_A \|\phi\|_0.$$

This implies that $\|T\| \in \mathcal{K}(U)'$.

Here we call, in general, $\mu = \lambda + i\nu$ a (complex-valued) Radon measure on U if λ and ν are real-valued Radon measures on U. For a Radon measure μ on U we define a positive measure $|\mu|$ by

$$|\mu|(B) = \sup\left\{ \sum_{j=1}^{\infty} |\mu(B_j)|;\ B_j \text{ are Borel subsets of } B \right.$$

$$\left. \text{such that } B_j \cap B_k = \varnothing\ (j \ne k) \text{ and } B = \bigcup_{j=1}^{\infty} B_j \right\},$$

where B is a Borel subset of U. We also get a distribution $T_\mu \in \mathcal{K}(U)'$ determined by

$$T_\mu(\phi) = \int_U \phi\, d\mu \quad (\phi \in \mathcal{K}(U)).$$

We see the converse by the following theorem due to Riesz.

(3.1.13) Theorem. *For an arbitrary $T \in \mathcal{K}(U)'$ there exists a unique Radon measure μ on U satisfying the conditions:*

(i) $T(\phi) = \int_U \phi\, d\mu \quad (\phi \in \mathcal{D}(U))$.

(ii) $\|T\|(\phi) = \int_U \phi\, d|\mu| \quad (\phi \in \mathcal{D}(U))$.

Thus we may consider T and $\|T\|$ to be Radon measures on U and call $\|T\|$ the total variation measure of T.

If a distribution T on U satisfies that $T(\phi) \ge 0$ for $\phi \in \mathcal{D}(U)$ with $\phi \ge 0$, T is called a **positive distribution** and written as $T \ge 0$.

(3.1.14) Lemma. *If a mapping $T : \mathcal{D}(U) \to \mathbf{C}$ satisfies*

$$T(\phi) \ge 0 \ \textit{for}\ \phi \in \mathcal{D}(U),\ \phi \ge 0,$$

then $T \in \mathcal{K}(U)'$ and is a positive Radon measure on U.

Proof. Let $A \subset U$ be an arbitrary compact subset and take a real $\phi_A \in \mathcal{D}(U)$ so that $\phi_A | A \equiv 1$. Then for any real $\phi \in \mathcal{D}_A(U)$, $\|\phi\|_0 \phi_A \pm \phi \ge 0$, so that

$$0 \le T(\|\phi\|_0 \phi_A \pm \phi) = \|\phi\|_0 T(\phi).$$

It follows that $|T(\phi)| \le T(\phi_A)\|\phi\|_0$, and hence $T \in K(U)'$. We easily see that T gives rise to a positive Radon measure. In fact, for any $\phi \in K(U)$ with $\phi \ge 0$, $\phi * \chi_\varepsilon \in D(U)$ and $\phi * \chi_\varepsilon \ge 0$. Since $\lim_{\varepsilon \to 0} \phi * \chi_\varepsilon = \phi$ in the sense of the topology of $K(U)$, we deduce that $T(\phi) = \lim_{\varepsilon \to 0} T(\phi * \chi_\varepsilon) \ge 0$. *Q.E.D.*

If a function $f \in L_{\mathrm{loc}}(U)$ is non-negative at almost all points of U with respect to the Lebesgue measure, then $[f] \ge 0$, and the converse holds, too. The following lemma will be clear by the definition and the proof of Lemma (3.1.8).

(3.1.15) Lemma. *The following conditions are mutually equivalent for $T \in D(U)'$.*

(i) $T \ge 0$.

(ii) *For any $\varepsilon > 0$, $T_\varepsilon(x) \ge 0$ $(x \in U_\varepsilon)$.*

(iii) *Every point $x \in U$ has a neighborhood $V \subset U$ such that $T|V \ge 0$.*

Now, let $T \in K(U)'$. Then T is identified with a Radon measure μ on U by Theorem (3.1.13). Therefore we can set

$$T(\phi) = \int_U \phi \, d\mu \tag{3.1.16}$$

for Borel measurable functions ϕ on U, provided that the above right hand side makes sense. Moreover, by using (3.1.16), we define the restriction $T|A$ of T over a Borel subset $A \subset U$ by

$$(T|A)(\phi) = \int_A \phi \, d\mu$$

for Borel measurable functions ϕ on A.

(f) Space of differentials. Let U be an open subset of $\mathbf{R}^m$ as before and $k \in \mathbf{Z}^+$ $(k \le m)$. We denote by $C^k(U)$ the complex vector space of all differential k-forms with continuous coefficients

$$\sum_{J \in \{m;\, k\}} \phi_J dx^J \quad (\phi_J \in C(U)).$$

Note that $C^0(U) = C(U)$. In what follows, we write $\sum' \phi_J dx^J$ for $\sum_{J \in \{m;\, k\}} \phi_J dx^J$. We define the support $\operatorname{supp} \phi$ of $\phi \in C^k(U)$ by

$$\operatorname{supp} \phi = \overline{\{x \in U;\, \phi(x) \ne 0\}}.$$

We denote by $K^k(U)$ the vector subspace of all $\phi \in C^k(U)$ with compact $\operatorname{supp} \phi$, and by $K^k_A(U)$ that of those $\phi \in K^k(U)$ with $\operatorname{supp} \phi \subset A$ for a subset A of U. Let $E^k(U)$ denote the complex vector space of all differential k-forms with C^∞-coefficients. Then $E^k(U)$ is a vector subspace of $C^k(U)$. Set

$$D^k(U) = K^k(U) \cap E^k(U),$$
$$D^k_A(U) = K^k_A(U) \cap E^k(U)$$

for a subset $A \subset U$. We define the seminorm $\| \phi \|_0$ of $\phi = \sum' \phi_J dx^J \in K^k(U)$ as follows:

$$\| \phi \|_0 = \sup\{ |\phi_J(x)| ;\, x \in U,\ J \in \{m;\, k\}\}.$$

For $\phi = \sum' \phi_J dx^J \in D^k(U)$ we set

$$\| \phi \|_l = \sup\{ |D^\alpha \phi_J(x)| ;\, x \in U,\ J \in \{m;\, k\},\ |\alpha| \le l\}$$
$$(l = 0, 1, 2, \ldots).$$

Using these seminorms, we define topologies on $K^k(U)$ and $D^k(U)$ as we have done for $K(U)$ and $C(U)$ in (d), so that they form topological complex vector spaces. On the other hand, the vector spaces $K^k(U)$ and $D^k(U)$ are written as direct sums

$$K^k(U) = \bigoplus_{J \in \{m;\, k\}} K(U) dx^J,$$
$$D^k(U) = \bigoplus_{J \in \{m;\, k\}} D(U) dx^J.$$

By making use of the topologies of $K(U)$ and $D(U)$ defined in (d) we naturally get topologies on $K^k(U)$ and $D^k(U)$ of the above finite direct sums. It is clear that these topologies coincide with each other. From now on, we consider $K^k(U)$ and $D^k(U)$ to be endowed with the above topologies.

(g) Currents. We call a continuous linear functional $T: D^k(U) \to \mathbf{C}$ a k-**dimensional current** on U; that is, T satisfies the following (cf. (3.1.4)):

(3.1.17) For an arbitrary compact subset $A \subset U$, there is a constant $C_A \ge 0$ and $l_A \in \mathbf{Z}^+$ such that

$$|T(\phi)| \le C_A \| \phi \|_{l_A} \quad (\phi \in D^k_A(U)).$$

If $l_A \le l$ with some $l \in \mathbf{Z}$, T is said to be of **order** $\le l$. If T is of order $\le l$ and not of order $\le l-1$, T is said to be of order l. For a convenience, we sometimes write

$$T(\phi) = \langle T, \phi \rangle$$

in the form of duality. Set

$$D'_k(U) = \{T : D^k(U) \to \mathbf{C};\ \text{current}\},$$

which is, of course, a complex vector space. Set similarly

$$K'_k(U) = \{T : K^k(U) \to \mathbf{C};\ \text{continuous linear functional}\}.$$

Then the restriction $T|D^k(U)$ of $T\in K_k'(U)$ is a current of order 0. By this restriction, the space $K_k'(U)$ is linearly isomorphic to the vector space of all currents of order 0 on U. We consider $K_k'(U)$ to be a vector subspace of $D_k'(U)$. Note that $K_0'(U)=K(U)'$, $D_0'=D(U)'$ $(k=0)$.

The mapping

$$\iota\colon \phi\in C^0(U)\to\phi\, dx^1\wedge\cdots\wedge dx^m\in C^m(U)$$

is a linear isomorphism and $\iota(D^0(U))=D^m(U)$. We denote the restriction $\iota|D^0(U)$ by the same ι. Then the mapping

$$\iota\colon D^0(U)\to D^m(U)$$

and its dual

$$\iota^*\colon D_m'(U)\to D_0'(U)$$

are linear isomorphisms, by which we identify $D_0'(U)$ with $D_m'(U)$ unless confusion occurs. Since $\iota^*(K_m'(U))=K_0'(U)$, $K_0'(U)$ is identified with $K_m'(U)$.

We denote by $L^k_{\mathrm{loc}}(U)$ the complex vector space of all differential k-forms on U with coefficients in $L_{\mathrm{loc}}(U)$. We get the following inclusions as complex vector spaces:

$$E^k(U)\subset C^k(U)\subset L^k_{\mathrm{loc}}(U).$$

Every $\omega\in L^{m-k}_{\mathrm{loc}}(U)$ defines a current

$$[\omega]\colon \phi\in D^k(U)\to\int_U\omega\wedge\phi\in\mathbf{C}.$$

Then $[\omega]$ is of order 0 and the linear mapping $\omega\in L^{m-k}_{\mathrm{loc}}(U)\to[\omega]\in K_k'(U)$ is injective. Regarding $L^{m-k}_{\mathrm{loc}}(U)$ as a complex vector subspace of $K_k'(U)$, we have the following inclusions:

$$E^{m-k}(U)\subset C^{m-k}(U)\subset L^{m-k}_{\mathrm{loc}}(U)\subset K_k'(U)\subset D_k'(U).$$

Let $V\subset U$ be an open subset. By extending $\phi\in D^k(V)$ to be 0 on $U-V$, we have that $D^k(V)\subset D^k(U)$. Then we define the restriction $T|V$ of $D_k'(U)$ over V by

$$T|V(\phi)=T(\phi)\quad(\phi\in D^k(V)\subset D^k(U)).$$

Let $S\in D_k'(U)$. If $T=S$, then $T|V=S|V$ for any open subset $V\subset U$. The converse holds, too:

(3.1.18) Proposition. *Let T, $S\in D_k'(U)$. Then $T=S$ if and only if for any point $x\in U$ there is a neighborhood V of x such that $T|V=S|V$.*

The proof is similar to that of Proposition (3.1.9). Similarly to Proposition (3.1.10), the following holds.

(3.1.19) Proposition. *A current $T \in D'_k(U)$ is of order $\leq l$ if and only if for any point $x \in U$ there is a neighborhood V of x such that $T|V$ is of order $\leq l$.*

We define the **support** supp T of $T \in D'_k$ as follows:

$$W = \{x \in U\,;\, x \text{ has a neighborhood } V \text{ such that } T|V = 0\},$$

$$\operatorname{supp} T = U - W.$$

(h) Operations for currents. Let $k,\ p \in \mathbf{Z}^+$ such that $k + p = m$. Because of a reason explained later we set

$$D'_k(U) = D'^p(U).$$

For $T \in D'^p(U)$ and $\alpha \in E^q(U)$ we define $T \wedge \alpha \in D'^{p+q}(U)$ by

$$\langle T \wedge \alpha,\ \phi\rangle = \langle T,\ \alpha \wedge \phi\rangle \quad (\phi \in D^{m-p-q}(U)).$$

We set

$$\alpha \wedge T = (-1)^{pq} T \wedge \alpha.$$

For instance, if $\omega \in L^p_{\mathrm{loc}}(U)$, then $[\omega] \in D'^p(U)$ and

$$[\omega] \wedge \alpha = [\omega \wedge \alpha].$$

In the case of $\alpha \in E^0(U)$ we write $\alpha \wedge T = \alpha T$.

For $T \in D'^p(U)$ we set

$$dT : \phi \in D^{m-p-1}(U) \to (-1)^{p-1} T(d\phi) \in \mathbf{C}.$$

Then $dT \in D'^{p+1}(U)$ and it is called the exterior differential of T. If $dT = 0$, then T is called a **closed current.** The following lemma states that the closedness of currents is a local property.

(3.1.20) Lemma. *A current $T \in D'^p(U)$ is closed if and only if for any point $x \in U$ there is a neighborhood V of x such that $T|V$ is closed.*

Proof. The "only if" part is clear. Assume that every point of U has such a neighborhood. Then there is a locally finite open covering $\{V_\nu\}$ of U such that $d(T|V_\nu) = 0$, and the partition of unity $\{c_\nu\}$ subordinated to $\{V_\nu\}$. For an arbitrary $\phi \in D^{m-p-1}(U)$ we have

$$dT(\phi) = (-1)^{p-1} T(d\phi) = (-1)^{p-1} T\left(d \sum_\nu c_\nu \phi\right)$$

$$= (-1)^{p-1} T\left[\sum_\nu d(c_\nu \phi)\right] = \sum_\nu (-1)^{p-1} T(d(c_\nu \phi))$$
$$= \sum_\nu (-1)^{p-1} d(T|V_\nu)(c_\nu \phi) = 0.$$

Therefore T is closed. *Q.E.D.*

We see by Stokes' theorem that

$$d[\omega] = [d\omega]$$

for $\omega \in E^p(U)$. Let $T \in D'^p(U)$ and $\alpha \in E^q(U)$. Then

$$d(T \wedge \alpha) = dT \wedge \alpha + (-1)^p T \wedge d\alpha.$$

By definition we get

$$\text{supp } dT \subset \text{supp } T$$

for $T \in D'^p(U)$.

Let $T \in D'^p(U)$. For $J \in \{m; p\}$ we define $T_J \in D'_m(U)$ by

$$T_J: \phi\, dx^1 \wedge \cdots \wedge dx^m \in D^m(U) \to \delta(J, \hat{J}) T(\phi\, dx^{\hat{J}}) \in \mathbf{C},$$

where $\delta(J, \hat{J})$ stands for the signature of the permutation $\{J, \hat{J}\} \to \{1, 2, \ldots, m\}$. Then T is written as

$$T = \sum_{J \in \{m; p\}} T_J dx^J. \tag{3.1.21}$$

By the identifications, $D'^0(U) = D'_m(U) = D(U)'$, we have $T_J \in D(U)'$. Thus every current $T \in D'^p(U)$ is considered a differential p-form with distribution coefficients. This is the reason why we write $D'_k(U) = D'^p(U)$ $(p + k = m)$. For $\phi = \sum_{I \in \{m; k\}} \phi_I dx^I \in D^k(U)$ $(\phi_I \in D(U))$ we have

$$T(\phi) = \sum_{J \in \{m; p\}} T_J(dx^J \wedge \phi) = \sum_{J \in \{m; p\}} T_J\left[dx^J \wedge \phi_{\hat{J}} dx^{\hat{J}}\right]$$
$$= \sum_{J \in \{m; p\}} T_J\left[\phi_{\hat{J}} \delta(J, \hat{J}) dx^1 \wedge \cdots \wedge dx^m\right]$$
$$= \sum_{J \in \{m; p\}} \delta(J, \hat{J}) T_J(\phi_{\hat{J}}).$$

By a simple computation we see that

$$dT = \sum_{J \in \{m; p\}} \sum_{j=1}^{m} \frac{\partial T_J}{\partial x^j} dx^j \wedge dx^J. \tag{3.1.22}$$

For $T \in K_k'(U)$ we have as in (3.1.21)

$$T = \sum_{J \in \{m;p\}} T_J dx^J, \quad T_J \in K(U)' = K_m'(U). \tag{3.1.23}$$

Therefore we also write $K_k'(U) = K'^p(U)$ $(k+p=m)$. On the other hand there is a reason for the notation $K_k'(U)$ (or $D_k'(U)$) by the following example.

(3.1.24) *Example*. Let X be a closed oriented k-dimensional submanifold of U. We set

$$T_X\colon \phi \in K^k(U) \to \int_X \phi \in \mathbf{C}.$$

In the case of $k=0$, we define

$$T_X(\phi) = \sum_{x \in X} \phi(x) \quad (\phi \in K^0(U) = K(U)).$$

Then T_X is a current of order 0, and closed by Stokes' theorem. If there is no confusion, we simply denote T_X by X.

Let $T = \sum' T_J dx^J \in K'^p(U)$. Then the coefficients T_J are considered to be Radon measures on U by Theorem (3.1.13). We define the total variation $\|T\|$ of T by $\|T\| = \sum' \|T_J\|$. Using the identification $K(U)' = K_m'(U)$, we know that all Borel measurable functions are T_J-measurable. If a Borel measurable function ϕ is T_J-integrable, we can set by (3.1.16)

$$T_J(\phi) = T_J(\phi\, dx^1 \wedge \cdots \wedge dx^m) \in \mathbf{C}.$$

We denote by $B^k(U)$ the complex vector space of all differential k-forms of which coefficients are Borel measurable functions. If $\phi \in B^k(U)$ is $\|T\|$-integrable, then it is integrable with respect to all T_J, so that the value $T(\phi)$ of T for ϕ is defined by

$$T\colon \phi \in B^k(U) \to \sum' T_J(dx^J \wedge \phi) \in \mathbf{C}. \tag{3.1.25}$$

Therefore we can define the restriction $T|A$ of T over a Borel subset $A \subset U$ by

$$T|A\colon \phi \in B^k(U) \to T(\chi_A \phi) \in \mathbf{C}, \tag{3.1.26}$$

where χ_A denotes the characteristic function of A.

(i) Convergence of currents. Let $T, T_j \in D'^p(U)$ $(j = 1, 2, \ldots)$. If $\lim_{j\to\infty} T_j(\phi) = T(\phi)$ for any $\phi \in D^k(U)$ $(k+p=m)$, we say that the sequence $\{T_j\}_{j=1}^\infty$ of currents converges to T, and write $\lim_{j\to\infty} T_j = T$. If we put as in (3.1.21)

$$T = \sum' T_J dx^J, \quad T_J \in D(U)',$$

$$T_j = \sum' T_{jJ} dx^J, \quad T_{jJ} \in D(U)',$$

we easily see that

(3.1.27) $$\lim_{j\to\infty} T_j = T \Longleftrightarrow \lim_{j\to\infty} T_{jJ} = T_J \text{ for all } J.$$

By this and Lemma (3.1.8) we have the following lemma.

(3.1.28) Lemma. *A sequence $\{T_j\}_{j=1}^{\infty} \subset D'^p(U)$ converges to $T \in D'^p(U)$ if and only if for any point $x \in U$ there is a neighborhood V such that $\lim_{j\to\infty} T_j|V = T|V$.*

We similarly have the following theorem by (3.1.27) and Theorem (3.1.7).

(3.1.29) Theorem. *Let $\{T_j\} \subset D'^p(U)$ be a sequence of currents and $k + p = m$. If the limit $\lim_{j\to\infty} T_j(\phi)$ exists for any $\phi \in D^k(U)$, then there is a current $T \in D'^p(U)$ such that $\lim_{j\to\infty} T_j = T$.*

Let $T = \sum' T_J dx^J \in D'^p(U)$. We define the smoothing T_ε $(\varepsilon > 0)$ of T by

$$T_\varepsilon = \sum{}' (T_J)_\varepsilon dx^J \in D'^p(U_\varepsilon).$$

As $(T_J)_\varepsilon$ are identified with C^∞-functions on U_ε, so is T_ε with an element of $E^p(U_\varepsilon)$. For any relatively compact open subset $V \subset U$ we have

$$\lim_{\varepsilon\to 0} T_\varepsilon|V = T|V.$$

We easily see that

(3.1.30) $$dT_\varepsilon = (dT)_\varepsilon.$$

Let $V \subset \mathbf{R}^n$ be an open set and $f : U \to V$ a differentiable mapping (C^∞-mapping). If f is proper, then the pull-back

$$f^* : D^k(V) \to D^k(U)$$

is a continuous linear mapping. Setting $(f_*T)(\phi) = T(f^*\phi)$, we have a continuous linear mapping between spaces of currents

(3.1.31) $$f_* : T \in D'_k(U) \to f_*T \in D'_k(V).$$

Then the following holds:

(3.1.32) $$d(f_*T) = f_*(dT).$$

If $f : V \to U$ is a diffeomorphism, then f_* is a continuous isomorphism.

(j) Currents on differentiable manifolds. Let M be an m-dimensional differentiable manifold satisfying the second countability axiom. It will be clear that the function spaces appeared above, such as $L^k_{\text{loc}}(U)$, $B^k(U)$, $C^k(U)$, $E^k(U)$, $K^k(U)$ and $D^k(U)$, can be defined on M. We denote them by

$$L^k_{\text{loc}}(M),\ B^k(M),\ C^k(M),\ E^k(M),\ K^k(M),\ D^k(M).$$

We are going to define topologies on $K^k(M)$ and $D^k(M)$. Take a locally finite covering $\{U_\lambda(x^1_\lambda, \ldots, x^m_\lambda)\}_{\lambda=1}^\infty$ of M consisting of local coordinate neighborhood systems $U_\lambda(x^1_\lambda, \ldots, x^m_\lambda)$, such that there are relatively compact open subsets $V_\lambda \subset U_\lambda$ with $\bigcup_\lambda V_\lambda = M$. For $\alpha = (\alpha_1, \ldots, \alpha_m) \in (\mathbf{Z}^+)^m$ we put

$$D^\alpha_\lambda = \left[\frac{\partial}{\partial x^1_\lambda}\right]^{\alpha_1} \cdots \left[\frac{\partial}{\partial x^m_\lambda}\right]^{\alpha_m}$$

on U_λ. Let $\phi \in D^k(M)$ and write $\phi = \sum' \phi_{\lambda J} dx^J_\lambda$. Then we define seminorms

$$\| \phi \|_l = \sup\{ |D^\alpha_\lambda \phi_{\lambda J}(x)|;\ x \in V_\lambda,\ \lambda \in \mathbf{N},\ J \in \{m; k\},\ |\alpha| \le l \},$$

where $l = 0, 1, \ldots$ We say that a sequence $\{\phi_j\}_{j=1}^\infty \subset D^k(M)$ converges to $\phi \in D^k(M)$ if the following conditions are satisfied:

(i) There exists a compact subset $A \subset M$ such that

$$\text{supp}\ \phi \subset A,\quad \text{supp}\ \phi_j \subset A \quad (j = 1, 2, \ldots).$$

(ii) For every $l \in \mathbf{Z}^+$, $\lim_{j\to\infty} \| \phi_j - \phi \|_l = 0$.

This defines a topology on $D^k(M)$, so that $D^k(M)$ becomes a topological complex vector space. It is easily checked that the above topology is independent of the choice of the pair $(\{U_\lambda\}, \{V_\lambda\})$ of coverings of M.

A continuous linear functional $T: D^k(M) \to \mathbf{C}$ is called a k-dimensional current on M. We denote by $D'_k(M) = D'^p(M)$ $(k + p = m)$ the complex vector space of all k-dimensional currents on M. In particular, elements of $D'_0(M)$ are called distributions on M. If M is orientable, fixing a volume form Ω on M we can identify $D^0(M)$ with $D^m(M)$ by

$$\phi \in D^0(M) \to \phi\,\Omega \in D^m(M).$$

Hence $D'^0(M)$ is identified with $D'^m(M)$. As in (3.1.21), we see that a current $T \in D'^p(M)$ is written on a local coordinate neighborhood U as a differential p-form with coefficients of distributions on U. All properties we described above, except for the convolution and the smoothing, hold for currents on M.

The seminorm $\|\cdot\|_0$ defines a topology on the vector space $K^k(M)$. Set

$$K'_k(M) = K'^p(M) = \{T: K^k(M) \to \mathbf{C};$$
$$\text{continuous linear functional}\}.$$

Then $K'^p(M)$ is identified with the subspace of D'^p consisting of currents of order 0.

3.2 Positive Currents

(a) Type of currents. We naturally identify $\mathbf{C}^m$ with $\mathbf{R}^{2m}$. As in the previous section, we have currents on an open subset $U \subset \mathbf{C}^m$ and moreover on a complex manifold M satisfying the second countability axiom. Note that a complex manifold carries the canonical orientation. As differential forms on M are decomposed into sums of those of type (p, q), we show here that currents on M are similarly decomposed. From now on we consider the case $U \subset \mathbf{C}^m$, but it will be clear that the notions described below are all well-defined on complex manifolds.

Let $z = (z^1, \dots, z^m) \in \mathbf{C}^m$ be the natural complex coordinate system and put $z^j = x^j + iy^j$ $(1 \le j \le m)$. Then $(x^1, y^1, \dots, x^m, y^m)$ forms the natural real coordinate system of $\mathbf{C}^m = \mathbf{R}^m$. We set

$$dz^j = dx^j + idy^j, \quad d\bar{z}^j = dx^j - idy^j,$$

$$\frac{\partial}{\partial z^j} = \frac{1}{2}\left(\frac{\partial}{\partial x^j} - i\frac{\partial}{\partial y^j}\right), \quad \frac{\partial}{\partial \bar{z}^j} = \frac{1}{2}\left(\frac{\partial}{\partial x^j} + i\frac{\partial}{\partial y^j}\right).$$

For an element $J = (j_1, \dots, j_k) \in \{m; k\}$ we set

$$dz^J = dz^{j_1} \wedge \cdots \wedge dz^{j_k},$$

$$d\bar{z}^J = d\bar{z}^{j_1} \wedge \cdots \wedge d\bar{z}^{j_k}.$$

Let $0 \le l \le 2m$ and $\phi \in C^l(U)$. Then ϕ is uniquely written as

$$\phi = \sum_{\substack{p,\, q \ge 0 \\ p+q=l}} \sum_{\substack{J \in \{m; p\} \\ K \in \{m; q\}}} \phi_{J\bar{K}} dz^J \wedge d\bar{z}^K \quad (\phi_{J\bar{K}} \in C(U)). \tag{3.2.1}$$

Let $p, q \ge 0$, $p + q = l$ and set

$$C^{(p, q)}(U) = \left\{ \sum_{\substack{J \in \{m; p\} \\ K \in \{m; q\}}} \phi_{J\bar{K}} dz^J \wedge d\bar{z}^K ;\ \phi_{J\bar{K}} \in C(U) \right\}.$$

A differential $(p+q)$-form belonging to $C^{(p, q)}$ is said to be of **type** (p, q). Moreover we set

$$K^{(p, q)}(U) = C^{(p, q)}(U) \cap K^{p+q}(U),$$

$$E^{(p, q)}(U) = C^{(p, q)}(U) \cap E^{p+q}(U),$$

$$D^{(p, q)}(U) = C^{(p, q)}(U) \cap D^{p+q}(U),$$

$$B^{(p, q)}(U) = \left\{ \sum_{\substack{J \in \{m; p\} \\ K \in \{m; q\}}} \phi_{J\bar{K}} dz^J \wedge d\bar{z}^K ;\ \phi_{J\bar{K}} \in B(U) \right\},$$

$$L_{\mathrm{loc}}^{(p,q)}(U)=\left\{\sum_{\substack{J\in\{m;p\}\\K\in\{m;q\}}}\phi_{J\bar{K}}dz^J\wedge d\bar{z}^K;\ \phi_{J\bar{K}}\in L_{\mathrm{loc}}(U)\right\}.$$

We also say that differential forms belonging to the above spaces are of type (p, q). Unless confusion occurs, we write for a simplicity

$$\sum_{\substack{J\in\{m;p\}\\K\in\{m;q\}}}\phi_{J\bar{K}}dz^J\wedge d\bar{z}^K=\sum_{\substack{|J|=p\\|K|=q}}{}'\phi_{J\bar{K}}dz^J\wedge d\bar{z}^K$$
$$=\sum{}'\phi_{J\bar{K}}dz^J\wedge d\bar{z}^K.$$

By (3.2.1) we have the following direct sums:

$$K^k(U)=\bigoplus_{\substack{p,\,q\ge 0\\p+q=k}}K^{(p,q)}(U),\quad E^k(U)=\bigoplus_{\substack{p,\,q\ge 0\\p+q=k}}E^{(p,q)}(U),$$
$$D^k(U)=\bigoplus_{\substack{p,\,q\ge 0\\p+q=k}}D^{(p,q)}(U),\quad B^k(U)=\bigoplus_{\substack{p,\,q\ge 0\\p+q=k}}B^{(p,q)}(U),$$
$$L_{\mathrm{loc}}^k(U)=\bigoplus_{\substack{p,\,q\ge 0\\p+q=k}}L_{\mathrm{loc}}^{(p,q)}(U),\quad 0\le k\le 2m.$$

For $\phi=\sum{}'\phi_{J\bar{K}}dz^J\wedge d\bar{z}^K\in E^{(p,q)}(U)$ we set

$$\partial\phi=\sum{}'\sum_{j=1}^{m}\frac{\partial\phi_{J\bar{K}}}{\partial z^j}dz^j\wedge dz^J\wedge d\bar{z}^K\in E^{(p+1,q)}(U),$$
$$\bar{\partial}\phi=\sum{}'\sum_{j=1}^{m}\frac{\partial\phi_{J\bar{K}}}{\partial \bar{z}^j}d\bar{z}^j\wedge dz^J\wedge d\bar{z}^K\in E^{(p,q+1)}(U).$$

We extend ∂ and $\bar{\partial}$ over $E^k(U)$ as linear mappings. Then we get the following identities.

(3.2.2) $$d=\partial+\bar{\partial},\ \partial\partial=\bar{\partial}\ \bar{\partial}=\partial\bar{\partial}+\bar{\partial}\partial=0.$$

Now we say that a current $T\in D'^l(U)$ is of **type** (p, q) $(p,\ q\ge 0,\ p+q=l)$ if $T(\phi)=0$ for all $\phi\in D^{(s,t)}(U)$ such that $s+t=2m-l$ and $(s,\ t)\ne(m-p,\ m-q)$. According to (3.1.21), $T\in D'^l(U)$ is of type (p, q) if and only if

(3.2.3) $$T=\sum_{\substack{|J|=p\\|K|=q}}{}'T_{J\bar{K}}dz^J\wedge d\bar{z}^K,\quad T_{J\bar{K}}\in D(U)'.$$

The space of all currents of type (p, q) is denoted by $D'^{(p,q)}(U)$ or by $D'_{(m-p,\,m-q)}(U)$. Set

$$K'^{(p,q)}(U) = K'_{(m-p,\, m-q)}(U) = K'^{p+q}(U) \cap D'^{(p,q)}(U).$$

Let $l + k = 2m$. Then we have

$$D'^{l}(U) = \bigoplus_{\substack{p,\, q \geq 0 \\ p+q=l}} D'^{(p,q)}(U) = \bigoplus_{\substack{s,\, t \geq 0 \\ s+t=k}} D'_{(s,t)}(U) = D'_k(U),$$

$$K'^{l}(U) = \bigoplus_{\substack{p,\, q \geq 0 \\ p+q=l}} K'^{(p,q)}(U) = \bigoplus_{\substack{s,\, t \geq 0 \\ s+t=k}} K'_{(s,t)}(U) = K'_k(U).$$

For $T \in D'^{l}(U) = D'_k(U)$ we set

$$\partial T : \phi \in D^{k-1}(U) \to (-1)^{l-1} T(\partial \phi) \in \mathbf{C},$$

$$\bar{\partial} T : \phi \in D^{k-1}(U) \to (-1)^{l-1} T(\bar{\partial} \phi) \in \mathbf{C}.$$

Then ∂T, $\bar{\partial} T \in D'^{l+1}$. If T is of type (p, q), then ∂T is of type $(p+1, q)$ and $\bar{\partial} T$ of $(p, q+1)$. If $\partial T = 0$ (resp. $\bar{\partial} T = 0$), T is said to be ∂-closed (resp. $\bar{\partial}$-closed). It follows from (3.2.2) that

(3.2.4) $$dT = \partial T + \bar{\partial} T, \quad \partial\partial T = \bar{\partial}\bar{\partial} T = (\partial\bar{\partial} + \bar{\partial}\partial) T = 0.$$

(3.2.5) *Example*. Let $\omega \in L^{(p,q)}_{\text{loc}}$ and put

$$[\omega] : \phi \in D^{2m-p-q}(U) \to \int \omega \wedge \phi \in \mathbf{C}.$$

Then $[\omega] \in K'^{(p,q)}(U)$.

(3.2.6) *Example*. Let M be a k-dimensional (closed) complex submanifold of U. Then M has the canonical orientation determined by the complex structure. As in Example (3.1.24), we set

$$M = T_M : \phi \in D^{2k}(U) \to \int_M \phi \in \mathbf{C}.$$

Then $M = T_M \in K'_{(k,k)}(U) = K'^{(m-k,\, m-k)}(U)$.

Now, let $\phi = \sum' \phi_{J\bar{K}}\, dz^J \wedge d\bar{z}^K \in D^{(p,q)}$. We set

$$\bar{\phi} = \sum' \bar{\phi}_{J\bar{K}}\, d\bar{z}^J \wedge dz^K \in D^{(q,p)}(U)$$

and linearly extend it over $D^k(U)$. We similarly define $\bar{\phi}$ for elements of E^k or L^k_{loc} and call $\bar{\phi}$ the **complex conjugate** of ϕ. If $\bar{\phi} = \phi$, ϕ is called a **real differential form.** In general, we have

(3.2.7) $$\phi = \overline{(\bar{\phi})}.$$

Let $T \in D'^{l}(U)$ and define $\bar{T}$ by

(3.2.8)
$$\bar{T}(\phi) = \overline{T(\bar{\phi})}.$$

If $T \in D'^{(p,q)}(U)$, $\bar{T} \in D'^{(q,p)}(U)$. If $T = \sum' T_{J\bar{K}} dz^J \wedge d\bar{z}^K \in D'^{(p,q)}(U)$, then

$$\bar{T} = \sum' \bar{T}_{J\bar{K}} d\bar{z}^J \wedge dz^K \in D'^{(q,p)}(U).$$

If $T = \bar{T}$, T is called a **real current.** The following is clear by definition.

(3.2.9) Lemma. *If $T \in D'^l(U)$ is of type (p, q), then so is the smoothing T_ε of T. If T is real, so is T_ε.*

Let $V \subset \mathbf{C}^n$ be an open subset and $f : U \to V$ an holomorphic mapping. Then the pull-back mapping $f^* : E^k(V) \to E^k(U)$ satisfies

(3.2.10)
$$\partial(f^*\phi) = f^*(\partial\phi),\ \bar{\partial}(f^*\phi) = f^*(\bar{\partial}\phi)\ \ (\phi \in E^k(V)),$$
$$f^*(E^{(p,q)}(V)) \subset E^{(p,q)}(U).$$

If $f : U \to V$ is proper, then as in (3.1.31) $f_* : D_k'(U) \to D_k'(V)$ is defined by $(f_*T)(\phi) = T(f^*\phi)$. We have by (3.2.10)

(3.2.11)
$$\partial(f_*T) = f_*(\partial T),\ \ \bar{\partial}(f_*T) = f_*(\bar{\partial}T)\ \ (T \in D_k'(U)),$$
$$f_*(D'_{(p,q)}(U)) \subset D'_{(p,q)}(V).$$

(b) Positive currents. Let $k + p = m$ with $k, p \in \mathbf{Z}^+$ and put

$$\sigma_k = \begin{cases} 2^{-k}, & \text{for } k \text{ even} \\ i2^{-k}, & \text{for } k \text{ odd.} \end{cases}$$

Note that

$$\frac{i}{2} dz^{j_1} \wedge d\bar{z}^{j_1} \wedge \frac{i}{2} dz^{j_2} \wedge d\bar{z}^{j_2} \wedge \cdots \wedge \frac{i}{2} dz^{j_k} \wedge d\bar{z}^{j_k} = \sigma_k dz^J \wedge d\bar{z}^J$$

for $J = (j_1, \ldots, j_k) \in \{m; k\}$. We say that a real current $T \in D'^{(p,p)}(U)$ is **positive** if $T \wedge (\sigma_k \eta \wedge \bar{\eta})$ are positive distributions for all $\eta \in E^{(k,0)}(U)$. If $T \in D'^{(p,p)}(U)$ is a positive current, then so is the restriction $T|V \in D'^{(p,p)}(V)$ of T over any open subset $V \subset U$. Conversely, as in Lemma (3.1.8), we have the following.

(3.2.12) Lemma. *A current $T \in D'^{(p,p)}(U)$ is positive if and only if for an arbitrary point $x \in U$ there is a neighborhood $V \subset U$ such that $T|V$ is positive.*

The following lemma will be clear.

(3.2.13) Lemma. *If $T \in D'^{(p,p)}(U)$ is positive, then so is the smoothing $T_\varepsilon \in D'^{(p,p)}(U_\varepsilon)$ $(\varepsilon > 0)$; the converse holds, too.*

(3.2.14) Theorem. *If a current* $T \in \mathcal{D}'^{(p,p)}(U)$ *is positive, then* $T \in \mathcal{K}'^{(p,p)}(U)$.

Proof. Let $W^{(p,p)}$ denote the vector subspace of $\mathcal{E}^{(p,p)}(\mathbf{C}^m)$ generated by $\{dz^J \wedge d\bar{z}^K; J, K \in \{m; p\}\}$ over $\mathbf{C}$. Then

$$W^{(p,p)} = \overset{p}{\wedge} W^{(1,1)} \quad (p\text{-the exterior power}). \tag{3.2.15}$$

Every $u \in W^{(1,1)}$ is written as

$$u = \sum_{1 \le j,\, l \le m} a_{j\bar{l}}\, dz^j \wedge d\bar{z}^l \quad (a_{j\bar{l}} \in \mathbf{C}).$$

Putting $b_{j\bar{l}} = (a_{j\bar{l}} + \bar{a}_{l\bar{j}})/2$ and $c_{j\bar{l}} = (a_{j\bar{l}} - \bar{a}_{l\bar{j}})/(2i)$, we have

$$a_{j\bar{l}} = b_{j\bar{l}} + ic_{j\bar{l}},\ b_{j\bar{l}} = \overline{b_{l\bar{j}}},\ c_{j\bar{l}} = \overline{c_{l\bar{j}}}.$$

Hence $W^{(1,1)}$ is generated by the elements of the form

$$u = \sum_{1 \le j,\, l \le m} a_{j\bar{l}}\, dz^j \wedge d\bar{z}^l \quad (a_{j\bar{l}} = \bar{a}_{l\bar{j}} \in \mathbf{C}). \tag{3.2.16}$$

Then the matrix $(a_{j\bar{l}})$ formed by the coefficients of (3.2.16) is hermitian. Therefore there is a unitary matrix $(d_{j\bar{l}})$ such that

$$^t\overline{(d_{j\bar{l}})}\,(a_{j\bar{l}})\,(d_{j\bar{l}}) = \begin{pmatrix} \lambda_1 & & O \\ & \ddots & \\ O & & \lambda_m \end{pmatrix} \quad (\lambda_j \in \mathbf{R}).$$

Thus the element u given in (3.2.16) is written as

$$u = i \sum_{l=1}^{m} \lambda_l \alpha^l \wedge \bar{\alpha}^l \quad (\lambda_l \in \mathbf{R}), \tag{3.2.17}$$

where $\alpha^l = \sum_{j=1}^{m} \beta^l_j dz^j (\beta^l_j \in \mathbf{C})$ with a certain regular matrix (β^l_j). It follows from (3.2.15) and (3.2.17) that

(3.2.18) $W^{(p,p)}$ is generated by the elements of the form $(i\alpha^1 \wedge \bar{\alpha}^1) \wedge \cdots \wedge (i\alpha^p \wedge \bar{\alpha}^p)$, where α^j are linear combinations of $dz^1, \ldots, dz^m$.

Since

$$W^{(m,m)} = \left\{ a\left[\frac{i}{2} dz^1 \wedge d\bar{z}^1\right] \wedge \cdots \wedge \left[\frac{i}{2} dz^m \wedge d\bar{z}^m\right]; a \in \mathbf{C} \right\},$$

we identify $W^{(m,m)}$ with $\mathbf{C}$. Then by the bilinear mapping

$$(\phi, \eta) \in W^{(p,p)} \times W^{(k,k)} \to \phi \wedge \eta \in \mathbf{C} \tag{3.2.19}$$

the above two spaces are mutually dual spaces. Take a base $\{\phi_1, \ldots, \phi_l\}$ $\left[l = \binom{m}{k}^2\right]$ of $W^{(k,k)}$ as in (3.2.18) and let $\{\phi_1^*, \ldots, \phi_l^*\}$ be the dual base of $W^{(p,p)}$. Then T is written as

$$T = \sum_{j=1}^{l} T_j \phi_j^*, \quad T_j \in D(U)'. \tag{3.2.20}$$

Since T is a positive current, $T \wedge \phi_j$ is a positive distribution. Hence we have

$$0 \leq T(f\phi_j) = T \wedge \phi_j(f) = \sum_{i=1}^{l} T_i \phi_i^* \wedge \phi_j(f) = T_j(f)$$

for all $f \in D(U)$ $(f \geq 0)$. We see by Lemma (3.2.14) that $T_j \in K(U)'$ and hence $T \in K'^{(p,p)}(U)$ by (3.2.20). *Q.E.D.*

(3.2.21) Theorem. *A current* $T = \sigma_p \sum' T_{J\bar{K}} dz^J \wedge d\bar{z}^K \in D'^{(p,p)}(U)$ *is positive if and only if the following two conditions are satisfied:*

(i) $\overline{T_{J\bar{K}}} = T_{K\bar{J}}$.

(ii) For any vector $(\xi^J) \in \mathbf{C}^{\binom{m}{p}}$, $\sum' T_{J\bar{K}} \xi^J \bar{\xi}^K$ is a positive distribution.

Proof. Condition (i) is necessary and sufficient for T to be real. Assume that T is a positive current. Take an arbitrary vector $(\xi^J) \in \mathbf{C}^{\binom{m}{p}}$ and put

$$\eta = \sum_{J \in \{m;p\}} \xi^J \delta(J, \hat{J}) dz^{\hat{J}} \in E^{(k,0)}(U),$$

where $\delta(J, \hat{J})$ stands for the signature of the permutation $\{J, \hat{J}\} \to \{1, 2, \ldots, m\}$. By the assumption, $\sigma_k T \wedge \eta \wedge \bar{\eta}$ is a positive distribution. Note that

$$\sigma_k T \wedge \eta \wedge \bar{\eta} = \sigma_k \sigma_p \sum' T_{J\bar{K}} dz^J \wedge d\bar{z}^K \wedge \eta \wedge \bar{\eta} \tag{3.2.22}$$

$$= \sigma_k \sigma_p \sum'_{\substack{|J|=p \\ |K|=p}} \sum'_{\substack{|J'|=p \\ |K'|=p}} T_{J\bar{K}} \xi^{J'} \bar{\xi}^{K'} \delta(J', \hat{J}') \delta(K', \hat{K}') dz^J \wedge d\bar{z}^K \wedge dz^{\hat{J}'} \wedge d\bar{z}^{\hat{K}'}$$

$$= \sigma_k \sigma_p \sum'_{\substack{|J|=p \\ |K|=p}} T_{J\bar{K}} \xi^J \bar{\xi}^K \delta(J, \hat{J}) \delta(K, \hat{K}) dz^J \wedge d\bar{z}^K \wedge dz^{\hat{J}} \wedge d\bar{z}^{\hat{K}}$$

$$= \sigma_k \sigma_p (-1)^{kp} \sum' T_{J\bar{K}} \xi^J \bar{\xi}^K \delta(J, \hat{J}) \delta(K, \hat{K}) dz^J \wedge dz^{\hat{J}} d\bar{z}^K \wedge d\bar{z}^{\hat{K}}$$

$$= \sigma_k \sigma_p (-1)^{kp} \left[\sum' T_{J\bar{K}} \xi^J \bar{\xi}^K\right] dz^1 \wedge \cdots \wedge dz^m \wedge d\bar{z}^1 \wedge \cdots \wedge d\bar{z}^m$$

$$= \sigma_k \sigma_p (-1)^{kp} (\sigma_m)^{-1} \left[\sum' T_{J\bar{K}} \xi^J \bar{\xi}^K\right] \bigwedge_{j=1}^{m} \frac{i}{2} dz^j \wedge d\bar{z}^j$$

Continued

$$= \left[\sum{}' T_{J\bar{K}} \xi^J \bar{\xi}^K \right] \bigwedge_{j=1}^{m} \frac{i}{2} dz^j \wedge d\bar{z}^j.$$

Thus $\sum' T_{J\bar{K}} \xi^J \bar{\xi}^K$ is a positive distribution. Conversely, assume that a real current T satisfies condition (ii). Then for $J, K \in \{m; p\}$, $T_{J\bar{K}} + T_{K\bar{J}}$ and $iT_{J\bar{K}} - iT_{K\bar{J}}$ are positive distributions. Hence $T_{J\bar{K}} \in \mathcal{K}(U)'$, so that $T \in \mathcal{K}'^{(p,p)}(U)$. Take an arbitrary $\eta \in \mathcal{E}^{(p,0)}(U)$ and $f \in \mathcal{D}(U)$ with $f \geq 0$. Take a relatively compact open subset $V \subset U$ such that $\operatorname{supp} f \subset V$, and put $\eta = \sum' \eta_J dz^J$. Approximate each η_J uniformly on V by a sequence of step functions

$$\eta_{\lambda J} = \sum_{\alpha} c_{\lambda J \alpha} \chi_{B_{\lambda J \alpha}}, \quad \lambda = 1, 2, \ldots,$$

where $\{B_{\lambda J \alpha}\}_{\alpha}$ is a finite family of Borel subsets of V such that $B_{\lambda J \alpha} \cap B_{\lambda J \beta} = \emptyset$ for $\alpha \neq \beta$. By Theorem (3.1.13) $T_{J\bar{K}}$ are considered Radon measures. Put $\eta_\lambda = \sum'_{|J|=p} \eta_{\lambda J} dz^J$. Then $\operatorname{supp} \eta_\lambda \subset V$ and $\eta_\lambda \to \eta$ uniformly as $\lambda \to \infty$. Therefore we obtain

$$(\sigma_k T \wedge \eta \wedge \bar{\eta})(f) = T(\sigma_k f \eta \wedge \bar{\eta}) = \lim_{\lambda \to \infty} T(\sigma f \eta_\lambda \wedge \bar{\eta}_\lambda).$$

It is sufficient to show that $\sigma_k T \wedge \chi_B \eta \wedge \bar{\eta}$ is a positive distribution for arbitrary Borel subset $B \subset U$ and $(\xi^J) \in \mathbf{C}^{\binom{m}{p}}$. In fact, we see by the computation of (3.2.22) that

$$(\sigma_k T \wedge \chi_B \eta \wedge \bar{\eta})(f) = \sum{}' T_{J\bar{K}} (\chi_B f) \xi^J \bar{\xi}^K \geq 0$$

for all $f \in \mathcal{D}(U)$ with $f \geq 0$. *Q.E.D.*

For a current $T = \sigma_p \sum' T_{J\bar{K}} dz^J \wedge d\bar{z}^K \in \mathcal{D}'^{(p,p)}(U)$, Trace $T = \sum'_{|J|=p} T_{J\bar{J}}$ is called the **trace** of T.

(3.2.23) Corollary. *If* $T = \sigma_p \sum' T_{J\bar{K}} dz^J \wedge d\bar{z}^K \in \mathcal{D}'^{(p,p)}(U)$ *is a positive current, then* Trace T *is a positive Radon measure and each* $T_{J\bar{K}}$ *is absolutely continuous with respect to* Trace T.

Proof. Take an arbitrary $f \in \mathcal{D}(U)$ with $f \geq 0$. It follows from Theorem (3.2.21) that the matrix $\left[T_{J\bar{K}}(f) \right]_{J, K \in \{m; p\}}$ is hermitian and semipositive definite. Schwarz's inequality implies

$$\sum_J{}' T_{J\bar{J}}(f) \geq 0,$$

$$|T_{J\bar{K}}(f)| \leq \text{Trace } T(f). \quad \textit{Q.E.D.}$$

(3.2.24) Lemma. *Let* $\omega \in E^{(1,1)}(U)$ *be a real differential form and* $\omega \geq 0$. *Let* $T \in D'^{(p,p)}(U)$ *be a positive current. Then* $T \wedge \omega^l$ $(1 \leq l \leq k (= m - p))$ *is positive, where* $\omega^l = \omega \wedge \cdots \wedge \omega$ *(l-times).*

Proof. By (3.2.16)~(3.2.18) we can write

$$\omega = \frac{i}{2} \sum_{j=1}^{m} a_j \alpha^j \wedge \bar{\alpha}^j \quad (a_j \in E(U),\ a_j \geq 0),$$

where α^j are linear combinations of $dz^1, \ldots, dz^m$. Then we have

$$\omega^l = l! \left(\frac{i}{2} \right)^l \sum_{1 \leq j_1 < \cdots < j_l \leq m} a_{j_1} a_{j_2} \cdots a_{j_l} \alpha^{j_1} \wedge \bar{\alpha}^{j_1} \wedge \cdots \wedge \alpha^{j_l} \wedge \bar{\alpha}^{j_l}$$

$$= l! \sigma_l \sum_{|J|=l}{}' a_J \alpha^J \wedge \bar{\alpha}^J,$$

where $a_J = a_{j_1} \cdots a_{j_l}$ and $\alpha^J = \alpha^{j_1} \wedge \cdots \wedge \alpha^{j_l}$ for $J = (j_1, \ldots, j_l) \in \{m; l\}$. For any $\eta \in E^{(m-p-l,0)}(U)$ we obtain

$$T \wedge \omega^l \wedge (\sigma_{m-p-l} \eta \wedge \bar{\eta}) = l! \sum_{|J|=l}{}' \sigma_l \sigma_{m-p-l} a_J T \wedge \alpha^J \wedge \bar{\alpha}^J \wedge \eta \wedge \bar{\eta}$$

$$= l! \sum_{|J|=l}{}' \sigma_l \sigma_{m-p-l} (-1)^{l(m-p-l)} \sigma_{m-p} a_J T \wedge (\alpha^J \wedge \eta) \wedge \overline{(\alpha^J \wedge \eta)}$$

$$= l! \sum_{|J|=l}{}' \left(\frac{1}{2} \right)^{l(m-p-l)} \sigma_k \, a_J T \wedge (\alpha^J \wedge \eta) \wedge \overline{(\alpha^J \wedge \eta)}.$$

Hence $T \wedge \omega^l$ is a positive current. *Q.E.D.*

Now we give the most important example of closed positive currents.

(3.2.25) *Example.* Let $X \subset U$ be a k-dimensional complex submanifold of U. As in Example (3.2.6), we get a closed current of order 0:

$$T_X \colon \phi \in K^{2k}(U) \to \int_X \phi \in \mathbf{C}.$$

Moreover, it is easy to see that T_X is positive. The closed positive current T_X is sometimes denoted simply by X, unless confusion occurs. It is noted that the same holds for a purely k-dimensional analytic subset of U which may have singularities (cf. Chapter IV for analytic subsets). Let $R(X)$ (resp. $S(X)$ denote the set of regular (resp. singular) points of X and set

$$T_X(\phi) = \langle X, \phi \rangle = \int_{R(X)} \phi \quad (\phi \in K^{2k}(U)).$$

We will prove in Chapter V, §1 that T_X is a positive current. Furthermore, it is known that T_X is closed. We will show the closedness of T_X for X analytic hypersurface in Chapter V, §1. There is an elementary but rather lengthy proof of this fact for general X (cf., e.g., Lelong [64]), but here we give a simple proof depending on Hironaka's resolution ([47]). That is, there are a k-dimensional complex manifold $\tilde{X}$ and a proper holomorphic mapping $\pi: \tilde{X} \to X$ such that $\pi|\tilde{X}-\pi^{-1}(S(X))$: $\tilde{X}-\pi^{-1}(S(X)) \to X-S(X)=R(X)$ is biholomorphic. Since $\pi^{-1}(S(X))$ is a proper analytic subset of $\tilde{X}$ and hence of measure 0 in $\tilde{X}$, we obtain

$$T_X(\phi) = \int_{\tilde{X}-\pi^{-1}(S(X))} \pi^* \phi = \int_{\tilde{X}} \pi^* \phi.$$

Hence it is clear that T_X defines a closed positive current on U.

Remark. Let M be a complex manifold satisfying the second countability axiom. The notion of currents described in §1, (j) is defined on M and the results of the present section for positive currents hold locally or globally on M.

(c) Lelong number. Let $(z^1, \dots, z^m)$ be the standard holomorphic coordinate system of $\mathbf{C}^m$ as above. For $z=(z^1, \dots, z^m)$ we set

$$\|z\|^2 = \sum_{j=1}^m |z^j|^2, \quad d^c = \frac{i}{4\pi}(\bar{\partial} - \partial),$$

$$\alpha = dd^c\|z\|^2 = \frac{i}{2\pi}\sum_{j=1}^m dz^j \wedge d\bar{z}^j,$$

$$\beta = dd^c \log\|z\|^2 \quad (z \neq 0),$$

$$\alpha^k = \overset{k}{\wedge}\alpha, \quad \beta^k = \overset{k}{\wedge}\beta, \quad \alpha^0 = 1, \quad \beta^0 = 1,$$

$$\eta = d^c \log\|z\|^2 \wedge \beta^{m-1}.$$

It is easy to see that

(3.2.26) $$d\alpha = 0, \quad d\beta = 0, \quad \beta^m = 0.$$

Set

$$B(z;r) = \{w \in \mathbf{C}^m;\ \|w-z\| < r\},$$

$$\Gamma(z;r) = \{w \in \mathbf{C}^m;\ \|w-z\| = r\},$$

$$B(r) = B(O;r), \quad \Gamma(r) = \Gamma(O;r).$$

One notes that

(3.2.27) $$\alpha^m = \frac{m!}{\pi^m}\bigwedge_{j=1}^m \frac{i}{2} dz^j \wedge d\bar{z}^j,$$

and as the volume of $B(1)$ is $\pi^m/m!$,

(3.2.28)
$$\int_{B(z;r)} \alpha^m = r^{2m}.$$

Since $d\|w\|^2 = \partial\|w\|^2 + \bar{\partial}\|w\|^2 = 0$ on $\Gamma(r)$, we have

(3.2.29)
$$\beta|\Gamma(r) = \frac{1}{r^2}\alpha|\Gamma(r).$$

Let $T \in D'^{(p,p)}(U)$ be a positive current on an open subset $U \subset \mathbf{C}^m$, $k+p=m$ and $B(z;R) \subset U$ $(R>0)$. For $0<r<R$ we set

$$n(z;r,T) = \frac{1}{r^{2k}} T(\chi_{B(z;r)}\alpha^k)(\geq 0), \quad n(r,T) = n(O;r,T).$$

Here the right-hand side of the first equality makes sense by Theorems (3.2.14) and (3.1.13) and is non-negative by Lemma (3.2.24).

Remark. We will use the above notation throughout the rest of this book.

The following is called Poincaré's lemma and known well.

(3.2.30) Lemma. *In general, let $\phi \in E^p(B(r))$ $(r>0)$ be a d-closed differential form Then there is a differential form $\psi \in E^{p-1}(B(r))$ such that $d\psi = \phi$.*

(3.2.31) Theorem. *Let $T \in D'^{(p,p)}(U)$ be a closed positive current, $z \in U$ and put*

$$R = \operatorname{dist}(z, \partial U) = \inf\{\|w-z\|; w \in \partial U\}.$$

Then $n(z;r,T)$ is an increasing function in $0<r<R$.

Proof. We may assume that $z=O$. Let $0<r_1<r_2<R$ and take $\varepsilon>0$ so that $\varepsilon < R-r_2$. Then the smoothing T_ε of T is a closed differential form of degree $2p$ with C^∞-coefficients on U_ε (cf. (3.1.30)). By Lemma (3.2.30) there is a differential form $S_\varepsilon \in E^{2p-1}(B(R-\varepsilon))$ such that $dS_\varepsilon = T_\varepsilon$. Then we have

(3.2.32)
$$\begin{aligned}
&n(r_2,T_\varepsilon) - n(r_1,T_\varepsilon)\\
&= \frac{1}{r_2^{2k}}\int_{B(r_2)} T_\varepsilon \wedge \alpha^k - \frac{1}{r_1^{2k}}\int_{B(r_1)} T_\varepsilon \wedge \alpha^k\\
&= \frac{1}{r_2^{2k}}\int_{\Gamma(r_2)} S_\varepsilon \wedge \alpha^k - \frac{1}{r_1^{2k}}\int_{\Gamma(r_1)} S_\varepsilon \wedge \alpha^k\\
&\qquad \text{(by Stokes' Theorem and (3.2.26))}\\
&= \int_{\Gamma(r_2)} S_\varepsilon \wedge \beta^k - \int_{\Gamma(r_1)} S_\varepsilon \wedge \beta^k \quad \text{(by (3.2.29))}\\
&= \int_{\overline{B(r_2)-B(r_1)}} T_\varepsilon \wedge \beta^k \quad \text{(by Stokes' Theorem and (3.2.26))}.
\end{aligned}$$

Let E be the set of $r \in [0, R]$ such that $\Gamma(r)$ has positive measure with respect to Trace T. Put

$$E_N = \left\{ r \in [0, R);\ \text{Trace}\ T(\Gamma(r)) > \frac{1}{N} \right\},\quad N = 1, 2, \ldots$$

Since Trace T is a Radon measure, any compact subset of $[0, R)$ contains at most finitely many elements of E_N, so that E_N is at most a countable set. Since $E = \bigcup_{N=1}^{\infty} E_N$, E is at most a countable set, too. We claim that

$$\lim_{\varepsilon \to 0} \int_{B(r)} T_\varepsilon \wedge \alpha^k = T(\chi_{B(r)} \alpha^k) \tag{3.2.33}$$

for an arbitrary $r \in (0, R) - E$. Note first that $T \wedge \alpha^k$ is a positive Radon measure. Since $r \notin E$, by using the outer regularity of Radon measure, we see that for an arbitrary $\delta > 0$, there is a positive number $\sigma > 0$ such that

$$T(\chi_{B(r+3\sigma) - \overline{B(r-3\sigma)}}\ \alpha^k) < \delta. \tag{3.2.34}$$

Define non-negative valued functions μ_1 and μ_2 by

$$\mu_1(t) = \begin{cases} 1, & t \le r - 2\sigma, \\ 1 - \{t - (r - 2\sigma)\}/\sigma, & r - 2\sigma \le t \le r - \sigma, \\ 0, & t \ge r - \sigma, \end{cases}$$

$$\mu_2(t) = 1 - \mu_1(t).$$

It follows from (3.2.34) that if $\varepsilon < \sigma$,

$$0 \le T(\mu_2\, \chi_{B(r)} \alpha^k) < \delta, \tag{3.2.35}$$

$$0 \le \int_{B(r)} \mu_2\, T_\varepsilon \wedge \alpha^k < T(\chi_{B(r+3\sigma) - \overline{B(r-3\sigma)}} \alpha^k) < \delta.$$

Since T_ε converges to T in the sense of $\|\cdot\|_0$ as $\varepsilon \to 0$, there is $\varepsilon_0 > 0$ such that for $\varepsilon < \varepsilon_0$

$$\left| \int \mu_1 T_\varepsilon \wedge \alpha^k - T(\mu_1 \alpha^k) \right| < \delta. \tag{3.2.36}$$

Hence it follows from (3.2.35) and (3.2.36) that

$$\left| \int_{B(r)} T_\varepsilon \wedge \alpha^k - T(\chi_{B(r)} \wedge \alpha^k) \right| < 3\delta$$

for $\varepsilon < \min\{\sigma, \varepsilon_0\}$. This shows (3.2.33). Suppose that $r_1, r_2 \notin E$. Then we see by (3.2.33) that

$$\lim_{\varepsilon \to 0} n(r_i, T_\varepsilon) = n(r_i, T), \quad i = 1, 2.$$

The same arguments as above with replacing α with β implies

$$\lim_{\varepsilon \to 0} \int_{B(r_2) - \overline{B(r_1)}} T_\varepsilon \wedge \beta^k = T(\chi_{B(r_2) - \overline{B(r_1)}} \beta^k).$$

Thus it follows from (3.2.32) that

$$n(r_2, T) - n(r_1, T) = T(\chi_{B(r_2) - \overline{B(r_1)}} \beta^k) = T \wedge \beta^k (\chi_{B(r_2) - \overline{B(r_1)}}).$$

On the other hand, since $\beta \geq 0$, we know by Lemma (3.2.24) that $T \wedge \beta^k$ is a positive Radon measure on $U - \{O\}$, so that $T \wedge \beta^k (\chi_{B(r_2) - \overline{B(r_1)}}) \geq 0$. Thus we see that $n(r_1, T) \leq n(r_2, T)$ for $r_1, r_2 \notin E$. Note that the inner regularity of Radon measure implies the left-continuity of $n(r, T)$. Therefore we have $n(r_1, T) \leq n(r_2, T)$ for all $r_1 < r_2$. *Q.E.D.*

Let T be as in Theorem (3.2.31). Then by that theorem, there exists the limit

$$\lim_{r \to 0} n(z; r, T) = L(z; T) \quad (z \in U),$$

which is called the **Lelong number** of T at $z \in U$.

(d) Lelong number for complex submanifolds. Let S be a closed k-dimensional complex submanifold of an open subset $U \subset \mathbf{C}^m$. As in Example (3.2.25), S defines a closed current $S \in D'^{(p,p)}(U)$ $(p = m - k)$ by

$$\phi \in D^{(k,k)}(U) \to \int_S \phi.$$

Moreover, S is positive and hence a positive closed current. For a compact subset $K \subset U$

(3.2.37)
$$\mathrm{Vol}(K \cap S) = \frac{\pi^k}{k!} \int_{K \cap S} \alpha^k,$$

where $\mathrm{Vol}(\cdot)$ denotes the euclidean volume and $\alpha = dd^c \|z\|^2$. In special, let V be a k-dimensional complex vector subspace of $\mathbf{C}^m$. We remark that

(3.2.38)
$$\int_{B(r) \cap V} \alpha^k = \frac{k!}{\pi^k} \mathrm{Vol}(B(r) \cap V) = r^{2k}.$$

(3.2.39) Proposition. *Let S be a k-dimensional complex submanifold of U ($\subset \mathbf{C}^m$). Then we have*

$$n(z;r,S)=\int_{S\cap B(z;r)}\alpha^k \geq L(z;S)=1$$

for $z\in S$, where $0<r<\operatorname{dist}(z,\partial U)$.

Proof. We may assume that $U=B(R)$ and $z=O$ and put $n(r,S)=n(O;r,S)$ as above. By Theorem (3.2.31) $n(r,S)$ is an increasing function in r. Here one can easily show this fact as follows. For almost all $r\in(0,R)$ with respect to the Lebesgue measure $\Gamma(r)\cap S$ are smooth differentiable submanifolds of $B(R)$, where $\Gamma(r)=\partial B(r)$. Taking such $r_1<r_2$ and putting $\beta=dd^c\log\|z\|^2$, we have

$$\begin{aligned} &n(r_2,S)-n(r_1,S)\\ &=\frac{1}{r^{2k}}\int_{\Gamma(r_2)\cap S}d^c\|z\|^2\wedge\alpha^{k-1}-\frac{1}{r^{2k}}\int_{\Gamma(r_1)\cap S}d^c\|z\|^2\wedge\alpha^{k-1}\\ &=\int_{\Gamma(r_2)\cap S}d^c\log\|z\|^2\wedge\beta^{k-1}-\int_{\Gamma(r_1)\cap S}d^c\log\|z\|^2\wedge\beta^{k-1}\\ &=\int_{(B(r_2)-\overline{B(r_1)})\cap S}\beta^k\geq 0. \end{aligned}$$

Since $n(r,S)$ is left-continuous, we see that $n(r,S)$ is increasing.

Let V be a complex vector subspace of $\mathbf{C}^m$ such that V is tangent to S at O. By a unitary transform, we may assume that

$$V=\{(z^1,\dots,z^k,0,\dots,0);\ z^j\in\mathbf{C}\}.$$

Then we may use $(z^1,\dots,z^k)$ as a holomorphic local coordinate system of S around the origin O. The holomorphic functions $z^j|S=F^j(z^1,\dots,z^k)$, $k+1<j\leq m$, defined around O, have zeros of order ≥ 2 at O. Therefore we have for small r (cf. (3.2.38))

$$n(r,S)=\frac{k!}{\pi^k r^{2k}}\int_{V\cap B(r)}\frac{i}{2}dz^1\wedge d\bar{z}^1\wedge\cdots\wedge\frac{i}{2}dz^k\wedge d\bar{z}^k+o(r),$$

so that

$$n(r,S)\geq\lim_{r\to 0}n(r,S)=L(O;S)=1.\quad Q.E.D.$$

We obtain the following corollary from (3.2.38).

(3.2.40) Corollary. *Let S be as above. Then*

$$\operatorname{Vol}(B(r)\cap S)\geq \frac{\pi^k}{k!}r^{2k}.$$

Remark. Let S be a purely k-dimensional analytic subset of U. Then S defines a closed positive current as in Example (3.2.25), and $L(z;S)\geq 1$ if and only if $z\in S$. Moreover, $L(z;S)\geq 2$ for singular points $z\in S$. In fact, $L(z;S)$ coincides with the so-called multiplicity of S at z (cf. Thie [117]). This fact will be proved for analytic hypersurfaces in Chapter V, §1.

3.3 Plurisubharmonic Functions

In the present section we show elementary facts on plurisubharmonic functions. Since the notion of plurisubharmonic function is broadly used in the complex analysis and geometry, we will discuss it fairly in detail. The results of this section will be used in Chapter V.

(a) Subharmonic functions. We begin with recalling subharmonic functions in one complex variable. Let U be an open subset of $\mathbf{C}$. We set

$$U_\varepsilon=\left\{z\in U;\ \inf_{w\notin U}|w-z|>\varepsilon\right\},\ \varepsilon>0.$$

(3.3.1) *Definition.* A function $u: U\to[-\infty,\infty)$ is said to be **subharmonic** if the following three conditions are satisfied.

(i) On every connected component of U, $u\not\equiv-\infty$:

(ii) The function u is upper semicontinuous; that is, $\{z\in U;\ u(z)<a\}$ are open subsets for all $a\in\mathbf{R}$:

(iii) For every $\varepsilon>0$ and $z\in U_\varepsilon$

$$u(z)\leq\frac{1}{2\pi}\int_0^{2\pi}u(z+\varepsilon^{i\theta})d\theta.$$

A function $u: U\to(-\infty,\infty)$ is said to be **harmonic** if both u and $-u$ are subharmonic.

(3.3.2) Theorem (maximum principle). *Let u be a non-constant subharmonic function in a domain U. Then u does not attain the maximum value inside U.*

Proof. Suppose that u attains the maximum value at $z_0\in U$. Take any $\varepsilon>0$ so that $z_0\in U_\varepsilon$. Then by definition

$$\frac{1}{2\pi}\int_0^{2\pi}\{u(z_0+\varepsilon e^{i\theta})-u(z)\}d\theta\geq 0.$$

Since $u(z_0)\geq u(z_0+\varepsilon^{i\theta})$, we conclude that $u(z_0+\varepsilon e^{i\theta})=u(z_0)$ for all most all θ. The upper semicontinuity of u implies that $u(z_0+\varepsilon e^{i\theta})=u(z_0)$ for all θ.

Therefore u identically takes the maximum value $u(z_0)$ in a neighborhood of z_0. Now, put $V = \{w \in U; u(w) \geq u(z_0)\}$. Then by the above argument we see that V is open. On the other hand, V is closed by the upper semicontinuity of u. Since U is connected, $V = U$. *Q.E.D.*

(3.3.3) Lemma. (i) *Let u_j be subharmonic functions in U and a_j non-negative constants $(j=1, 2)$. Then $a_1u_1 + a_2u_2$ is subharmonic in U, where we set $0u_j = 0$.*

(ii) *Let $\{u_\lambda\}(\lambda \in \Lambda)$ be a family of subharmonic functions in U. Put*

$$u(z) = \sup_{\lambda \in \Lambda} u_\lambda(z) \quad \text{for } z \in U.$$

If $u(z) < \infty$ for all $z \in U$ and u is upper semicontinuous (these are satisfied for finite set Λ), then u is subharmonic.

(iii) *Let $\{u_j\}_{j=1}^{\infty}$ be a monotone-increasing sequence of subharmonic functions $(u_j(z) \leq u_{j+1})$. Then the limit function $u(z) = \lim_{j \to \infty} u_j$ is subharmonic or identically $-\infty$ in each connected component of U.*

Proof. The statment (i) will be clear. We show (ii). It is trivial that $u \not\equiv -\infty$. It is also clear by definition that if Λ is finite, u is upper semicontinuous. Take arbitrary $\varepsilon > 0$ and $z \in U_\varepsilon$. Since u is bounded from above on an arbitrary compact subset K of U, u_λ are uniformly bounded from above on K. Thus we can apply Fatou's lemma to get

$$\begin{aligned} u(z) = \sup_{\lambda \in \Lambda} u_\lambda(z) &\leq \sup_{\lambda \in \Lambda} \frac{1}{2\pi} \int_0^{2\pi} u_\lambda(z + \varepsilon e^{i\theta})\, d\theta \\ &\leq \frac{1}{2\pi} \int_0^{2\pi} u(z + \varepsilon e^{i\theta})\, d\theta. \end{aligned}$$

Hence u is subharmonic. For (iii) we assume that U is connected and $u \not\equiv -\infty$. It is clear that u is upper semicontinuous, so that u_j are uniformly bounded from above. Again by Fatou's lemma we have

$$\begin{aligned} u(z) = \lim_{j \to \infty} u_j(z) &\leq \overline{\lim_{j \to \infty}} \frac{1}{2\pi} \int_0^{2\pi} u_j(z + \varepsilon e^{i\theta})\, d\theta \\ &\leq \frac{1}{2\pi} \int_0^{2\pi} \overline{\lim_{j \to \infty}}\, u_j(z + \varepsilon e^{i\theta})\, d\theta = \frac{1}{2\pi} \int_0^{2\pi} u(z + \varepsilon e^{i\theta})\, d\theta. \end{aligned}$$

Therefore u is subharmonic. *Q.E.D.*

(3.3.4) Theorem. *Let u be a subharmonic function in U and $P(t)$ a monotone increasing convex function in* **R**. *Then $P \circ u$ is subharmonic, where $P(-\infty) = \lim_{t \to -\infty} P(t) \in [-\infty, \infty)$.*

Proof. It is clear that $P \circ u \not\equiv -\infty$. It follows from the conditions that P is continuous, so that $P \circ u$ is upper semicontinuous. Take arbitrary $\varepsilon > 0$ and $z \in U_\varepsilon$. Since P is monotone increasing and convex, we obtain

$$P(u(z)) \le P\left[\frac{1}{2\pi}\int_0^{2\pi} u(z+\varepsilon e^{i\theta})d\theta\right]$$

$$\le \frac{1}{2\pi}\int_0^{2\pi} P(u(z+\varepsilon e^{i\theta}))d\theta.$$

Thus $P \circ u$ is subharmonic. *Q.E.D.*

(3.3.5) Theorem. *Let u be a subharmonic function in U. Then $u \in L_{\text{loc}}(U)$ and moreover $dd^c[u]$ is a positive current.*

Proof. For a simplicity we assume that U is a domain. Let V be a set of points $z \in U$ such that z carry neighborhoods on which u is integrable. Then V is obviously open. Since $u \not\equiv -\infty$, there is a point $z \in U$ with $u(z) \ne -\infty$. Take $\varepsilon > 0$ so that $z \in U$. Then we have

$$u(z) \le \frac{1}{2\pi}\int_0^{2\pi} u(z+te^{i\theta})d\theta$$

for $0 < t < \varepsilon$. Multiplying both the hands of the above inequality by t, we have

$$\frac{\varepsilon^2}{2}u(z) \le \frac{1}{2\pi}\int_0^{\varepsilon} dt \int_0^{2\pi} tu(z+te^{i\theta})d\theta. \tag{3.3.6}$$

On the other hand, as u is upper semicontinuous, u is bounded from above on $\Delta(z;\varepsilon)$. Hence u is integrable on $\Delta(z;\varepsilon)$ and $V \ne \emptyset$; moreover

$$V \supset \{z \in U;\, u(z) \ne -\infty\}. \tag{3.3.7}$$

We claim that $U - V$ is open. In fact we take $z \in U - V$ and $\varepsilon > 0$ so that $z \in U_{4\varepsilon}$. Then it follows that $u(w) = -\infty$ for all $w \in \Delta(z;\varepsilon)$. Otherwise, take a point $w \in \Delta(z;\varepsilon)$ with $u(w) \ne -\infty$. Since $w \in U_{2\varepsilon}$, the above arguments imply that $\Delta(w;2\varepsilon) \subset V$; especially, $z \in V$. This is a contradiction. Thus V is non-empty, open and closed in the domain U and hence $V = U$.

We next show that $dd^c[u] \ge 0$. Take an arbitrary $\phi \in \mathcal{D}(U)$ with $\phi \ge 0$ and then a positive number ε so that $\operatorname{supp}\phi \subset U_\varepsilon$. Since

$$2\pi u(z) \le \int_0^{2\pi} u(z+\varepsilon e^{i\theta})d\theta,$$

we have

$$2\pi \int_U u(z)\phi(z)\frac{i}{2}dz \wedge d\bar{z} \le \int_0^{2\pi} d\theta \int_U u(z+\varepsilon e^{i\theta})\phi(z)\frac{i}{2}dz \wedge d\bar{z}$$

Continued

$$= \int_0^{2\pi} d\theta \int_U \phi(z - \varepsilon e^{i\theta}) u(z) \frac{i}{2} dz \wedge d\bar{z}.$$

Therefore

$$\text{(3.3.8)} \qquad \frac{1}{\varepsilon^2} \int_U u(z) \left\{ \int_0^{2\pi} (\phi(z - \varepsilon e^{i\theta}) - \phi(z)) d\theta \right\} \frac{i}{2} dz \wedge d\bar{z} \geq 0.$$

Put $w = -\varepsilon e^{i\theta}$ in the Taylor expansion

$$\phi(z+w) = \phi(z) + \frac{\partial \phi}{\partial z}(z) w + \frac{\partial \phi}{\partial \bar{z}}(z) \bar{w}$$

$$+ \frac{\partial^2 \phi}{\partial z^2}(z) w^2 + 2 \frac{\partial^2 \phi}{\partial z \partial \bar{z}}(z) w \bar{w} + \frac{\partial^2 \phi}{\partial \bar{z}^2}(z) \bar{w}^2 + \text{(higher term)},$$

substitute it to (3.3.8) and let $\varepsilon \to 0$. Then we get

$$\int_U u(z) \frac{\partial^2 \phi}{\partial z \partial \bar{z}}(z) \frac{i}{2} dz \wedge d\bar{z} \geq 0.$$

That is, $dd^c[u]$ is a positive current. *Q.E.D.*

(3.3.9) Theorem. *Let $u: U \to [-\infty, \infty)$ be a upper semicontinuous function and $u \not\equiv -\infty$. Then the following three conditions are equivalent.*

(i) *u is subharmonic.*

(ii) *For arbitrary $\varepsilon > 0$ and $z \in U_\varepsilon$*

$$u(z) \leq \frac{1}{\pi \varepsilon^2} \int_{w \in \Delta(\varepsilon)} u(z+w) \frac{i}{2} dw \wedge d\bar{w}.$$

(iii) *Let $K \subset U$ be a compact subset and h a continuous function on K which is harmonic in the set K° of interior points of K. If $u \leq h$ on $\partial K = K - K^\circ$, then $u \leq h$ on K.*

Proof. The implication (i)⇒(ii) is already showed in (3.3.6). We show (ii)⇒(iii). Applying (i)⇒(ii) to the subharmonic functions h and $-h$, we have

$$h(z) = \frac{1}{\pi \varepsilon^2} \int_{w \in \Delta(\varepsilon)} h(z+w) \frac{i}{2} dw \wedge d\bar{w}.$$

for $z \in (K^\circ)_\varepsilon$. Hence we have

$$\text{(3.3.10)} \qquad u(z) - h(z) \leq \frac{1}{\pi \varepsilon^2} \int_{w \in \Delta(\varepsilon)} \{u(z+w) - h(z+w)\} \frac{i}{2} dw \wedge d\bar{w}.$$

In the same way as in the proof of Theorem (3.3.2), we get

$$\sup\{u(z)-h(z);\ z\in K^{\circ}\}\le \sup\{u(z)-h(z);\ z\in\partial K\}.$$

Therefore $u\le h$. We show (iii)$\Rightarrow$(i). Take arbitrary $\varepsilon>0$ and $w\in U_{\varepsilon}$. Since u is upper semicontinuous on $\partial\Delta(w;\varepsilon)$, there is a monotone decreasing sequence $\{a_j\}_{j=1}^{\infty}$ of continuous functions u_j which converges point-wise to u. Integrating a_j with respect to the Poisson kernel, we obtain harmonic function h_j; i.e.,

$$h_j(z)=\frac{1}{2\pi}\int_0^{2\pi}a_j(w+\varepsilon e^{i\theta})\frac{\varepsilon^2-|z-w|^2}{|\varepsilon e^{i\theta}-(z-w)|^2}d\theta.$$

As well known, h_j are continuous in $\overline{\Delta(w;\varepsilon)}$ and $h_j|\partial B(w;\varepsilon)=a_j$. Since $u\le a_j|\partial\Delta(w;\varepsilon)$, $u\le h_j$ on $\Delta(w;\varepsilon)$ by the assumption; especially,

$$u(w)\le h_j(w)=\frac{1}{2\pi}\int_0^{2\pi}a_j(w+\varepsilon e^{i\theta})d\theta.$$

Letting $j\to\infty$, we have by Lebesgue's convergence theorem

$$u(w)\le\frac{1}{2\pi}\int_0^{2\pi}u(w+\varepsilon e^{i\theta})d\theta.\quad Q.E.D.$$

(3.3.11) Corollary. *If $u: U\to[-\infty,\infty)$ is subharmonic, then so is u in any open subset of U. Conversely, if any point of U carries a neighborhood where u is subharmonic, then u is subharmonic in U.*

Proof. Assume that u is subharmonic in U and U' is an open subset of U. For the subharmonicity of the restriction $u|U'$ it is sufficient to check that $u|U'\not\equiv-\infty$, and this follows from Theorem (3.3.5). For the proof of the second half we are going to use Theorem (3.3.9), (iii). Let u be a function satisfying the above condition. Let $K\subset U$ be a compact subset and h a continuous function on K such that u is harmonic in K° and $u\le h$ on ∂K. Take $0<\varepsilon<\delta$ so that $K\subset U_{\delta}$ and u is subharmonic in $\Delta(z;\varepsilon)$ for all $z\in K$. Then

$$u(z)\le\frac{1}{2\pi}\int_0^{2\pi}u(z+\varepsilon' e^{i\theta})d\theta$$

for all $\varepsilon'\le\varepsilon$. Since h is harmonic in K°,

$$u(z)-h(z)\le\frac{1}{2\pi}\int_0^{2\pi}\{u(z+\varepsilon' e^{i\theta})-h(z+\varepsilon' e^{i\theta})\}d\theta \tag{3.3.12}$$

for all $z\in(K^{\circ})_{\varepsilon}$. In the proof of the maximum principle Theorem (3.3.2) we used only the property (3.3.12). Hence it follows that $u-h$ does not attain the maximum at any point of every connected component of K° if it is not constant there. Since $u-h\le 0$ on ∂K, $u-h\le 0$ on K. *Q.E.D.*

We see by the above corollary that the subharmonicity is a local property. Here we give an important example of subharmonic function.

(3.3.13) Corollary. *Let $F(z)\not\equiv 0$ be a holomorphic function in U. Then* $\log|F|$ *is subharmonic, and so is $|F|^a$ for any positive $a\in\mathbf{R}$.*

Proof. It is clear that $\log|F|$ is upper semicontinuous. Let $K\subset U$ be a compact subset and h a continuous function on K such that h is harmonic in K° and $\log|F|\le h$ on ∂K. Let M be the maximum of $\log|F|-h$ on K. Assuming $M>0$, we are going to obtain a contradiction. Let $w\in K$ be a point with $M=\log|F(w)|-h(w)$. Then, of course, $w\in K^\circ$ and $F(w)\ne 0$. Therefore $\log|F|$ is harmonic in a neighborhood of w. It follows from Theorem (3.3.2) that $\log|F|-h$ is harmonic in the neighborhood of w. We see that $\log|F|-h\equiv M$ in the connected component V of K° containing w. On the other hand, $\log|F|-h\le 0$ on $\partial V\subset\partial K$. This is a contradiction. Hence $\log|F|\le h$, and then Theorem (3.3.9), (iii) implies that $\log|F|$ is subharmonic. Note that $t\in\mathbf{R}\to e^{at}\in\mathbf{R}$ ($a>0$) is a convex function. By Theorem (3.3.4) $|F|^a$ is subharmonic. *Q.E.D.*

We show a preparatory integral formula:

(3.3.14) Lemma. *Let $0<R<\infty$ and u be a C^2-function in a neighborhood of $\overline{\Delta(R)}$. Then for $0\le r\le R$*

$$\frac{1}{2\pi}\int_0^{2\pi}u(Re^{i\theta})d\theta-\frac{1}{2\pi}\int_0^{2\pi}u(re^{i\theta})d\theta=2\int_r^R\frac{dt}{t}\int_{\Delta(t)}dd^cu.$$

Proof. Put $z=te^{i\theta}$. The form $d^c\log|z|^2$ restricted over $\Gamma(t)$ is $d\theta/2\pi$. Using Stokes' theorem, we have the following for $r\ge 0$:

$$\frac{1}{2\pi}\int_0^{2\pi}u(Re^{i\theta})d\theta-\frac{1}{2\pi}\int_0^{2\pi}u(re^{i\theta})d\theta$$
$$=\int_{\Gamma(R)}ud^c\log|z|^2-\int_{\Gamma(r)}ud^c\log|z|^2$$
$$=\int_{\Delta(R)-\Delta(r)}d(ud^c\log|z|^2)=\int_{\Delta(R)-\Delta(r)}du\wedge d^c\log|z|^2$$
$$=\int_{\Delta(R)-\Delta(r)}d\log|z|^2\wedge d^cu\quad(\text{for } du\wedge d^c\log|z|^2=d\log|z|^2\wedge d^cu)$$
$$=2\int_r^R\frac{dt}{t}\int_{\Gamma(t)}d^cu\quad(\text{by Fubini's Theorem})$$
$$=2\int_r^R\frac{dt}{t}\int_{\Delta(t)}dd^cu.$$

We see the case of $r=0$ by letting $r\to 0$ in the above formula. *Q.E.D.*

(3.3.15) Corollary. *Let u be a C^2-function in U. Then u is subharmonic if and only if Δu $(= 4\partial^2 u / \partial z \partial \bar{z}) \geq 0$; especially, u is harmonic if and only if $\Delta u = 0$.*

Proof. The "only if" part follows from Theorem (3.3.5). The "if" part follows from Theorem (3.3.14) and the identity $dd^c u = (\Delta u / 4\pi)(i/2) dz \wedge d\bar{z}$. *Q.E.D.*

Now we set $u_\varepsilon = u * \chi_\varepsilon$ $(\varepsilon > 0)$ for a function u as explained in §1, (c).

(3.3.16) Theorem. (i) *Let u be a subharmonic function in U. Then $u_\varepsilon(z)$ is a* subharmonic C^∞-function in U_ε and converges monotone decreasingly to $u(z)$ as $\varepsilon \to 0$.

(ii) *If u is harmonic in U, then u is C^∞ and $\Delta u = 0$.*

Proof. The assertion (ii) easily follows from (i). We show (i). Noting that $(dd^c[u])_\varepsilon = dd^c[u_\varepsilon]$, we infer from Theorem (3.3.5) and Lemma (3.2.13) that $dd^c u_\varepsilon \geq 0$. By Corollary (3.3.15) u_ε is subharmonic. We next show the monotone decreasing convergence of U_ε. Changing the parameter in the integration, we have

$$u_\varepsilon(z) = \int_{w \in \mathbf{C}} u(z + \varepsilon w)\, \chi(w) \frac{i}{2} dw \wedge d\bar{w} \tag{3.3.17}$$

$$= \int_0^1 \chi(t) t dt \int_0^{2\pi} u(z + \varepsilon t e^{i\theta}) d\theta.$$

If u is C^2, $dd^c u \geq 0$ by Theorem (3.3.5), and moreover by Lemma (3.3.14)

$$\int_0^{2\pi} u(z + \delta t e^{i\theta}) d\theta \leq \int_0^{2\pi} u(z + \varepsilon t e^{i\theta}) d\theta$$

for $0 < \delta < \varepsilon$, $z \in U_\varepsilon$ and $0 \leq t \leq 1$. Thus it follows from (3.3.17) that $u_\delta(z) \leq u_\varepsilon(z)$. For a general u, we consider $(u_\gamma)_\varepsilon$ with $\gamma, \varepsilon > 0$. Since u_γ is a C^∞-subharmonic function, we deduce from the above arguments that $(u_\gamma)_\delta \leq (u_\gamma)_\varepsilon$ for $0 < \delta < \varepsilon$. Since $u_\gamma \to u$ $(\gamma \to 0)$ in the sense of L_{loc}, we infer from (3.3.17) that for $\tau > 0$, $(u_\gamma)_\tau \to u_\tau$ $(\gamma \to 0)$. Therefore $u_\delta \leq u_\varepsilon$. It follows from (3.3.17) that

$$u_\varepsilon(z) \geq 2\pi u(z) \int_0^1 \chi(t) dt = u(z).$$

Hence $\lim_{\varepsilon \to 0} u_\varepsilon(z) \geq u(z)$. On the other hand, the upper semicontinuity of u implies that for arbitrary $z \in U$ and α there is $\beta > 0$ such that $u(w) < u(z) + \alpha$ for all $w \in \Delta(z; \beta)$. It follows again from (3.3.17) that $u_\varepsilon(z) \leq u(z) + \alpha$ for $\varepsilon < \beta$, so that $\lim_{\varepsilon \to 0} u_\varepsilon(z) \leq u(z) + \alpha$. Since $\alpha > 0$ is arbitrary, we have $\lim_{\varepsilon \to 0} u_\varepsilon(z) \leq u(z)$, and then $\lim_{\varepsilon \to 0} u_\varepsilon(z) = u(z)$. *Q.E.D.*

(3.3.18) Corollary. *Assume that $u \in L_{\text{loc}}(U)$ and $dd^c[u] \geq 0$. Then there is a plurisubharmonic function which is equal almost everywhere to u.*

Proof. It follows from the assumption and Corollary (3.2.13) that $dd^c[u_\gamma] = [dd^c u]_\gamma \geq 0$ $(\gamma > 0)$. Thus Corollary (3.3.15) implies that u_γ is plurisubharmonic in U_γ. By Lemma (3.3.16), $(u_\gamma)_\delta \leq (u_\gamma)_\varepsilon$ for $0 < \delta < \varepsilon$. Letting $\gamma \to 0$, we have that $(u_\delta) \leq (u_\varepsilon)$. Therefore the subharmonic functions u_ε converge monotone decreasingly to u in the sense of $L_{\mathrm{loc}}(U)$, as $\varepsilon \to 0$. Put $v(z) = \lim_{\varepsilon \to 0} u_\varepsilon(z)$. Then v is almost everywhere equal to u, and so $v \not\equiv -\infty$. Thus v is subharmonic by Lemma (3.3.3), (iii). *Q.E.D.*

(3.3.19) Theorem. *Let $f: U \to V$ be a holomorphic mapping between two domains U and V of* $\mathbf{C}$. *If u is a subharmonic function in V, then $f^* u = u \circ f$ is either identically $-\infty$ or subharmonic in U.*

Proof. We denote by z (resp. w) the coordinate of U (resp. V). If u is C^2, then so is $f^* u$ and

$$\frac{\partial^2 f^* u}{\partial z \partial \bar{z}} = \left| \frac{df}{dz} \right|^2 \frac{\partial^2 u}{\partial w \partial \bar{w}} \geq 0.$$

By Corollary (3.3.15) $f^* u$ is subharmonic. For a general u, we first consider u_ε, which is C^∞ and subharmonic. Therefore $f^* u_\varepsilon$ is subharmonic in $f^{-1}(V_\varepsilon)$. We infer from Lemmas (3.3.16) and (3.3.3), (iii) that $f^* u$ is either identically $-\infty$ or subharmonic. *Q.E.D.*

Now we prove famous **Jensen's formula** which has various applications.

(3.3.20) Lemma. *Let u be a C^2- or subharmonic function in a neighborhood of $\overline{\Delta(R)}$ $(0 \leq R < +\infty)$. Then we have*

$$\frac{1}{2\pi}\int_0^{2\pi} u(Re^{i\theta})d\theta - \frac{1}{2\pi}\int_0^{2\pi} u(re^{i\theta})d\theta = 2\int_r^R \langle dd^c[u], \chi_{\Delta(t)} \rangle \frac{dt}{t}$$

for $0 \leq r \leq R$, where the second term of the left hand side stands for $u(0)$ for $r = 0$.

Proof. When u is a C^2-function, this was proved in Lemma (3.3.14). Assume that u is subharmonic. Taking small $\varepsilon > 0$, we obtain from Lemma (3.3.14)

$$\text{(3.3.21)} \qquad \frac{1}{2\pi}\int_0^{2\pi} u_\varepsilon(Re^{i\theta})d\theta - \frac{1}{2\pi}\int_0^{2\pi} u_\varepsilon(re^{i\theta})d\theta$$
$$= 2\int_r^R \frac{dt}{t}\int_{\Delta(t)} dd^c u_\varepsilon = 2\int_r^R \langle dd^c[u], \chi_{\Delta(t)} \rangle \frac{dt}{t}.$$

Since u_ε is uniformly bounded from above on $\overline{\Delta(R)}$ and converge monotone decreasingly to u, we have by Lebesgue's monotone convergence theorem

$$\text{(3.3.22)} \qquad \lim_{\varepsilon \to 0} \frac{1}{2\pi}\int_0^{2\pi} u_\varepsilon(Re^{i\theta})d\theta = \frac{1}{2\pi}\int_0^{2\pi} u(Re^{i\theta})d\theta,$$

$$\lim_{\varepsilon \to 0} \frac{1}{2\pi} \int_0^{2\pi} u_\varepsilon(re^{i\theta})d\theta = \frac{1}{2\pi} \int_0^{2\pi} u(re^{i\theta})d\theta.$$

For a sufficiently small $\delta > 0$, $dd^c[u]$ is a positive current (distribution in the present case), so that there is a positive Radon measure μ such that

$$<dd^c[u], \phi> = \int \phi \, d\mu \quad (\phi \in D(\Delta(R+\delta))) \text{ (cf. §1)}.$$

We have

$$0 \le \int_{\Delta(t)} dd^c u_\varepsilon \le <dd^c[u], \chi_{\Delta(R+\delta)}> \quad (0 \le t \le R). \tag{3.3.23}$$

In fact,

$$\int_{\Delta(t)} dd^c u_\varepsilon = \int_{z \in \Delta(t)} \left\{ \int_{w \in \mathbf{C}} \chi_\varepsilon(w-z) d\mu(w) \right\} \frac{i}{2} dz \wedge d\bar{z}$$

$$= \int_{w \in \mathbf{C}} d\mu(w) \int_{z \in \Delta(t)} \chi_\varepsilon(w-z) \frac{i}{2} dz \wedge d\bar{z} \le \int_{w \in \Delta(R+\delta)} d\mu(w)$$

$$\le <dd^c[u], \chi_{\Delta(R+\delta)}>.$$

As in (3.2.33), we see that

$$\lim_{\varepsilon \to 0} \int_{\Delta(t)} dd^c u_\varepsilon \le <dd^c[u], \chi_{\Delta(t)}>$$

for all but countably many $t \in [0, R]$. Combining this with (3.3.23), we may apply Lebesgue's convergence theorem to obtain

$$\lim_{\varepsilon \to 0} \int_r^R \frac{dt}{t} \int_{\Delta(t)} dd^c u_\varepsilon = \int_r^R <dd^c[u], \chi_{\Delta(t)}> \frac{dt}{t}.$$

Our assertion follows from this, (3.3.21) and (3.3.22). *Q.E.D.*

(3.3.24) Corollary. *Let u be a subharmonic function on U.*

(i) *For an arbitrary $0 < \delta < \varepsilon$ and $z \in U_\varepsilon$*

$$u(z) \le \frac{1}{2\pi} \int_0^{2\pi} u(z+\delta e^{i\theta})d\theta \le \frac{1}{2\pi} \int_0^{2\pi} u(z+\varepsilon e^{i\theta})d\theta.$$

Especially, u is integrable over the boundary circle of any closed disk in U.

(ii) $u(z) = \lim_{\varepsilon \to 0} \frac{1}{2\pi} \int_0^{2\pi} u(z+\varepsilon e^{i\theta})d\theta \ (z \in U).$

(iii) $u(z) = \overline{\lim_{w \to z}}\, u(w) \quad (z \in U).$

Proof. The inequalities of (i) follows from $dd^c[u] \geq 0$ and Jensen's formula, Lemma (3.3.20). Since $u \in L_{\mathrm{loc}}$ (Theorem (3.3.5)), for arbitrary $\varepsilon > 0$ and $z \in U_\varepsilon$ there is $\delta \in (0, \varepsilon)$ such that

$$\left| \frac{1}{2\pi} \int_0^{2\pi} u(z + \delta e^{i\theta}) d\theta \right| < \infty.$$

Hence we see by the inequality obtained just above that u is integrable over $\partial\Delta(z; \varepsilon)$. The assertion (ii) follows from (i) and the upper semicontinuity of u. We show (iii). By the definition of upper semicontinuity, $u(z) \geq \overline{\lim}_{w \to z} u(w)$. The converse immediately follows from (ii) and Fatou's lemma. *Q.E.D.*

We next show an extension theorem of Riemann for subharmonic functions.

(3.3.25) Theorem. *Let u be a subharmonic function in $\Delta^*(1)$ which is bounded from above around the origin. Then u is uniquely extended over $\Delta(1)$ as a subharmonic function.*

Proof. For an arbitrary $z \in \Delta(1)$ we put

$$v(z) = \overline{\lim_{\substack{w \to z \\ w \neq 0}}} u(z).$$

It follows from Corollary (3.3.24) that $v | \Delta^*(1) = u$ and $-\infty \leq v(z) < \infty$. It is obvious that $v \not\equiv -\infty$ and is upper semicontinuous. Suppose that v is not subharmonic in $\Delta(1)$. Then by Theorem (3.3.9), (iii) there are a relatively compact subdomain V of $\Delta(1)$ and a continuous function h such that h is harmonic in V, $v + h$ attains the maximum value L at some point $w \in V$ and $L > \max\{v(z) + h(z);\ z \in \partial V\}$. Necessarily, $w = 0$. Take $R > 0$ so that $\overline{\Delta(R)} \subset V$. Since $v + h$ is upper semicontinuous on $\partial\Delta(1)$, it attains the maximum value M there. Furthermore, since $v + h$ is subharmonic in $V - \{0\}$, $M < L$ by Theorem (3.3.2). Take $\varepsilon > 0$ with $\varepsilon < L - M$. Taking an arbitrary $r \in (0, R)$, we put

$$h_r(z) = \frac{\varepsilon}{\log \frac{r}{R}} \log \left\{ \left(\frac{r}{R} \right)^{L/\varepsilon} \frac{|z|}{r} \right\}$$

for $r \leq |z| \leq R$. Then

$$h_r(z) = \begin{cases} L - \varepsilon & \text{for } |z| = R, \\ L & \text{for } |z| = r. \end{cases}$$

Note that $v + h$ is subharmonic in a neighborhood of $\{r \leq |z| \leq R\}$, on which boundary $v + h \leq h_r$. Hence we have

$$v(z)+h(z) \leq h_r(z), \quad r \leq |z| \leq R.$$

For any fixed $z \in \overline{\Delta(R)}-\{0\}$, $h_r(z) \to L-\varepsilon$ as $r \to 0$, and henceforth $v(z)+h(z) \leq L-\varepsilon$. Therefore it follows that $v(0)+h(0) \leq L-\varepsilon$. This is a contradiction. The uniqueness follows from Corollary (3.3.24), (ii). *Q.E.D.*

(b) Plurisubharmonic functions. We consider the case of several complex variables. In this paragraph we denote by U an open subset of $\mathbf{C}^m$.

(3.3.26) *Definition.* A function $u: U \to [-\infty, \infty)$ is said to be **plurisubharmonic** if the following two conditions are satisfied:

(i) u is not identically $-\infty$ in any connected component of U, and upper semicontinuous.

(ii) For arbitrary $z \in U$ and $a \in \mathbf{C}^m$, the function in one variable

$$\zeta \in \mathbf{C} \to u(z+\zeta a) \in [-\infty, \infty)$$

is either identically $-\infty$ or subharmonic in every connected component of the open set $\{\zeta \in \mathbf{C};\ z+\zeta a \in U\}$.

Almost all the results proved in (a) for subharmonic functions are valid for plurisubharmonic functions. We simply give remarks and omit the proofs for those results which are obvious by the definitions and the results in (a).

(3.3.27) *Remark.* Theorem (3.3.2), Lemma (3.3.3) and Theorem (3.3.4) hold for plurisubharmonic functions.

(3.3.28) Theorem. *Let u be a plurisubharmonic function in U. Then $u \in L_{\mathrm{loc}}(U)$ and the current $dd^c[u]$ is positive and of type* $(1, 1)$.

Proof. As in (a) we set

$$U_\varepsilon = \{z \in U;\ \inf_{w \notin U} \|z-w\| > \varepsilon\}$$

for $\varepsilon > 0$. For $z \in U$ and $a \in \Gamma(1)$ we have

$$u(z) \leq \frac{1}{2\pi}\int_0^{2\pi} u(z+te^{i\theta}a)d\theta,\ 0 \leq t \leq \varepsilon. \tag{3.3.29}$$

Let $\rho: \Gamma(1) \to \mathbf{P}^{m-1}(\mathbf{C})$ be the Hopf fibering and ω the Fubini-Study-Kähler form on $\mathbf{P}^{m-1}(\mathbf{C})$. Then

$$u(z) = \int_{\mathbf{P}^{m-1}(\mathbf{C})} u(z)\omega^{m-1} \tag{3.3.30}$$

$$\leq \int_{[a] \in \mathbf{P}^{m-1}(\mathbf{C})} \left\{ \frac{1}{2\pi}\int_0^{2\pi} u(z+te^{i\theta}a)d\theta \right\} \omega^{m-1}$$

Continued

$$= \int_{x \in \mathbf{P}^{m-1}(\mathbf{C})} \left\{ \int_{w \in \rho^{-1}(x)} u(z + tw) d^c \log \|w\|^2 \right\} \omega^{m-1}$$

$$= \int_{w \in \Gamma(t)} u(z + w)\eta(w)$$

(cf. §2, (c)). Multiplying $2mt^{2m-1}$ to (3.3.30) and integrating it in $t \in [0, \varepsilon]$, we obtain

(3.3.31) $$\int_0^{\varepsilon} u(z) 2mt^{2m-1} dt \le \int_0^{\varepsilon} 2mt^{2m-1} dt \int_{w \in \Gamma(t)} u(z+w)\eta(w)$$

$$= \int_{w \in B(z;\, \varepsilon)} u(z+w)\alpha^m.$$

Hence we have

(3.3.32) $$u(z) \le \frac{1}{\varepsilon^{2m}} \int_{w \in B(z;\, \varepsilon)} u(z+w)\alpha^m.$$

Since $u \not\equiv -\infty$ in any connected component of U, we infer as in the proof of Theorem (3.3.5) that $u \in L_{\text{loc}}(U)$. Now we consider the current $dd^c[u]$. Note by definition that

$$dd^c[u] = \frac{i}{2\pi} \sum_{j,\, k=1}^{m} \frac{\partial^2 [u]}{\partial z^j \partial \bar{z}^k} dz^j \wedge d\bar{z}^k.$$

By Theorem (3.2.21) it is sufficient to show that for an arbitrary vector $a = (a^1, \ldots, a^m) \in \mathbf{C}^m$ the distribution

$$\sum_{j,k} \frac{\partial^2 [u]}{\partial z^j \partial \bar{z}^k} a^j \bar{a}^k$$

is positive. We may assume that $a \in \Gamma(1)$. It is obvious that the plurisubharmonicity is invariant under unitary transformations, so that we may assume $a = (1, 0, \ldots, 0)$. Take an arbitrary $w = (w^j) \in U$ and $\varepsilon > 0$ so that

$$D = \{z = (z^j) \in \mathbf{C}^m;\ |z^j - w^j| < \varepsilon,\ 1 \le j \le m\} \subset U.$$

Put

$$D' = \{z' = (z^2, \ldots, z^m);\ |z^j - w^j| < \varepsilon,\ 2 \le j \le m\}.$$

Take $\phi \in \mathcal{D}(D)$ with $\phi \ge 0$. Then

(3.3.33) $$\left\langle \frac{\partial^2 [u]}{\partial z^1 \partial \bar{z}^1}, \phi \right\rangle = \left\langle [u], \frac{\partial^2 \phi}{\partial z^1 \partial \bar{z}^1} \right\rangle$$

$$= \int_D u \frac{\partial^2 \phi}{\partial z^1 \partial \bar{z}^1} \frac{\pi^m}{m!} \alpha^m$$

$$= \int_{z' \in D'} \left\{ \int_{\{|z^1 - w^1| < \varepsilon\}} u(z^1, z') \frac{\partial^2 \phi}{\partial z^1 \partial \bar{z}^1} \pi \alpha(z^1) \right\} \frac{\pi^{m-1}}{m!} \alpha^{m-1}(z').$$

For almost all $z \in D'$, $u(z^1, z')$ are locally integrable in z^1 and hence not identically $-\infty$. Thus $u(z^1, z')$ are subharmonic in z^1 for almost all $z' \in D'$. For such z' we see by Theorem (3.3.5)

$$\int u(z^1, z') \frac{\partial^2 \phi}{\partial z^1 \partial \bar{z}^1}(z^1, z') \pi \alpha(z^1) \geq 0.$$

Combining this with (3.3.33), we obtain $<\partial^2[u]/\partial z^1 \partial \bar{z}^1, \phi> \geq 0$. *Q.E.D.*

(3.3.34) *Remark.* The local integrability of plurisubharmonic function proved above implies that Corollary (3.3.11) holds for plurisubharmonic functions.

(3.3.35) *Remark.* Let F be a holomorphic function in U which is not identically 0 in any connected component of U. Then, as in Corollary (3.3.13), $\log |F|$ is plurisubharmonic and so is $|F|^\alpha$ for $\alpha > 0$.

(3.3.36) Lemma. *Let u be a C^2-function in U. Then u is plurisubharmonic if and only if $dd^c u \geq 0$.*

Proof. The "only if" part is clear by Theorem (3.3.28). For the "if" part we take an arbitrary complex line L in $\mathbf{C}^m$. Then

$$dd^c(u | L \cap U) = (dd^c u) | L \cap U.$$

Thus $dd^c(u|L \cap U) \geq 0$. By Corollary (3.3.15) $u|L \cap U$ is subharmonic. Therefore u is plurisubharmonic. *Q.E.D.*

(3.3.37) Lemma. *Let u be a plurisubharmonic function in U. For $\varepsilon > 0$ the smoothing u_ε is C^∞ plurisubharmonic function. Moreover, $u \leq u_\delta \leq u_\varepsilon$ for $0 < \delta < \varepsilon$ and $\lim_{\varepsilon \to 0} u_\varepsilon = u$.*

Proof. Since $u \in L_{\mathrm{loc}}(U)$ (Theorem (3.3.28)), $u_\varepsilon \in E(U_\varepsilon)$. Since $dd^c[u_\varepsilon] = (dd^c[u])_\varepsilon$, $dd^c[u_\varepsilon] \geq 0$. Then Lemma (3.3.34) implies that u_ε is plurisubharmonic. By definition we have

$$u_\varepsilon(z) = \int_{w \in \mathbf{C}^m} u(w) \chi_\varepsilon(w - z) \frac{\pi^m}{m!} \alpha^m$$

$$= \int_{w \in \mathbf{C}^m} u(z + \varepsilon w) \chi(w) \frac{\pi^m}{m!} \alpha^m.$$

Making use of the Hopf fibering as in (3.3.30) and (3.3.31) and of Lemma (3.3.16), we see that $u \le u_\delta \le u_\varepsilon$ for $0 < \delta < \varepsilon$ and $\lim_{\varepsilon \to 0} u_\varepsilon = u$. *Q.E.D.*

(3.3.38) *Remark.* By the above Lemma (3.3.37), Corollary (3.3.18) and Theorem (3.3.19) hold for plurisubharmonic functions.

(3.3.39) Lemma (Jensen's formula). *Let u be a C^2-function or a plurisubharmonic function on a neighborhood of $\overline{B(R)}$ $(R > 0)$. Then*

$$\int_{\Gamma(R)} u\eta - \int_{\Gamma(r)} u\eta = 2\int_r^R \langle dd^c[u] \wedge \alpha^{m-1}, \chi_{B(t)} \rangle \frac{dt}{t^{2m-1}}$$

for $0 \le r \le R$, where for $r = 0$ we put

$$\int_{\Gamma(0)} u\eta = u(O).$$

Proof. We first consider the case where u is of class C^2. Denoted by I, the left-hand side of the given equation, we have by Stokes' theorem

$$I = \int_{B(R)-B(r)} d(u \wedge \eta)$$

$$= \int_{B(R)-B(r)} du \wedge d^c \log \|z\|^2 \wedge \beta^{m-1} \quad (\text{by } d\eta = 0)$$

$$= 2\int_{B(R)-B(r)} d\log \|z\|^2 \wedge d^c u \wedge \beta^{m-1}$$

$$(\text{by } du \wedge d^c \log \|z\|^2 \wedge \beta = d\log \|z\|^2 \wedge d^c u \wedge \beta^{m-1})$$

$$= 2\int_r^R \frac{dt}{t} \int_{\Gamma(t)} d^c u \wedge \beta^{m-1} \quad (\text{by Fubini's theorem})$$

$$= 2\int_r^R \frac{dt}{t^{2m-1}} \int_{\Gamma(t)} d^c u \wedge \alpha^{m-1} \quad (\text{by (3.2.29)})$$

$$= 2\int_r^R \frac{dt}{t^{2m-1}} \int_{B(t)} dd^c u \wedge \alpha^{m-1} \quad (\text{by Stokes' theorem}).$$

For general u we take the smoothing u_ε and apply the above arguments. Then, as in the proof of Lemma (3.3.20), we have the required equality by letting $\varepsilon \to 0$. *Q.E.D.*

As in Corollary (3.3.24), we obtain the following.

(3.3.40) Corollary. *Let u be a plurisubharmonic function in U.*

(i) *For $0 \le \delta \le \varepsilon$ and $z \in U_\varepsilon$*

$$u(z) \leq \int_{w\in\Gamma(\delta)} u(z+w)\eta(w) \leq \int_{w\in\Gamma(\varepsilon)} u(z+w)\eta(w).$$

In special, u is integrable over spheres in U.

(ii) $u(z) \leq \lim_{\varepsilon\to 0} \int_{w\in\Gamma(\delta)} u(z+w)\eta(w)$ *for* $z\in U$.

(iii) $u(z) = \overline{\lim_{w\to z}}\, u(w)$ *for* $z\in U$.

Now we prove an extension theorem of Riemann type for plurisubharmonic functions.

(3.3.41) Theorem. *Let A be a nowhere dense analytic subset of U and u a plurisubharmonic function in* $U-A$. *If u is locally bounded around all points of A, then u is uniquely extended over U as plurisubharmonic function.*

Proof. The uniqueness is clear by Lemma (3.3.37). We put

$$v(z) = \overline{\lim_{\substack{w\to z\\ w\in U-A}}}\; u(w)$$

for $z\in U$. Then Corollary (3.3.40), (iii) implies that $v\,|\,(U-A) = u$. The function v is upper semicontinuous and $v\not\equiv -\infty$ in any connected component of U. Take arbitrarily a point $z\in U$ and a vector $a\in \mathbf{C}^m$ with $\|a\| = 1$. Then we show that the function

$$\zeta \to v(z+\zeta a)$$

is either identically $-\infty$ or subharmonic in the connected component V of the open subset $\{\zeta\in\mathbf{C};\, z+\zeta a\in U\}$ which contains z. We may assume that $u\not\equiv-\infty$ in V. It is sufficient to show that

$$v(z) \leq \frac{1}{2\pi}\int_0^{2\pi} v(z+\varepsilon e^{i\theta}a)d\theta$$

for all small $\varepsilon > 0$. There is a sequence $\{z_j\}\subset U-A$ $(j = 1, 2, \ldots)$ such that $\lim_{j\to\infty} z_j = z$ and $\lim_{j\to\infty} v(z_j) = v(z)$. Take a sequence $\{a_j\}$ of vectors $a_j\in\Gamma(1)$ so that $\lim_{j\to\infty} a_j = a$ and the sets

$$S_j = \{\zeta\in W;\, z_j+\zeta a_j\in A\}$$

are discrete. By Theorem (3.3.25) and the assumption we may suppose that $v(z_j+\zeta a_j)$ are plurisubharmonic functions in V. Hence we have

$$v(z_j) \leq \frac{1}{2\pi}\int_0^{2\pi} v(z_j+\varepsilon e^{i\theta}a_j)d\theta.$$

It follows from Fatou's lemma that

$$v(z) \le \frac{1}{2\pi}\int_0^{2\pi} \overline{\lim_{j\to\infty}}\, v(z_j + \varepsilon e^{i\theta} a_j) d\theta$$
$$\le \frac{1}{2\pi}\int_0^{2\pi} v(z + \varepsilon e^{i\theta} a) d\theta \quad Q.E.D.$$

Next we prove an extension theorem of Hartogs type.

(3.3.42) Theorem. *Let A be an analytic subset of U and u a plurisubharmonic function in $U - A$. If* $\operatorname{codim} A \ge 2$, *then u is uniquely extended to a plurisubharmonic function in U.*

Proof. We first show that for $z \in U$ and $a \in \mathbf{C}^m - \{O\}$ such that

(3.3.43) $$\{z + \zeta a;\, \zeta \in \mathbf{C}\} \cap U \not\subset A,$$

the function

$$\zeta \to u(z + \zeta a)$$

extends to a subharmonic function or identically $-\infty$ in each connected component of $\{\zeta \in \mathbf{C};\, z + \zeta a \in U\}$. We may assume that $z \in A$ and deal with the problem around $\zeta = 0$. Suppose that $u(z + \zeta a) \not\equiv -\infty$ there. By (3.3.43) there is a constant $\delta > 0$ such that

$$\{z + \zeta a;\, |\zeta| \le \delta\} \cap A = \{z\}.$$

Since $\operatorname{codim} A \ge 2$, there is $b \in \mathbf{C}^m - \{O\}$ such that z is an isolated point of

$$\{z + \zeta a + \xi b;\, \zeta, \xi \in \mathbf{C}\} \cap A.$$

The case of $m = 2$ is clear, since A is discrete. In general, if A is discrete, then this is obvious. Otherwise, we show this fact by the induction on m. Suppose that it holds up to $m - 1$ ($m \ge 3$). There is a complex hyperplane H through $\{z,\, z + a\}$ such that $A \not\subset H$. Otherwise, we have a contradiction:

$$A \subset \bigcap_{z,\, z+a \in H} H = \{z + \zeta a;\, \zeta \in \mathbf{C}\}.$$

The set $A \cap H$ is a proper analytic subset of H and $\operatorname{codim}_H A \cap H \ge 2$. By the induction hypothesis there is $b \in \mathbf{C}^m - \{O\}$ such that $z + b \in H$ and $z \in H$ is isolated in

$$\{z + \zeta a + \xi b;\, \zeta, \xi \in \mathbf{C}\} \cap (A \cap H).$$

Therefore z is an isolated point of

$$\{z + \zeta a + \xi b;\, \zeta, \xi \in \mathbf{C}\} \cap A.$$

Taking δ smaller if necessary, we have some $\varepsilon > 0$ such that

$$\{z+\zeta a+\xi b;\, |\zeta| \le \delta,\, 0<|\xi| \le \varepsilon\} \cap A = \emptyset,$$

$$\{z+\zeta a+\xi b;\, 0<|\zeta| \le \delta,\, |\xi| \le \varepsilon\} \cap A = \emptyset.$$

For an arbitrary $0 < |\zeta| < \delta$ we have

$$u(z+\zeta a) \le \frac{1}{2\pi}\int_0^{2\pi} u(z+\zeta a+\varepsilon e^{i\theta} b)d\theta$$

$$\le \max\{u(z+\zeta a+\xi b);\, |\zeta| \le \delta,\, |\xi| = \varepsilon\}.$$

Thus $u(z+\zeta a)$ is bounded from above in $\{\zeta \in \mathbf{C};\, 0 < |\zeta| < \delta\}$, and hence by Theorem (3.3.25) it extends to a subharmonic function around 0.

We next show that u is bounded from above in a neighborhood of every point $w \in A$. Note that this and Theorem (3.3.42) complete the proof. By a parallel translation we may assume that $w = 0$. By making use of $\operatorname{codim} A \ge 2$ and the above argument there is a holomorphic coordinate system $(z^1, \ldots, z^m)$ such that $\Delta(\delta^1) \times \cdots \times \Delta(\delta^m) \subset U$ and $\pi^{-1}(O) = 0$, where δ^j are positive constants and

$$\pi\colon (z^j) \in \{(z^j) \in A;\, |z^j| < \delta^j\} \to z' = (z^3, \ldots, z^m) \in \prod_{j=3}^{m} \Delta(\delta^j)$$

is the projection. Moreover, we may assume that π is proper (cf. Chapter IV, §1). We put

$$E = \left\{ (z^j) \in \mathbf{C}^m;\, |z^1| = \delta^1,\, |z^j| \le \frac{\delta^j}{2},\, 2 \le j \le m \right\}.$$

Since π is proper, $E \cap A = \emptyset$. Noting that E is compact, we put

$$L = \max_{z \in E} u(z).$$

Take an arbitrary point $w = (w^j) \in \Delta(\delta^1) \times \Delta(\delta^2/2) \times \cdots \times \Delta(\delta^m/2) - A$. Since π is proper,

$$\{(\zeta, w^2, \ldots, w^m);\, |\zeta| \le \delta^1\} \not\subset A.$$

Therefore $u(\zeta, w^2, \ldots, w^m)$ is extended to a subharmonic function in a neighborhood of $\{|\zeta| \le \delta^1\}$. Then we have

$$u(w) \le \max_{|\zeta| = \delta^1} u(\zeta, w^2, \ldots, w^m) \le L. \quad Q.E.D.$$

(3.3.44) Corollary. *Let U and A be as in Theorem (3.3.42). Let F be a holomorphic function in $U - A$. Then there is a unique holomorphic function $\tilde{F}$ on U such that $\tilde{F}|(U-A) = F$.*

The proof is clear by Theorem (3.3.42) and Riemann's extension theorem.

Notes

The theory of currents was initiated by de Rham (cf. [95]). The theory of positive currents is due to Lelong [63]. The notion of Lelong number $L(z;T)$ for a closed positive current T is very important. Siu [105] proved that $L(z;T)$ is independent of the choice of holomorphic local coordinate system (cf, also, Demailly [24]). Moreover, Siu [105] proved that for an arbitrary constant $c>0$ the subset $\{z;L(z;T)\geq c\}$ forms an analytic subset of codimension $\geq p$ (cf, also, Demailly [24]). Several extension theorems for closed positive currents have been obtained (cf. e.g., [108, 69, 104]).

It is known that the extension Theorems (3.3.41) and (3.3.42) for plurisubharmonic functions are also valid over normal complex spaces (Grauert-Remmert [36]). The potential theory of plurisubharmonic functions related to the complex Monge-Ampére equation $\overset{m}{\wedge}\, dd^c u = \phi$ was lately explored by Bedford-Taylor [5, 6]. There are also some works on this subject from the viewpoint of probability theory ([91, 31]).

CHAPTER IV

Meromorphic Mappings

4.1 Analytic Subsets

In this section we state well-known fundamental facts on analytic subsets without proofs, which are already used frequently in this book. For general references, we refer to [72] and a part of [18]. Let M be an m-dimensional complex manifold. For $x \in M$ we put

$$F_{M,x} = \{a_U : U \to \mathbf{C}, \text{ holomorphic in neighborhoods } U \text{ of } x\}.$$

We introduce an equivalence relation in $F_{M,x}$ as follows. We say that $a_U, b_V \in F_{M,x}$ is equivalent and write $a_U \sim b_V$, if there is a neighborhood $W \subset U \cap V$ of x such that $a_U | W = b_V | W$. Then we set

$$\gamma_x : F_{M,x} \to F_{M,x}/\sim = O_{M,x}$$

and call $O_{M,x}$ the **local ring of holomorphic functions** at x over M. An element of $\gamma_x(a_U) \in O_{M,x}$ is called a **germ of holomorphic function** at x over M. Then $O_{M,x}$ naturally forms a ring with 1 by

$$\gamma_x(a_U) + \gamma_x(b_V) = \gamma_x(a_U | U \cap V + b_V | U \cap V),$$

$$\gamma_x(a_U)\gamma_x(b_V) = \gamma_x((a_U | U \cap V)(b_V | U \cap V)).$$

An element of the ring is called a **unit** if it has an inverse element, and we denote by $O^*_{M,x}$ the set of all units of $O_{M,x}$; i.e.,

$$O^*_{M,x} = \{\gamma_x(a_U) \in O_{M,x};\ a_U(x) \neq 0\}.$$

For $f, g \in O_{M,x}$ with $g \neq 0$, we say that g divides f and write $g | f$, if there is a non-zero element $h \in O_{M,x}$ such that $f = gh$; moreover, g is called a **divisor** of f in this case. Then $O_{M,x}$ is an **integral domain**; i.e., there is no non-zero element of $O_{M,x}$

which divides 0. A non-zero element $f \in \mathcal{O}_{M,x}$ is said to be **irreducible** if there is no divisor of f other than one equivalent to itself or unit; otherwise, it is said to be **reducible.** Let $f_1, f_2, \ldots \in \mathcal{O}_{M,x}$. If there is no common divisor other than unit for all $f_1, f_2, \ldots$, then $f_1, f_2, \ldots$ are said to be mutually irreducible. The following is an important fact.

(4.1.1) Theorem. (i)(unique factorization) *Every element $f \in \mathcal{O}_{M,x}$ is written as a product of finitely many irreducible elements. Moreover, if $f = f_1 \cdots f_p = g_1 \cdots g_q$ with f_j, g_k irreducible, then $p = q$ and there are a permutation σ of $\{1, \ldots, p\}$ and units $u_j \in \mathcal{O}^*_{M,x}$ such that $f_j = u_j g_{\sigma(j)}$ for all j.*

(ii) *$\mathcal{O}_{M,x}$ is a Noetherian ring.*

(iii) *If $\gamma_x(a_U)$ and $\gamma_x(b_U)$ are mutually irreducible, there is a neighborhood $W \subset U \cap V$ of x such that $\gamma_y(a_U)$ and $\gamma_y(b_V)$ are mutually irreducible for all $y \in W$.*

A subset X of M is called an **analytic subset** if for any $x \in M$ there are a neighborhood U of x and holomorphic functions $f_1, \ldots, f_l$ such that $X \cap U = \{z \in U; f_1(z) = \cdots = f_l(z) = 0\}$. The following are clear by the definition and Theorem (4.1.1), (ii).

(4.1.2) Lemma. (i) *Analytic subsets are closed.*

(ii) *M itself and the empty set $\emptyset$ are analytic subsets.*

(iii) *Let X be an analytic subset of M and U an open subset of M. Then $X \cap U$ is an analytic subset of U.*

(iv) *If an analytic subset X of M contains an interior point, then $X = M$.*

(v) *Let $\{X_\lambda\}_{\lambda \in \Lambda}$ be a family of analytic subsets of M. Then $\bigcap_{\lambda \in \Lambda} X_\lambda$ is an analytic subset of M, and so is $\bigcup_{\lambda \in \Lambda} X_\lambda$ if $\{X_\lambda\}_{\lambda \in \Lambda}$ is locally finite; i.e., for any compact subset K of M, $\{\lambda \in \Lambda; \bar{X}_\lambda \cap K \neq \emptyset\}$ is finite.*

Let X be an analytic subset of M. Then X is said to be **reducible** if there are non-empty distinct analytic subset X_1 and X_2 of M such that $X = X_1 \cup X_2$ and $X \neq X_j$ for $j = 1, 2$; otherwise, X is said to be **irreducible.** Let $x \in X$. We say that X is **locally reducible** at x if for any neighborhood U of x there is a neighborhood $V \subset U$ of x such that $X \cap V$ is reducible; otherwise, X is said to be **locally irreducible** at x. The point x is called a **regular** (**non-singular**) **point** of X if there is a neighborhood U such that $X \cap U$ is a complex submanifold. We denote by $R(X)$ the set of all regular points of X and set $S(X) = X - R(X)$. A point of $S(X)$ is called a **singular point** of X, and if $S(X) = \emptyset$, then X is called a **non-singular (smooth)** analytic subset.

(4.1.3) Lemma. (i) $R(M) = M$.

(ii) *A non-singular analytic subset $X \subset M$ is irreducible if and only if X is connected.*

(iii) *Let $X \subset M$ be an analytic subset. Then X is locally irreducible at all $x \in R(X)$.*

The following are known important facts.

(4.1.4) Theorem. *Let X be an analytic subset of M.*

(i) *$R(X)$ is a dense open subset of X.*

(ii) *$S(X)$ is an analytic subset of M and $S(X) \underset{\neq}{\subset} X$.*

(iii) *Let $R(X) = \bigcup_{\lambda \in \Lambda} X_\lambda$ be the decomposition of $R(X)$ into connected components. Then X_λ are locally closed complex submanifolds of M and the closures $\overline{X}_\lambda$ are irreducible analytic subsets of M.*

(iv) *$X = \bigcup_{\lambda \in \Lambda} \overline{X}_\lambda$ and $\overline{X}_\lambda \neq \overline{X}_\mu$ for $\lambda \neq \mu$. The family $\{\overline{X}_\lambda\}$ is locally finite.*

(v) *Let $Y \subset M$ be any irreducible analytic subset such that $Y \subset X$. Then there is some $\overline{X}_\lambda$ such that $Y \subset \overline{X}_\lambda$.*

Each of the above $\overline{X}_\lambda$ is called an **irreducible component** of X and $X = \bigcup_{\lambda \in \Lambda} \overline{X}_\lambda$ is called the **irreducible decomposition** of X. The irreducible decomposition is unique in the sense that if $X = \bigcup_{\gamma \in \Gamma} Y_\gamma$ with irreducible analytic subsets $Y_\gamma \subset M$, then there is a bijection $\sigma: \Lambda \to \Gamma$ such that $\overline{X}_\lambda = Y_{\sigma(\lambda)}$. The following corollary is clear.

(4.1.5) Corollary. *Let $X \subset M$ be an analytic subset. Then X is irreducible if and only if $R(X)$ is connected. Moreover, if X is irreducible, then X is connected and X is the only one irreducible component of X.*

The following is a local version of Theorem (4.1.4) and elementary fact in the local theory of analytic subsets.

(4.1.6) Theorem. *Let $X \subset M$ be an analytic subset and $x \in X$. Then there is a neighborhood U of x satisfying the following:*

(i) *Let $X \cap U = \bigcup_{j=1}^{l} A_j$ be the irreducible decomposition of $X \cap U$. Then $l < \infty$ and $x \in A_j$ for all j.*

(ii) *For an arbitrary neighborhood V of x, there is a neighborhood $W \subset V \cap U$ of x such that $X \cap W = \bigcup_{j=1}^{l} A_j \cap W$ is the irreducible decomposition of the analytic subset $X \cap W$ of W.*

Each of the above A_j is called a **local irreducible component** of X at x and $X \cap U = \bigcup_{j=1}^{l} A_j$ is called the **local irreducible decomposition** of X at x. These two notions are uniquely determined in the sense of above (ii). The number l is determined only by X and x, and X is locally irreducible at x if and only if $l=1$.

Let $x \in X$ and define the **local codimension** of X at x as follows. In the case where X is locally irreducible at x, we set

$$\begin{gathered}\operatorname{codim}_{M,x} X = \sup\{k;\text{ there is a } k\text{-dimensional} \\ \text{complex submanifold } N \text{ containing } x \text{ such that} \\ x \text{ is an isolated point of } N \cap X\}.\end{gathered}$$

In general, we take a neighborhood U of x as in Theorem (4.1.6), (ii), so that $X \cap U = \bigcup_{j=1}^{l} A_j$ is the local irreducible decomposition of X at x. Then we set

$$\operatorname{codim}_{M,x} X = \inf_{1 \le j \le l} \operatorname{codim}_{U,x} A_j.$$

This is independent of the choice of U by Theorem (4.1.6), (ii). We set the **local dimension** $\dim_x X$ of X at x by

$$\dim_x X = \dim M - \operatorname{codim}_{M,x} X.$$

In case $x \in R(X)$, there is a neighborhood U of x such that $X \cap U$ is a closed complex submanifold of U and $\dim_x X$ coincides the complex dimension of $X \cap U$. Furthermore, if $N \subset M$ is a closed complex submanifold containing X, then

$$\dim N - \operatorname{codim}_{N,x} X = \dim M - \operatorname{codim}_{M,x} X$$

for $x \in X$. Thus we see that $\dim_x X$ is determined only by X and x. We define the **dimension** of X by

$$\dim X = \sup\{\dim_x X;\, x \in X\}.$$

For the sake of convenience, we set

$$\dim \emptyset = -1.$$

In general, X is said to be of **pure dimension** if $\dim_x X = \dim X$ for all $x \in X$.

(4.1.7) Theorem. *Let $U \subset \mathbf{C}^m$ be a neighborhood of the origin O and $X \subset U$ an analytic subset containing O such that $\dim_O X = n$. Then, after a linear change of coordinates, there are $r_j > 0$, $j = 1, 2, \ldots, m$ such that the projection*

$$\pi\colon (z^j) \in \left[\prod_{j=1}^{m} \Delta(r_j)\right] \cap X \to (z^1, \ldots, z^n) \in \prod_{j=1}^{n} \Delta(r_j)$$

is surjective, proper and of finite fibers.

(4.1.8) Theorem. *Let $X \subset M$ be an analytic subset and $x \in X$. Assume that X is* *locally irreducible at x. Then there is a neighborhood W of x in X such that* $\dim_x X = \dim_y X$ *for all* $y \in W$.

We immediately have the following by Theorem (4.1.4), (i) , Corollary (4.1.5) and Theorem (4.1.8).

(4.1.9) Corollary. *An irreducible analytic subset of M is of pure dimension.*

(4.1.10) Corollary. *The function* $x \in X \to \dim_x X \in \mathbf{Z}$ *is upper semicontinuous.*

(4.1.11) Theorem. *Let X and Y be analytic subsets of M.*

(i) *Assume that X is irreducible. If* $X \underset{\neq}{\supset} Y$, *then* $\dim X > \dim Y$; *especially* $\dim S(X) < \dim X$.

(ii) *Let* $X = \bigcup_{\lambda \in \Lambda} X_\lambda$ *be the irreducible decomposition. If* $X_\lambda \not\subset Y$ *for all* λ, *then* $X - Y = X - (X \cap Y)$ *is a dense, locally connected, open subset of X. In special, if X is irreducible,* $X - Y$ *is connected.*

(4.1.12) Lemma. *Let* M_j $(j = 1, 2)$ *be complex submanifolds and* $X_j \subset M_j$ *analytic subsets. Then* $X_1 \times X_2$ *is an analytic subset of* $M_1 \times M_2$.

This is clear by definition. The following theorem is not trivial but will be frequently used.

(4.1.13) Theorem (Remmert-Stein). *Let* $Y \subset M$ *and* $X \subset M - Y$ *be analytic subsets. Let* $X = \bigcup_{\lambda \in \Lambda} X_\lambda$ *be the irreducible decomposition. If* $\dim X_\lambda > \dim Y$ *for all* λ, *then the closure* $\bar{X}$ *in M is an analytic subset of M and* $\bar{X} = \bigcup_{\lambda \in \Lambda} \bar{X}_\lambda$ *the irreducible decomposition of* $\bar{X}$.

4.2 Divisors and Meromorphic Functions

In this section we denote by M an m-dimensional complex manifold. A subset $X \subset M$ is called an **analytic hypersurface** if for any $x \in M$ there are a neighborhood U of x and a holomorphic function f such that $X \cap U = \{z \in U; f(z) = 0\}$. Of course, X is an analytic subset of M in the sense of §1. The following lemma is not trivial but fundamental.

(4.2.1) Lemma. *Let* $X \subset M$ *be an analytic hypersurface. Then X is of pure dimension* $m - 1$. *Conversely, any analytic subset X of pure dimension* $m - 1$ *is an analytic hypersurface.*

Let $X \subset M$ be an analytic subset. We set

$$I_{X,x} = \{\gamma_x(a_U) \in O_{M,x};\ a_U | X = 0\}$$

for $x \in M$. Then $I_{X,x}$ forms an ideal of $O_{M,x}$. Since $O_{M,x}$ is Noetherian (Theorem (4.1.1), (ii)), $I_{X,x}$ is finitely generated. Hence there are a neighborhood U of x and finitely many holomorphic functions $F_1, \ldots, F_l$ on U such that

$$(4.2.2) \qquad I_{X,x} = \sum_{i=1}^{l} O_{M,x}\gamma_x(F_i),$$

$$X \cap U = \{F_1 = \cdots = F_l = 0\}.$$

Those $F_1, \ldots, F_l$ are called the **defining functions** of X at x and $F_1 = \cdots = F_l = 0$ is called the **defining equation** of X at x. It is known by the famous coherence theorem that there is a neighborhood $V \subset U$ of x such that

$$I_{X,y} = \sum_{i=1}^{l} O_{M,y}\gamma_y(F_i)$$

for all $y \in V$ (cf. [18] and [90], VIII). We denote by $O_{X,x}$ the quotient ring of $O_{M,x}$ by $I_{X,x}$ and the quotient mapping by

$$f \in O_{M,x} \to f|X \in O_{X,x} = O_{M,x}/I_{X,x}.$$

It follows from Theorem (4.1.4), Corollary (4.1.5) and Theorem (4.1.7) that X is locally irreducible if and only if $O_{X,x}$ is an integral domain. The following is important.

(4.2.3) Theorem. *Let $X \subset M$ be an analytic hypersurface and $x \in X$.*

(i) *$I_{M,x}$ is a principal ideal; i.e., there is an element $f \in I_{M,x}$ such that $I_{X,x} = O_{M,x} \cdot f$.*

(ii) *If $g \in I_{X,x}$ satisfies $I_{X,x} = O_{M,x} \cdot g$, then there is an element $u \in O^*_{M,x}$ such that $g = uf$.*

(iii) *Let $\gamma_x(a_U) \in I_{X,x}$ satisfy $I_{X,x} = O_{M,x} \cdot \gamma_x(a_U)$, then there is a neighborhood $V \subset U$ of x such that $I_{X,y} = O_{M,y} \cdot \gamma_y(a_U)$ for any $y \in V$.*

A formal sum $D = \sum_{\lambda \in \Lambda} \nu_\lambda X_\lambda$ of a locally finite family $\{X_\lambda\}_{\lambda \in \Lambda}$ with coefficients $\nu_\lambda \in \mathbf{Z}$ is called a **divisor** on M. Without loss of generality, we may assume that X_λ are irreducible and mutually distinct, and that any $\nu_\lambda \neq 0$. Then we define the **support** supp D of the divisor D by

$$\operatorname{supp} D = \bigcup_{\lambda \in \Lambda} X_\lambda$$

Note that supp D is an analytic hypersurface. In the case of all $\nu_\lambda \geq 0$, D is called an **effective divisor.** Every X_λ is called an **irreducible component** of the divisor D. Especially, an irreducible hypersurface is also called an **irreducible divisor.** The family of all divisors on M forms a $\mathbf{Z}$-module, which is denoted by Div(M). Let $U \subset M$ be an open subset. Then we define the restriction divisor $D|U$ of D over U by $D|U = \sum_\lambda \nu_\lambda (X_\lambda \cap U)$.

We then define meromorphic functions in M. Let $X \subset M$ be an analytic hypersurface. A holomorphic function $f_{M-X}: M-X \to \mathbf{C}$ is said to have **at most poles** on X if

(4.2.4) for an arbitrary $x \in M$, there are a neighborhood $U \subset M$ of x and a holomorphic function a_U defined in U such that $a_U f_{M-X}$ is holomorphic in U.

This condition (4.2.4) is essential only for $x \in X$; i.e., for $x \in M-X$ we may take $U = M-X$ and $a_U = 1$. Moreover, if f_{M-X} can not holomorphically extends over any neighborhood of any point of X, then f_{M-X} is said to have **poles** on X.

Let f_{M-X} a holomorphic function on M–X such that it has at most poles on X. By definition, for every $x \in M$ there are a neighborhood U_x and a holomorphic function $a_x: U_x \to \mathbf{C}$ such that $a_x f_{M-X}$ is holomorphic in U_x. Put $b_x = a_x f_{M-X}$. By Theorem (4.1.1), (i) we may assume that $\gamma_x(a_x)$ and $\gamma_x(b_x)$ are coprime. Taking U_x smaller if necessary, we may assume by Theorem (4.1.1), (ii) that $\gamma_y(a_x)$ and $\gamma_y(b_x)$ are mutually prime for all $y \in U_x$. Therefore we see that there are a neighborhood U_x and holomorphic functions a_x and b_x such that

(4.2.5) (i) $a_x f_{M-X} = b_x$ in $U_x - X$,

(ii) $\gamma_y(a_x)$ and $\gamma_y(b_x)$ are coprime for all $y \in U_x$.

We see by Theorem (4.2.3) that such $\gamma_x(a_x)$ and $\gamma_x(b_x)$ are unique up to the multiplications of units. Therefore we obtain an analytic hypersurface Z_∞ (resp. Z_0) of M by defining locally $Z_\infty = \{a_x = 0\}$ (resp. $Z_0 = \{b_x = 0\}$) in U_x. Note that $Z_\infty \subset X$, and that Z_∞ and Z_0 have no common irreducible component. We put

$$Z = Z_\infty \cup Z_0 = \bigcup_\mu Z_\mu,$$

where Z_μ are irreducible components of Z. Then the function f_{M-X} on $M-X$ uniquely extends to a holomorphic function f on $M - Z_\infty$ which has poles on Z_∞. Let $x \in Z_\mu \cap R(Z)$ be an arbitrary point and $(U, \phi, \Delta(1)^m)$ be a holomorphic local coordinate system around x such that $\phi(x) = O$ and

$$U \cap Z = \{(z^1, ..., z^m) \in \Delta(1)^m; z^1 = 0\},$$

where $\phi = (z^1, ..., z^m)$. We consider $f = f(z, w)$ to be a holomorphic function in one variable $z \in \Delta^*(1)$ with parameter $w \in \Delta(1)^{m-1}$. Then we have the Laurent expansion

$$f(z, w) = \sum_{j \geq k}^{\infty} A_j(w) z^j, \tag{4.2.6}$$

where $k \in \mathbf{Z}$ and A_j are holomorphic functions with $A_k \not\equiv 0$ (cf. (2.6.7)). The integer k is determined independently from $(U, \phi, \Delta(1)^m)$, so that it is denoted by $k(x)$. It is clear that $k(x)$ is locally constant in $x \in R(Z)$. Since $Z_\mu \cap R(Z)$ is connected (cf.

Corollary (4.1.5) and Theorem (4.1.11)), $k(x)$ is constant in $Z_\mu \cap R(Z)$. We denote this integer by $m(Z_\mu; f_{M-X})$. For a convenience, we define $m(\Xi; f_{M-X}) = 0$ for any irreducible analytic hypersurface $\Xi \subset M$ other than Z_μ. We set

$$(f_{M-X})_0 = \sum_{m(Z_\mu; f_{M-X})>0} m(Z_\mu; f_{M-X}) Z_\mu,$$

$$(f_{M-X})_\infty = \sum_{m(Z_\mu; f_{M-X})<0} -m(Z_\mu; f_{M-X}) Z_\mu.$$

We call $(f_{M-X})_0$ (resp. $(f_{M-X})_\infty$) the **zero divisor** (resp. **polar divisor**) of f_{M-X}. Moreover, we define the **divisor** of f_{M-X} by

$$(f_{M-X}) = (f_{M-X})_0 - (f_{M-X})_\infty$$
$$= \sum m(Z_\mu; f_{M-X}) Z_\mu.$$

For a_x, b_x and U_x in (4.2.5), we have

(4.2.7)
$$(f_{M-X})_0 | U_x = (a_x),$$
$$(f_{M-X})_\infty | U_x = (b_x),$$
$$(f_{M-X}) | U_x = (a_x) - (b_x).$$

Let f_{M-X} (resp. g_{M-Y}) be a holomorphic function on $M-X$ (resp. $M-Y$) which has at most poles on X (resp. Y). These are said to be equivalent (written as $f_{M-X} \sim g_{M-Y}$) if there exits a non-empty open subset $U \subset M-(X \cup Y)$ such that $f_{M-X}|U = g_{M-Y}|U$. This is clearly an equivalence relation and the equivalence class is called a **meromorphic function** on M. All meromorphic functions on M naturally form a field, called the **meromorphic function field** of M, which is denoted by Mer(M). The meromorphic function field Mer(M) includes the ring of holomorphic functions on M. The equivalence class of f_{M-X} is denoted by $[f_{M-X}]$. If $[f_{M-X}] = [g_{M-Y}]$, then $(f_{M-X})_0 = (g_{M-Y})_0$ and $(f_{M-X})_\infty = (g_{M-Y})_\infty$. Hence we set

$$(\phi)_0 = (f_{M-X})_0, \quad (\phi)_\infty = (f_{M-X})_\infty, \quad (\phi) = (\phi)_0 - (\phi)_\infty,$$
$$m(\Xi; \phi) = m(\Xi; f_{M-X})$$

for $\phi = [f_{M-X}]$ and an irreducible analytic hypersurface Ξ of M. The following easily follows from (4.2.5) and (4.2.7).

(4.2.8) Lemma. *Let ϕ, $\psi \in$ Mer(M). Then $(\phi) = (\psi)$ if and only if ϕ/ψ is a nowhere vanishing holomorphic function on M.*

(4.2.9) Theorem. *Let $\{U_\lambda\}_{\lambda \in \Lambda}$ be an open covering of M.*

(i) *Let D_λ be divisors on U_λ such that $D_\lambda | (U_\lambda \cap U_\mu) = D_\mu | (U_\lambda \cap U_\mu)$ for* *all* λ *and* μ *with* $U_\lambda \cap U_\mu \neq \varnothing$. *Then there exists a unique divisor* D *on* M *such that* $D|U_\lambda = D_\lambda$ *for all* $\lambda \in \Lambda$.

(ii) *Let* $\phi_\lambda \in \mathrm{Mer}(U_\lambda)$ *be meromorphic functions such that if* $U_\lambda \cap U_\mu \neq \varnothing$, *then* $\phi_\lambda / \phi_\mu |(U_\lambda \cap U_\mu)$ *are nowhere zero holomorphic functions in* $U_\lambda \cap U_\mu$. *Then there exists a unique divisor* D *on* M *such that* $D|U_\lambda = (\phi_\lambda)$.

(iii) *For any divisor* $D \in \mathrm{Div}(M)$, *there are an open covering* $\{V_\lambda\}$ *of* M *and* $\psi_\lambda \in \mathrm{Mer}(V_\lambda)$ *such that* $D|V_\lambda = (\psi_\lambda)$.

Proof. (i) By the condition there is an analytic hypersurface X of M with $X \cap U_\lambda = \operatorname{supp} D_\lambda$. Let $X = \bigcup_\alpha X_\alpha$ be the decomposition into irreducible components. For every X_α we take a point $x \in R(X) \cap X_\alpha$. Then there is a divisor D_λ with $x \in \operatorname{supp} D_\lambda$. The divisor is written uniquely up to the order as

$$D_\lambda = \sum \nu_{\lambda\beta} X_{\lambda\beta}, \quad \nu_{\lambda\beta} \in \mathbf{Z} - \{0\},$$

where $X_{\lambda\beta}$ are distinct irreducible analytic hypersurfaces of U_λ. Since $x \in R(X) \cap U_\lambda = R(\operatorname{supp} D_\lambda)$, there is a unique $X_{\lambda'\beta'}$ with $x \in R(X_{\lambda'\beta'})$. Put

$$\nu_\alpha = \nu_{\lambda'\beta'}.$$

Then it follows that this ν_α is independent of the choice of U_λ. Therefore the desired divisor D is given by

$$D = \sum_\alpha \nu_\alpha X_\alpha.$$

(ii) Put $D_\lambda = (\phi_\lambda) \in \mathrm{Div}(U_\lambda)$. By Lemma (4.2.8), $D_\lambda|(U_\lambda \cap U_\mu) = D_\mu|(U_\lambda \cap U_\mu)$. Hence we have such D by (i).

(iii) For an arbitrary point $x \in M$, there are holomorphic functions $a_1, \ldots, a_l$ in a neighborhood U_x of x such that $\gamma_x(a_i)$ are irreducible and mutually coprime and $a_1 \cdots a_l = 0$ is the defining equation of $\operatorname{supp} D$ in U_x. Taking U_x smaller if necessary, we may assume that the analytic hypersurfaces $X_i = \{y \in U_x;\ a_i(x) = 0\}$ $(1 \leq i \leq l)$ in U_x are distinct and irreducible. Then we may write uniquely

$$D|U_x = \sum_{i=1}^{l} \nu_i X_i, \quad \nu_i \in \mathbf{Z}.$$

Putting $\phi_x = a_1^{\nu_1} \cdots a_l^{\nu_l}$, we have that $(\phi_x) = D|U_x$. *Q.E.D.*

(4.2.10) Theorem. *Let* S *be an analytic subset of* M *with* $\operatorname{codim} S \geq 2$. *Let* ϕ *be a meromorphic function on* $M - S$. *Then* ϕ *uniquely extends to a meromorphic function on* M.

Proof. Put $X = \operatorname{supp}(\phi)_\infty \subset M - S$. It follows from Theorem (4.1.13) that X is an analytic hypersurface of M. For simplicity we put

$$M' = M - \overline{X}, \quad S' = M' \cap S.$$

Then S' is an analytic subset of M', $\operatorname{codim} S' \geq 2$ and $(M-S)-X = M'-S'$. The restriction $\phi|(M-S)-X$ is holomorphic. Hence by Corollary (3.3.44) there is a unique holomorphic extension $f_{M'}$ of $\phi|(M-S)-X$ over M'. We show that $f_{M'}$ has at most poles on $\overline{X}$. Let $y \in \overline{X}$ be an arbitrary point. Then there is a holomorphic function g in a neighborhood U of y such that g is a defining function of $\overline{X}$ at y. Taking U smaller, we may assume that g is a defining function of $\overline{X}$ at all points $z \in \overline{X} \cap U$, and that $\overline{U}$ is compact. Let $\overline{X} = \cup \overline{X}_{\mu}$ be the irreducible decomposition of $\overline{X}$ and put $X_{\mu} = \overline{X}_{\mu} \cap X$. Then $X = \cup X_{\mu}$ is the irreducible decomposition of X by Theorem (4.1.11). For any $z \in R(X_{\mu}) \cap U$, $g^{k(\mu)}\phi|(M-S)-X$ is holomorphic in a neighborhood of z, where $k(\mu) = m(X_{\mu}; f_{(M-S)-X})$. Put $k = \max\{k(\mu); X_{\mu} \cap \overline{U} \neq \emptyset\}$. Then $g^k\phi|(M-S)-X$ is holomorphic in $U-S$ and so is $g^k f_{M'}$. Since $\operatorname{codim} S \geq 2$, $g^k f_{M'}$ is holomorphic in U by Corollary (3.3.44). *Q.E.D.*

Let $\mathbf{L} \xrightarrow{\pi} M$ be a holomorphic line bundle over M, $(\{U_{\lambda}\}, \{s_{\lambda}\})$ a local trivialization of $\mathbf{L}$ and $(\{U_{\lambda}\}, \{T_{\lambda\mu}\})$ the subordinated system of holomorphic transition functions. Let $\sigma = \{\sigma_{\lambda}\} \in \Gamma(M, \mathbf{L}) - \{O\}$ be a holomorphic section, where $\sigma | U_{\lambda} = \sigma_{\lambda} s_{\lambda}$. Then we have

$$\sigma_{\lambda} = T_{\lambda\mu}\sigma_{\mu}, \quad \text{in } U_{\lambda} \cap U_{\mu}. \tag{4.2.11}$$

Since $T_{\lambda\mu}$ are non-vanishing holomorphic functions, it follows from Theorem (4.2.9), (ii) that there is an effective divisor D on M with $D | U_{\lambda} = (\sigma)$. We set

$$D = (\sigma).$$

$$|\mathbf{L}| = \{(\sigma); \sigma \in \Gamma(M, \mathbf{L}) - \{O\}\} \subset \operatorname{Div}(M).$$

Assume that M is compact. Then it is known that $\Gamma(M, \mathbf{L})$ is of finite dimension (cf. [121]). We have the natural identification

$$|\mathbf{L}| = P(\Gamma(M, \mathbf{L})). \tag{4.2.12}$$

Hence $|\mathbf{L}|$ carries the structure of the complex projective space of dimension $\dim \Gamma(M, \mathbf{L}) - 1$. More generally, a collection $\sigma = \{\sigma_{\lambda}\}$ of meromorphic functions σ_{λ} in U_{λ} satisfying the relation (4.2.11) is called a **meromorphic section** of $\mathbf{L}$. In the same way as above, we have a divisor (σ) determined by the meromorphic section σ, which is effective if and only if σ is a holomorphic section. We denote by $\Gamma_{\text{mer}}(M; \mathbf{L})$ the vector space of all meromorphic sections of $\mathbf{L}$.

Now let $D \in \operatorname{Div}(M)$. Then we deduce from Theorem (4.2.9), (iii) that there are an open covering $\{U_{\lambda}\}$ and meromorphic functions ϕ_{λ} on U_{λ} such that $D | U_{\lambda} = (\phi_{\lambda})$. We put

$$T_{\lambda\mu} = \frac{\phi_\lambda}{\phi_\mu}$$

in non-empty $U_\lambda \cap U_\mu$. Then $T_{\lambda\mu}$ are non-vanishing and satisfy the cocycle condition (2.1.3). Then we have a holomorphic line bundle over M denoted by $[D]$ such that $[D]$ is trivial over U_λ, $\{\phi_\lambda\}$ forms a meromorphic section ϕ and $(\{U_\lambda\}, \{T_{\lambda\mu}\})$ is the system of holomorphic transition functions (cf. Chapter II, §2). If we start from another $\{U'_\lambda\}$ and $\{\phi'_\lambda\}$ with $D \,|\, U'_\lambda = (\phi'_\lambda)$, then we easily verify that the resulted holomorphic line bundle is isomorphic to the above one; in this sense, $[D]$ is uniquely determined by D. If D is effective, then the above ϕ is a holomorphic section of $[D]$. By the construction we have

(4.2.13) $$(\phi) = D, \quad [-D] = [D]^*,$$

$$[D_1 + D_2] \cong [D_1] \otimes [D_2] \quad (D_1, D_2 \in \mathrm{Div}(M)),$$

$$\sigma \in \Gamma_{\mathrm{mer}}(M, \mathbf{L}) \Rightarrow [(\sigma)] \cong \mathbf{L}.$$

The following fact will be used in the next chapter.

(4.2.14) Theorem. *Let M be $\Delta(1)^m$ or $\mathbf{C}^m$, and $D \in \mathrm{Div}(M)$. Then there is a meromorphic function $\Psi \in \mathrm{Mer}(M)$ such that $(\Psi) = D$.*

Proof. Let $(z^1, \ldots, z^m)$ be the natural complex coordinate system of M. We give a proof depending on two facts:

(4.2.15) (i) A holomorphic function F on M is expanded in a convergent power series

$$f(z^1, \ldots, z^m) = \sum a_{\nu_1 \cdots \nu_m} (z^1)^{\nu_1} \cdots (z^m)^{\nu_m}, \quad a_{\nu_1 \cdots \nu_m} \in \mathbf{C}.$$

(ii) For $b \in E^{(0,1)}(M)$ with $\bar{\partial} b = 0$ and a relatively compact domain U in M, there is a C^∞-function u in U such that $\bar{\partial} u = b$.

The proofs of these facts are not difficult; especially, (i) immediately follows from the Cauchy integral formula. For (ii) we refer to Hörmander [48], Chapter II, §3.

We deal here with the case of $M = \mathbf{C}^m$. The proof for the polydisc $M = \Delta(1)^m$ is totally similar to this case. By definition, there are a locally finite open covering $\{U_i\}_{i=1}^\infty$ of $\mathbf{C}^m$ and meromorphic functions ϕ_i such that $(\phi_i) = D \,|\, U_i$. We may assume that U_i are polydiscs in $\mathbf{C}^m$. Since non-empty $U_i \cap U_j$ are simply connected and ϕ_i/ϕ_j are non-vanishing holomorphic functions in $U_i \cap U_j$ $(\neq \emptyset)$, $\log(\phi_i/\phi_j)$ give rise to 1-valued holomorphic functions. Take a partition $\{c_i\}$ of unit subordinated to $\{U_i\}$. We put

$$a_i = \sum_k c_k \log \frac{\phi_i}{\phi_k}.$$

Then a direct calculation yields

$$a_i - a_j = \log \frac{\phi_i}{\phi_j} + 2\pi i c_{ij} \text{ in } U_i \cap U_j \neq \emptyset,$$

where $c_{ij} \in \mathbf{Z}$. Hence $\bar{\partial} a_i = \bar{\partial} a_j$ in $U_i \cap U_j$ ($\neq \emptyset$), so that a 1-form $a \in E^{(0,1)}(\mathbf{C}^m)$ is defined by

$$b = \bar{\partial} a_i \text{ in } U_i.$$

Then $\bar{\partial} b = 0$. By making use of (4.2.15), (ii) with $U = \Delta(n)^m$ ($n = 1, 2, \ldots$), we have C^∞-functions $u_{(n)}$ defined in $\Delta(n)^m$ such that

$$\bar{\partial} u_{(n)} = b.$$

Putting $u_{(n)i} = u_{(n)} - a_i$ in $U_i \cap \Delta(n)^m$, we get

$$\bar{\partial} u_{(n)i} = 0,$$

$$u_{(n)i} - u_{(n)j} = \log \frac{\phi_i}{\phi_j} + 2\pi i c_{ij} \text{ in } U_i \cap U_j \cap \Delta(n)^m \neq \emptyset.$$

Hence $u_{(n)i}$ are holomorphic, and the meromorphic functions $e^{-u_{(n)i}}\phi_i$ on $U_i \cap \Delta(n)^m$ satisfy

$$e^{-u_{(n)i}}\phi_i = e^{-u_{(n)j}}\phi_j \text{ in } U_i \cap U_j \cap \Delta(n)^m \neq \emptyset.$$

Therefore there are meromorphic functions $\psi_{(n)}$ on $\Delta(n)^m$ such that $\psi_{(n)} | U_i \cap \Delta(n)^m = e^{-u_i}\phi_i$. It follows that

$$(\psi_{(n)}) = D | \Delta(n)^m, \quad n = 1, 2, \ldots$$

Note that $\psi_{(n+1)}/\psi_{(n)}$ are non-vanishing in $\Delta(n)^m$. It follows from (4.2.15), (i) that there are polynomials $P_{(n)}$ such that

$$\left| \frac{\psi_{(n+1)}}{\psi_{(n)}} e^{P_{(n)}} - 1 \right| < \frac{1}{2^n} \quad \text{in } \Delta(n-1)^m.$$

The infinite products

$$\Psi = \psi_{(1)} \prod_{n=1}^{\infty} \frac{\psi_{(n+1)}}{\psi_{(n)}} e^{P_{(n)}}.$$

converges to a meromorphic function on $\mathbf{C}^m$ and satisfies $(\Psi) = D$. *Q.E.D.*

4.3 Holomorphic Mappings

In the present section, we state without proofs, well-known fundamental facts on holomorphic mappings between complex manifolds. Let M and N be complex manifolds of dimensions m and n, respectively. Let $f: M \to N$ be a holomorphic mapping. Moreover, let $X \subset M$ be an analytic subset and denote by $f|X$ the restriction of f over X. The following will be clear by the definition:

(4.3.1) For an arbitrary $x \in X$, $(f|X)^{-1}((f|X)(x)) = f^{-1}(x) \cap X$ is an analytic subset of M.

We call the analytic subset $(f|X)^{-1}((f|X)(x))$ the **fiber** of $f|X$ at $y = f(x)$. We define the **rank** $\mathrm{rank}_x f|X$ at $x \in X$ by

$$\mathrm{rank}_x f|X = \dim_x X - \dim_x (f|X)^{-1}((f|X)(x)).$$

If X is irreducible, then the rank $\mathrm{rank}\, f|X$ of $f|X$ is defined by

$$\mathrm{rank}\, f|X = \sup\{\mathrm{rank}_x f|X\,;\, x \in X\}.$$

In general, let $X = \bigcup_{\lambda \in \Lambda} X_\lambda$ be the irreducible decomposition of X. Then set

$$\mathrm{rank}\, f|X = \sup\{\mathrm{rank}\, f|X_\lambda;\, \lambda \in \Lambda\}.$$

We set

$$E(f|X) = \{x \in X\,;\, \mathrm{rank}_x f|X < \mathrm{rank}\, f|X\},$$

which is called the **degeneracy locus** of $f|X$.

(4.3.2) Theorem. (i) *The mapping* $x \in X \to \dim_x (f|X)^{-1}(f|X(x)) \in \mathbf{Z}^+$ *is upper semicontinuous. Hence, If* X *is of pure dimension, then the mapping* $x \in X \to \mathrm{rank}_x f|X \in \mathbf{Z}^+$ *is lower semicontinuous.*

(ii) *Let* $(f|R(X))_{*x}: \mathbf{T}(R(X))_x \to \mathbf{T}(N)_{f(x)}$ *be the holomorphic differential. Then* $\mathrm{rank}\, f|X = \sup\{\mathrm{rank}\,(f|R(X))_{*x};\, x \in R(X)\}$.

(iii) *If* $(f|X)(X) = N$, *then* $\mathrm{rank}\, f|X = \dim N$.

(iv) $E(f|X)$ *is an analytic subset of* M *and* $E(f|X) \underset{\neq}{\subset} X$.

The following theorem due to Remmert [93] is called the proper mapping theorem and it is very useful.

(4.3.3) Theorem. *Let* $f|X: X \to N$ *be a proper mapping. Then the following hold:*

(i) *For any analytic subset* Y *of* X, $f(Y)$ *is an analytic subset of* N *and* $\dim f(Y) = \mathrm{rank}\, f|Y$.

(ii) *If* X *is irreducible, then so is* $f(X)$.

(4.3.4) Theorem. *Let* X *be of pure dimension. Then* $f|X: X \to N$ *is an open* *mapping if and only if* $\operatorname{rank}_x f|X = \dim N$ *for all* $x \in X$.

The mapping $f\,|X$ naturally induces a ring homomorphism:

$$(f|X)^*: \gamma_{f(x)}(a_U) \in O_{N, f(x)} \to \gamma_x((a_U \circ f)_{f^{-1}(U)})|X \in O_{X, x}.$$

In general, let R be a commutative ring with 1 and S be a subring of R containing 1. Then R is said to be integral over S if for an arbitrary $a \in R$, there are finitely many elements $h_1, \ldots, h_l$ of S such that

$$a^l + h_1 a^{l-1} + \cdots + h_{l-1} a + h_l = 0.$$

(4.3.5) Theorem. *A point* $x \in X$ *is an isolated point of* $(f|X)^{-1}(f(x))$ *if and only if* $O_{X, x}$ *is integral over* $(f|X)^*(O_{N, f(x)})$.

4.4 Meromorphic Mappings

We are going to define a meromorphic mapping and to discuss its general properties. Let M and N be complex manifolds of dimensions m and n, respectively, as in §3. The following theorem is due to Remmert [94].

(4.4.1) Theorem. *Let* $f: M \to N$ *be a mapping and* $G(f) = \{(x, f(x)); x \in M\} \subset M \times N$ *the graph of f. Then f is holomorphic if and only if* $G(f)$ *is an analytic subset of pure dimension m of* $M \times N$.

Proof. The "only if" part is clear. We show the "if" part. The projections $p: (x, y) \in M \times N \to x \in M$ and $q: (x, y) \in M \times N \to y \in N$ are holomorphic. Put

$$X = G(f), \ g = p|X.$$

For every $x \in M$, $g^{-1}(g(x)) = \{x\}$. Therefore

$$\operatorname{rank}_x g = \dim_x X - \dim_x g^{-1}(g(x)) = m.$$

By Theorem (4.3.4), $g: X \to M$ is an open mapping. Since g is surjective, $g^{-1}: M \to X$ is continuous. Hence, $f = q \circ g^{-1}: M \to N$ is continuous. Put

$$C = \{w \in R(X); \operatorname{rank} g_{*w} < m\} \ (\underset{\neq}{\subset} R(X)).$$

Then C is an analytic subset of $R(X)$. Note that $g: X \to M$ is holomorphic and homeomorphic. Hence $g(S(X))$ is an analytic subset of M and so is $g(C)$ of $M - g(S(X))$ (Theorem (4.3.3)). By the inverse function theorem we see that g^{-1} is holomorphic in $(M - g(S(X))) - g(C)$, and hence so is f. Since f is continuous, we infer from Riemann's extension theorem that f is holomorphic in M. *Q.E.D.*

Let $p: M \times N \to M$ and $q: M \times N \to N$ be the projections as in the above proof. Let W be an open dense subset of M and $f_W: W \to N$ a holomorphic mapping. We say that f_W is **meromorphic with respect to** M if the following conditions are satisfied:

(4.4.2) (i) The closure $\overline{G(f_W)}$ of the graph $G(f_W)$ of f in $M \times N$ is an analytic subset of $M \times N$.

(ii) the mapping $p|\overline{G(f_W)}: \overline{G(f_W)} \to M$ is proper.

For a convenience, we put

$$Mer(M, N) = \{f_W: W \to N; W \subset M \text{ is a dense open subset and}$$
$$f_W \text{ is holomorphic and meromorphic with respect to } M\}.$$

We define an equivalence relation $f_W \sim g_V$ for $f_W, g_V \in Mer(M, N)$ by $\overline{G(f_W)} = \overline{G(g_V)}$. This is clearly an equivalence relation and the equivalence class of f_W is denoted by $[f_W]$. We take the quotient

$$\mathrm{Mer}(M, N) = Mer(M, N)/\sim .$$

We call an element $f = [f_W] \in \mathrm{Mer}(M, N)$ a **meromorphic mapping** (determined by f_W) from M into N and write

$$f: M \to N.$$

We set

$$G(f) = \overline{G(f_W)},$$

which is called the **graph** of f. Let $M' \subset M$ be an open subset and $f_{W \cap M'} = f_W | W \cap M'$. Then $\overline{G(f_{W \cap M'})} = G(f) \cap p^{-1}(M')$. Hence $f_{W \cap M'}$ is meromorphic with respect to M', and hence determines an element $f|M' = [f_{W \cap M'}]$ called the restriction of the meromorphic mapping over M' of f. For a subset $A \subset M$ (resp. $B \subset N$) we define the image (resp. inverse image) $f(A)$ (resp. $f^{-1}(B)$) of A (B) by

$$f(A) = q((p|G(f))^{-1}(A)),$$
$$(\text{resp. } f^{-1}(B) = p((q|G(f))^{-1}(B))).$$

(4.4.3) Lemma. *Let W be a dense open subset of M and $f_W: W \to N$ a holomorphic mapping. Then f_W is meromorphic with respect to M if and only if every point of M has a neighborhood U in M such that $f_W | U \cap W \to N$ is meromorphic with respect to U.*

This says that condition (4.4.2) is a local property over M and the proof will be clear.

(4.4.4) Lemma. *Let $f = [f_W]: M \to N$ be a meromorphic mapping.*

(i) *$f(z)$ is a compact analytic subset of N, and $f(A)$ is compact if so is $A \subset M$.*

(ii) *Let X be an analytic subset of N. Then $f^{-1}(X)$ is an analytic subset of M.*

Proof. Condition (4.4.2) immediately implies that $(p|G(f))^{-1}(x) = p^{-1}(x) \cap G(f)$ is a compact analytic subset. Then it follows from Theorem (4.3.3) that $f(x) = q((p|G(f))^{-1}(x))$ is a compact analytic subset. The rest will be clear by the above argument. *Q.E.D.*

(4.4.5) Lemma. *Let $f = [f_W] \in \mathrm{Mer}(M, N)$ be a meromorphic mapping. Then we have the following.*

(i) *$G(f_W) = G(f) \cap p^{-1}(W)$; especially, $G(f_W)$ is a dense open subset of $G(f)$.*

(ii) *$G(f_W)$ is an m-dimensional closed complex submanifold of $W \times N$ and $p|G(f_W)\colon G(f_W) \to W$ is biholomorphic.*

(iii) *$p(G(f)) = M$.*

(iv) *$G(f)$ is irreducible and* $\dim G(f) = m$.

(v) *$f(x) = f_W(x)$ for $x \in W$.*

(vi) *$(p|G(f))^{-1}(x) = G(f) \cap p^{-1}(x)$ $(x \in M)$ is connected.*

Proof. (i) It is clear by definition that $G(f_W) \subset G(f) \cap p^{-1}(W)$. Take $(x, y) \in G(f) \cap p^{-1}(W)$. By the definition of $G(f)$ there is a sequence $\{(x_\nu, y_\nu)\}_{\nu=1}^{\infty} \subset G(f_W)$ such that $\lim x_\nu = x$ and $\lim y_\nu = y$. Then $y_\nu = f_W(x_\nu)$ ((i)) and $y_\nu = f_W(x_\nu)$. Since $x \in W$, $y = f_W(x)$, so that $(x, y) \in G(f_W)$.

(ii) This is trivial.

(iii) Since $p|G(f)\colon G(f) \to M$ is proper, $p(G(f))$ is a closed subset. On the other hand, $p(G(f)) \subset W$ and W is dense in M. Hence $p(G(f)) = M$.

(iv) Let $G(f) = \bigcup_{\lambda \in \Lambda} G_\lambda$ be the decomposition into irreducible components. Since $p|G(f)\colon G(f) \to M$ is proper, so is every $p|G_\lambda\colon G_\lambda \to M$. It follows from Theorem (4.3.3) that $p(G_\lambda)$ is an irreducible analytic subset of M. Since $M = p(G(f)) = \bigcup_{\lambda \in \Lambda} p(G_\lambda)$ and M is irreducible, there is some G_{λ_0} such that $p(G_{\lambda_0}) = M$. Then $G_{\lambda_0} \supset G(f_W)$ by (i), and hence $G_{\lambda_0} \supset \overline{G(f_W)} = G(f)$. Thus $G_{\lambda_0} = G(f)$, and $\dim G(f) = \dim R(G(f)) = \dim G(f_W) = m$.

(v) This immediately follows from (i).

(vi) Suppose that there is a point $x \in M$ such that $(p|G(f))^{-1}(x)$ is not connected. Then there are two open subsets U_1 and U_2 of $M \times N$ such that

$$U \cap U_2 = \varnothing, \ U_1 \cup U_2 \supset (p|G(f))^{-1}(x).$$

Since $p|G(f)$ is proper, we may assume that U_i are of the following form:

$$U_i = W' \times V_i, \ i = 1, 2,$$

where W' is a neighborhood of x, $V_i \subset N$ are open subsets and $\bar{V}_1 \cap \bar{V}_2 = \emptyset$. It follows that

$$G(f) \cap (U_1 \cup U_2) = (G(f) \cap U_1) \cup (G(f) \cap U_2).$$

Each $G(f) \cap U_i$ is a purely m-dimensional analytic subset of U_i. Since $G(f) \cap U_i \cap G(f_W) \neq \emptyset$, the rank of $p|G(f) \cap U_i$ is m over a non-empty subset. We see by Theorem (4.3.4) that $p|G(f)(G(f) \cap U_i)$ contains an interior point. On the other hand, the holomorphic mapping

$$p|G(f) \cap U_i : G(f) \cap U_i \to W'$$

is proper. Therefore $(p|G(f) \cap U_i)(G(f) \cap U_i) = W'$. Note that $W \cap W'$ is not empty. Therefore, for $x' \in W \cap W'$, $(p|G(f))^{-1}(x)$ contains at least one point in each $G(f) \cap U_i$ $(i = 1, 2)$. This contradicts to (v). *Q.E.D.*

We next show that "meromorphic mapping" is an extended notion of "holomorphic mapping".

(4.4.6) Theorem. *Let $f : M \to N$ be a meromorphic mapping. Then f is represented by a holomorphic mapping $f_M : M \to N$ if and only if $f(x)$ consists of one point for any $x \in M$.*

Proof. The "only if" part is obvious. To show the converse, we put

$$f_M : x \in M \to f(x) \in N.$$

Then $G(f_M) = G(f)$. We deduce from Theorem (4.4.5), (iv) and Theorem (4.4.1) that f_M is holomorphic, so that $f = [f_M]$. *Q.E.D.*

If a meromorphic mapping $f : M \to N$ is represented by a holomorphic mapping $f_M : M \to N$, we say f to be holomorphic and identify f with f_M. If a meromorphic mapping $f : M \to N$ is not holomorphic, then by Theorem (4.4.6) there necessarily exists a point $x \in M$ such that $f(x)$ is not a point. We investigate such a point for a while. We denote simply by E the degeneracy locus $E(p|G(f))$ of $p|G(f) : G(f) \to M$.

(4.4.7) Lemma. (i) *E coincides with the set of points $(x, y) \in G(f)$ such that (x, y) are not isolated points of $(p|G(f))^{-1}((p|G(f))(x, y))$.*

(ii) *$G(f) - E$ is connected and dense in $G(f)$.*

(iii) *$p' = p|(G(f) - E) : G(f) - E \to M$ is an injective open mapping.*

Proof. (i) This is clear by the definition of $f(x)$ and Lemma (4.4.5), (vi).

(ii) By Theorem (4.3.2), (iv) E is an analytic subset of $M \times N$, and $E \neq G(f)$. It follows from Theorem (4.1.11), (ii) and Lemma (4.4.5), (iv) that $G(f) - E$ is a connected , open dense subset of $G(f)$.

(iii) Since $G(f)$ is irreducible (Lemma (4.4.5), (iv)), $G(f)$ is purely m-

dimensional by Corollary (4.1.9), and so is $G(f)-E$ by Theorem (4.1.11), (ii). Take a point $x \in M$. By Lemma (4.4.5), (vi), one and only one of the following two cases occurs:

(a) $(p|G(f))^{-1}(x)$ consists of one point and hence $(p|G(f))^{-1}(x) \cap E = \varnothing$,

(b) $(p|G(f))^{-1}(x)$ is a connected positive dimensional compact analytic subset of $M \times N$ and hence $(p|G(f))^{-1}(x) \subset E$.

Therefore we see that $p' \colon G(f)-E \to M$ is injective and that p' has rank m at all points of $G(f)-E$. It follows from Theorem (4.3.4) that p' is an open mapping. *Q.E.D.*

We set

$$I(f) = p(E(p|G(f))),$$

which is called the **indeterminacy locus** of $f \in \mathrm{Mer}(M, N)$. The next theorem will explain why $I(f)$ is so called.

(4.4.8) Theorem. *Let $f: M \to N$ be a meromorphic mapping. Then the following hold.*

(i) *$I(f)$ is an analytic subset of M, and* $\dim I(f) \le m-2$.

(ii) *$(p|G(f))^{-1}(I(f)) = E(p|G(f))$ and $f(x)$ consists of one point for $x \in M - I(f)$.*

(iii) *If $x \in I(f)$, then $f(x)$ is a positive dimensional, compact, connected analytic subset of N.*

(iv) *If we put $f_{M-I(f)} \colon x \in M - I(f) \to f(x) \in N$, then $f_{M-I(f)} \in Mer(M, N)$ and $f = [f_{M-I(f)}]$.*

(v) *If $f = [f_W]$ with $f_W \in Mer(M, N)$, then $W \subset M - I(f)$ and $f_W = f_{M-I(f)}|W$.*

(vi) *Let X be an analytic hypersurface of N such that any irreducible component of X is not contained in $f(I(f))$ and $f(M) \cap X \neq \varnothing$. Then $f^{-1}(X)$ is an analytic hypersurface of M.*

Proof. (i) Put $E = E(p|G(f))$. Since $p|G(f) \colon G(f) \to M$ is proper, so is $p|E \colon E \to M$. Then it follows from Theorem (4.3.3), (i) that $p(E) = I(f)$ is an analytic subset of M, and $\dim I(f) = \operatorname{rank} p|E$. Let $R(E) = \bigcup_{\lambda \in \Lambda} E'_\lambda$ be the decomposition into connected components. By Theorem (4.1.4), the closures $E_\lambda = \overline{E'_\lambda}$ of E'_λ give rise to the irreducible components of E. By Theorems (4.1.4) and (4.3.2), (i), we have

$$\operatorname{rank} p|E_\lambda = \sup\{\operatorname{rank}_{(x,y)} p|E_\lambda;\ (x, y) \in E'_\lambda\}. \tag{4.4.9}$$

Take an arbitrary point $(x, y) \in E'_\lambda$. Then, by Lemma (4.4.7), (i), there is a positive dimensional irreducible component C of $(p|G(f))^{-1}(x)$ passing through

$(x, y) \in E'_\lambda$. Hence $C \subset E_\lambda$. Thus we see that $\dim_{(x, y)}(p|E_\lambda)^{-1}(x) \geq 1$. We obtain

$$\begin{aligned}\operatorname{rank}_{(x, y)} p|E_\lambda &= \dim_{(x, y)} E_\lambda - \dim_{(x, y)}(p|E_\lambda)^{-1}(x) \\ &\leq \dim E_\lambda - 1 \leq \dim E - 1 \\ &\leq \dim G(f) - 2 = m - 2.\end{aligned}$$

Thus $\operatorname{rank} p|E_\lambda \leq m-2$ for all $\lambda \in \Lambda$, so that $\dim I(f) = \operatorname{rank}(p|E) \leq m-2$.

(ii) This is easily deduced from the facts (a) and (b) in the proof of Lemma (4.4.7), (iii).

(iii) Since $(p|G(f))^{-1}(x)$ is a compact connected analytic subset, so is $q((p|G(f))^{-1}(x)) = f(x)$.

(iv) The graph $G(f_{M-I(f)})$ of $f_{M-I(f)}$ is given by

$$G(f_{M-I(f)}) = G(f) - E = G(f) \cap (M' \times N),$$

where $M' = M - I(f)$. Hence we see that $G(f_{M-I(f)})$ is a purely m-dimensional analytic subset of $M' \times N$. By Theorem (4.4.1), $f_{M-I(f)}$ is a holomorphic mapping. It is clear that $f_{M-I(f)} \in Mer(M, N)$ and $f = [f_{M-I(f)}]$.

(v) This is trivial.

(vi) We may assume that X is irreducible. We know by Lemma (4.4.4), (ii) that $f^{-1}(X)$ is an analytic subset of M. Since E contains no irreducible component of $(q|G(f))^{-1}(X)$, $(q|G(f))^{-1}(X) = \overline{(q|G(f))^{-1}(X) - E}$. Thus $f^{-1}(X) = \overline{f_{M-I(f)}^{-1}(X)}$, which is a purely $(m-1)$-dimensional analytic subset of M; that is, it is an analytic hypersurface of M. *Q.E.D.*

(4.4.10) Theorem. *Let L, M, and N be complex manifolds of dimension l, m and n, respectively. Let $f: L \to M$ and $g: M \to N$ be meromorphic mappings and put*

$$W = (f|L - I(f))^{-1}(M - I(g)).$$

Assume that $W \neq \emptyset$. Then W is a dense open subset of L, and the holomorphic mapping $h_W = (g|M - I(g)) \circ (f|W)\colon W \to N$ is meromorphic with respect to L.

Proof. Note that $W = (L - I(f)) - (f|L - I(f))^{-1}(I(g))$ $(\neq \emptyset)$. It follows from Theorem (4.1.11), (ii) that $L - I(f)$ is a connected open dense subset of L, so that W is a connected open dense subset of $L - I(f)$. Thus W is a connected open dense subset of L. Let

$$p: L \times M \to L, \quad q: L \times M \to M,$$

$$r: M \times N \to M, \quad s: L \times N \to N$$

be the natural projections. By Lemma (4.1.12), $G(f) \times G(g) \subset L \times M \times M \times N$ is an analytic subset. Put

$$\Delta = \{(y, y) \in M \times M;\ y \in M\},$$

$$H = G(f) \times G(g) \cap (L \times \Delta \times N).$$

Note that H is an analytic subset and $N \neq \emptyset$. In fact, $(x, f(x), f(x), h_W(x)) \in H$ for $x \in W$. Let

$$t: L \times \Delta \times N \to L \times N$$

be the natural projection. Then we claim that $t|H: H \to L \times N$ is proper. Take compact subsets $K_1 \subset L$ and $K_2 \subset N$. Then we have

$$(t|H)^{-1}(K_1 \times K_2) = \{(x, y, y, z) \in L \times \Delta \times N;$$

$$x \in K_1,\ y \in f(x),\ z \in g(y) \cap K_2\}$$

$$\subset \left[(p|G(f))^{-1}(K_1) \cap (q|G(f))((p|G(f))^{-1}(K_1)) \right] \times K_2.$$

Since $(p|G(f))^{-1}(K_1)$ is compact, so is $(t|H)^{-1}(K_1 \times K_2)$. Here we put $X = (t|H)(H)$. We see by Theorem (4.3.3) that X is an analytic subset of $L \times N$ and that the graph $G(h_W)$ of h_W is contained in X. Let A be an arbitrary subset of L. Then we get

$$(4.4.11) \qquad (s|X)^{-1}(A) = \{(x, z) \in L \times N;\ x \in A,\ z \in g(f(x))\},$$

for "$(x, z) \in H$" $\Longleftrightarrow$ "there is $y \in f(x)$ such that $z \in g(y)$" $\Longleftrightarrow$ "$z \in g(f(x))$". We are going to show that $s|X: X \to L$ is a proper surjective mapping. Let $K \subset L$ be a compact subset. It follows from (4.4.10) that $(s|X)^{-1}(K) = \{(x, z) \in K \times N;\ z \in g(f(x))\} \subset K \times g(f(K))$. Since $g(f(K))$ is compact, so is $(s|X)^{-1}(K)$. The surjectivity is clear by (4.4.10). We obtain again from (4.4.10)

$$(s|X)^{-1}(W) = \{(x, z) \in L \times M;\ x \in W,\ z \in g(f(x))\}$$

$$= \{(x, z) \in L \times M;\ x \in W,\ z = h_W(x)\} = G(h_W).$$

Especially, $G(h_W)$ is a connected open subset of X and $G(h_W) \subset R(X)$. There is a unique connected component X_0' of $R(X)$ such that $X_0' \supset G(h_W)$. It follows from Theorem (4.1.3) that $X_0 = \overline{X_0'}$ is the only one irreducible component of X containing $G(h_W)$. The mapping $s|X_0: X_0 \to L$ is proper. To finish the proof, we are going to show $\overline{G(h_W)} = X_0$. Since $p|G(f): G(f) \to L$ is proper, we deduce from Theorem (4.3.3) that $p|G(f)((q|G(f))^{-1}(I(g)))$ is an analytic subset of L. Moreover, we have

$$W = p|G(f)(G(f|L - I(f)) \cap (q|G(f))^{-1}(M - I(g)))$$

$$= p|G(f)(G(f|L - I(f)) - G(f|L - I(f)) \cap (q|G(f))^{-1}(I(g)))$$

$$\supset (L - I(f)) - p|G(f)((q|G(f))^{-1}(I(g)))$$
$$= L - (I(f) \cup p|G(f)((q|G(f))^{-1}(I(g))).$$

Therefore we see that

$$G(h_W) = (s|X_0)^{-1}(W)$$
$$= X_0 - (s|X_0)^{-1}(I(f) \cup p|G(f)((q|G(f))^{-1}(I(g)))).$$

Thus Theorem (4.1.11), (ii) implies that $G(h_W)$ is an open dense subset of X_0, so that $\overline{G(h_W)} = X_0$. *Q.E.D.*

The meromorphic mapping $h = [h_W] \in \mathrm{Mer}(L, N)$ obtained in the above Theorem (4.4.10) is called the **composition of the meromorphic mappings** f and g, and written as $h = f \circ g$.

We define the **rank of a meromorphic mapping** $f: M \to N$ by

$$\operatorname{rank} f = \operatorname{rank} q|G(f),$$

where $G(f) \subset M \times N$ is the graph of f and $q: M \times N \to N$ is the natural projection. Noting that $G(f)$ is irreducible and hence of pure dimension, we have

(4.4.12) $$\operatorname{rank} f = \operatorname{rank} q|G(f) - p^{-1}(I(f)) = \operatorname{rank} f|M - I(f),$$

where $p: M \times N \to M$ denotes the natural projection.

(4.4.13) *Example.* While we will explain meromorphic mappings from M into the projective space $\mathbf{P}^n(\mathbf{C})$ in §5, (e) of this chapter, we give here some simple examples showing that the notion is natural. Let ρ: $\mathbf{C}^{n+1} - \{O\} \to \mathbf{P}^n(\mathbf{C})$ be the Hopf fibering (Chapter I, §8). Then ρ is meromorphic with respect to $\mathbf{C}^{n+1}$ and defines a meromorphic mapping $[\rho]$: $\mathbf{C}^{n+1} \to \mathbf{P}^n(\mathbf{C})$. To see this, we let $(z^0, \ldots, z^n)$ be the standard complex coordinate system of $\mathbf{C}^{n+1}$. Consider an analytic subset $X \subset \mathbf{C}^{n+1} \times \mathbf{P}^n(\mathbf{C})$ defined by

(4.4.14) $$X = \{((z^j), [w^k]) \in \mathbf{C}^{n+1} \times \mathbf{P}^n(\mathbf{C}); \; z^j w^k - z^k w^j = 0 \text{ for all } j \text{ and } k\}.$$

Then it follows that $G([\rho])$ is the irreducible component of X containing the graph $G(\rho)$ of ρ. Let $p: \mathbf{C}^{n+1} \times \mathbf{P}^n(\mathbf{C}) \to \mathbf{C}^{n+1}$ and $q: \mathbf{C}^{n+1} \times \mathbf{P}^n(\mathbf{C}) \to \mathbf{P}^n(\mathbf{C})$ be the natural projections, as above. Then $p|G(f)$ is proper. Therefore ρ defines a meromorphic mapping $[\rho]$: $\mathbf{C}^{n+1} \to \mathbf{P}^n(\mathbf{C})$. It easily follows from (4.4.14) that the graph $G([\rho])$ is a complex submanifold of $\mathbf{C}^{n+1} \times \mathbf{P}^n(\mathbf{C})$. We see that $(p|G([\rho]))^{-1}(O)$ is an analytic hypersurface of $G([\rho])$, $[\rho](O) = q((p|G([\rho]))^{-1}(O)) = \mathbf{P}^n(\mathbf{C})$, and that

$$q|(p|G([\rho]))^{-1}(O): (p|G([\rho]))^{-1}(O) \to \mathbf{P}^n(\mathbf{C})$$

is biholomorphic. Put

$$\tilde{\mathbf{C}}^{n+1} = G([\rho]), \ \tilde{p} = p|\tilde{\mathbf{C}}^{n+1}.$$

Then

$$\tilde{p}|\tilde{\mathbf{C}}^{n+1} - \tilde{p}^{-1}(O): \tilde{\mathbf{C}}^{n+1^{n+1}} - \tilde{p}^{-1}(O) \to \mathbf{C}^{n+1} - \{O\}$$

is biholomorphic and the inverse

$$(\tilde{p}|\tilde{\mathbf{C}}^{n+1} - \tilde{p}^{-1}(O))^{-1}: \mathbf{C}^{n+1} - \{O\} \to \tilde{\mathbf{C}}^{n+1}$$

is meromorphic with respect to $\mathbf{C}^{n+1}$. The holomorphic mapping

$$\tilde{p}: \tilde{\mathbf{C}}^{n+1} \to \mathbf{C}^{n+1} \tag{4.4.15}$$

is called a **monoidal transformation** or a **blowing up**.

4.5 Meromorphic Functions and Meromorphic Mappings

(a) Pullback of holomorphic functions and plurisubharmonic functions. Let M and N be complex manifolds and $f: M \to N$ a meromorphic mapping. For a holomorphic function a on N, the pullback $(f_{M-I(f)})^* a = a \circ f_{M-I(f)}$ is a holomorphic function on $M - I(f)$. By Theorem (4.4.8), (i), $\dim I(f) \leq \dim M - 2$, and then by Corollary (3.3.43), $(f_{M-I(f)})^* a$ is uniquely extended to a holomorphic function on M, which is written as $f^* a$ and called the **pullback of a by the meromorphic mapping f.**

Let u be a plurisubharmonic function on N. By making use of Theorem (3.3.42), we define the pullback $f^* u$ in the same way as above, which is a plurisubharmonic function on M.

(b) Pullback of meromorphic functions. Let $f: M \to N$ be as in (a). Let $X \subset N$ be an analytic hypersurface and α_{N-X} a holomorphic function on $N - X$ which has at most poles on X. We assume that $f(M) \not\subset X$. Then we see by Theorem (4.4.8), (vi) that $Y = f^{-1}(X)$ is an analytic hypersurface of M. We consider the holomorphic function

$$\phi = \alpha_{N-X} \circ f|(M - (I(f) \cup Y))$$

on $M - (I(f) \cup Y)$. It is easy to check that ϕ has at most poles on $Y' = Y - I(f)$. Since $\operatorname{codim} I(f) \geq 2$ (Theorem (4.4.8), (i)), Theorem (4.2.10) implies that ϕ is extended to a holomorphic function $f^* \alpha_{N-X}$ on $N - Y$ which has at most poles on Y. If $[\alpha_{N-X}] = [\alpha'_{X'}]$ and if $f(M) \not\subset X$ and $f(M) \subset X'$, then we easily see that $[f^* \alpha_{N-X}] = [f^* \alpha'_{N-X'}]$. Hence we define the **pullback of a meromorphic function $\alpha = [\alpha_{N-X}] \in \mathrm{Mer}(N)$** by f as

$$f^* \alpha = [f^* \alpha_{N-X}].$$

If rank $f = \dim N$, then rank $f|M - I(f) = \dim N$ by (4.4.2). It follows from Theorem (4.3.2), (ii) that $f(M)$ contains a non-empty open subset of N, and the converse holds, too. In this case, f is said to be **dominant** and $f^*\alpha$ is defined for any $\alpha \in \mathrm{Mer}(N)$; furthermore, if $f^*\alpha = 0$, then $\alpha = 0$. Thus we obtain a field homomorphism

$$f^*: \alpha \in \mathrm{Mer}(N) \to f^*\alpha \in \mathrm{Mer}(M).$$

(c) Pullback of divisors. Let $f: M \to N$ be a meromorphic mapping as above and $D \in \mathrm{Div}(N)$. We define the pullback f^*D of D by f as follows. We first consider the case where f is holomorphic. By Theorem (4.2.9), (iii), there are an open covering $\{U_i\}$ of N and $\phi_i \in \mathrm{Mer}(U_i)$ such that $D | U_i = (\phi_i)$. Put $V_i = f^{-1}(U_i)$ and $\psi = f^*\phi_i$. Then $\{V_i\}$ is an open covering of M, and by Lemma (4.2.8) we see that $\psi_i | / \psi_j (V_i \cap V_j)$ are nowhere vanishing on $V_i \cap V_j \neq \emptyset$. We deduce from Theorem (4.2.9), (ii) that there is a unique divisor $E \in \mathrm{Div}(M)$ such that $E | U_i = (\psi_i)$ for all i. It is clear that E is independent of the choices of $\{U_i\}$ and $\{\phi_i\}$ and determined only by D. Hence we have the pullback of D by f:

$$f^*D = E.$$

In general, $I(f)$ may be non-empty. By the above argument we have

$$E = (f|(M - I(f)))^* D \in \mathrm{Div}(M - I(f)).$$

Let $\operatorname{supp} E = \bigcup_\alpha Y_\alpha$ be the irreducible decomposition. Then there are $\nu_\alpha \in \mathbf{Z} - \{0\}$ such that

$$E = \sum \nu_\alpha Y_\alpha.$$

By Theorem (4.1.13) $\overline{\operatorname{supp} E} = \bigcup_\alpha \bar{Y}_\alpha$ is an analytic hypersurface of M. We set

$$f^*D = \sum \nu_\alpha \bar{Y}_\alpha \in \mathrm{Div}(M),$$

which is called the **pullback of the divisor D by the meromorphic mapping f.**

(d) Pullback of holomorphic differential forms. We keep the above notation. Let $\omega \in \Gamma(N, \overset{k}{\wedge} \mathbf{T}(N))$. Then $(f|M - I(f))^*\omega \in \Gamma(M - I(f), \overset{k}{\wedge} \mathbf{T}(M))$. For an arbitrary point $x \in I(f)$, we take a holomorphic local coordinate system $(U, \phi, \Delta(1)^m)$ ($m = \dim M$) around x and put $\phi = (z^1, \dots, z^m)$. Then we have

$$((f|M - (f))^*\omega) | U - I(f) = \sum_{J \in \{m;\, k\}} a_J dz^J,$$

where a_J are holomorphic functions on $U - I(f)$. By Corollary (3.3.44), all a_J

uniquely extends to holomorphic functions on U, so that $(f|M-I(f))^*\omega$ uniquely extends to a holomorphic differential k-form defined on the whole M. We denote it by $f^*\omega$ and call it the **pullback of the holomorphic k-form ω by the meromorphic mapping f.** Hence we obtain a linear homomorphism

$$f^*\colon \omega\in\Gamma(N, \overset{k}{\wedge}\mathbf{T}(N))\to f^*\omega\in\Gamma(M, \overset{k}{\wedge}\mathbf{T}(M)).$$

Assume that $\dim M=\dim N=m$. Let $k=m$. Then we have

$$f^*\colon \omega\in\Gamma(N, \mathbf{K}(N))\to f^*\omega\in\Gamma(M, \mathbf{K}(M)).$$

In the same way as above, we can define

$$f^*\colon \omega\in\Gamma(N, \mathbf{K}(N)^l)\to f^*\omega\in\Gamma(M, \mathbf{K}(M)^l)$$

for all $l\in\mathbf{Z}$.

(4.5.1) Lemma. *Assume that $\Gamma(N, \mathbf{K}(N)^l)\neq\{O\}$ and $l\neq 0$. Then the linear homomorphism $f^*\colon\Gamma(N, \mathbf{K}(N)^l)\to\Gamma(M, \mathbf{K}(M)^l)$ is non-zero if and only if* $\operatorname{rank} f=m$, *and in this case f^* is injective.*

Proof. The first half is clear by Theorem (4.3.2), (ii) and (4.4.12). If $\operatorname{rank} f=m$, then it follows from Theorem (4.3.2), (ii) and (4.4.12) that $f(M-I(f))$ contains a non-empty open subset of N, so that f^* is injective. *Q.E.D.*

(e) Examples of meromorphic mappings. In Example (4.4.13) we gave an example of a monoidal transformation $\tilde{p}\colon\tilde{\mathbf{C}}^m\to\mathbf{C}^m$. Here we discuss more general mappings. Let M and N be m-dimensional complex manifolds. A holomorphic mapping $f\colon M\to N$ is called a **proper modification** or **blowing up** if f satisfied the following conditions:

(i) f is proper and surjective.

(ii) There is an analytic subset $A\underset{\neq}{\subset}N$ such that $f|M-f^{-1}(A)\colon M-f^{-1}(A)\to N-A$ is biholomorphic.

(4.5.2) Theorem. *Let $f\colon M\to N$ be a proper modification and $A\underset{\neq}{\subset}N$ as above. Then the holomorphic mapping $(f|M-f^{-1}(A))^{-1}\colon N-A\to M-f^{-1}(A)$ is meromorphic with respect to N and so defines a meromorphic mapping $f^{-1}\colon N\to M$, which is independent of the choice of A as far as the above condition (ii) is satisfied.*

Proof. Put $g=(f|M-f^{-1}(A))^{-1}$. Then we get

$$G(g)=\{(y, g(y))\in N\times M;\, y\in N-A\}$$
$$=\{(f(x), x)\in N\times M;\, x\in M-f^{-1}(A)\}.$$

The permutation $\sigma: (x, y) \in M \times N \to (y, x) \in N \times M$ is biholomorphic, and $G(g) = \sigma(G(f) - A \times N)$. Therefore we have

$$\overline{G(g)} = \overline{\sigma(G(f) - A \times N)}$$
$$= \overline{\sigma(G(f) - G(f) \cap (A \times N))} = \sigma(G(f)).$$

Thus $\overline{G(g)}$ is an analytic subset of $N \times M$. Let $p : N \times M \to N$ be the natural projection. We claim that $p | \overline{G(g)}$ is proper. Take a compact subset $K \subset N$. Then

$$(p | \overline{G(g)})^{-1}(K) = \{(y, x) \in \overline{G(g)};\ y \in K\}$$
$$= \sigma\{(x, y) \in G(f);\ y \in K\} = \sigma\{(x, f(x)) \in M \times N;\ f(x) \in K\}$$
$$\subset K \times f^{-1}(K).$$

Since $f^{-1}(K)$ is compact, so is $(p | \overline{G(g)})(K)$. The meromorphic mapping $[g]: N \to M$ satisfies

$$G([g]) = \sigma(G(f))$$

and hence is independent of the choice of A. *Q.E.D.*

Now we study what is a meromorphic mappings $f : M \to \mathbf{P}^N(\mathbf{C})$. Let $\rho: \mathbf{C}^{N+1} - \{O\} \to \mathbf{P}^N(\mathbf{C})$ be the Hopf fibering and $z = (z^0, \ldots, z^N)$ the natural holomorphic coordinate system of $\mathbf{C}^{N+1}$.

(4.5.3) Theorem. *Let $F = (\phi^0, \ldots, \phi^N): M \to \mathbf{C}^{N+1}$ be a holomorphic mapping, and put $A = F^{-1}(O)$. Assume that $A \neq M$. Then $f_{M-A} = \rho \circ F \mid M - A \to \mathbf{P}^N(\mathbf{C})$ is meromorphic with respect to M.*

Proof. Let $\mathbf{H}_0 \to \mathbf{P}^N(\mathbf{C})$ be the hyperplane bundle. Then z^j $(0 \le j \le N)$ form a base of $\Gamma(\mathbf{P}^N(\mathbf{C}), \mathbf{H}_0)$ (cf. Chapter II, §1). Put

$$G^* = \{(x, \rho(z)) \in M \times \mathbf{P}^N(\mathbf{C});\ \phi^i(x) z^j - \phi^j(x) z^i = 0,\ 0 \le i,\ j \le N\}.$$

Then G^* is an analytic subset of $M \times \mathbf{P}^N(\mathbf{C})$ and

$$G(f_{M-A}) = G^* \cap ((M - A) \times \mathbf{P}^N(\mathbf{C})). \tag{4.5.4}$$

By Theorem (4.1.11), (ii), $M - A$ is connected and open, so that $G(f_{M-A})$ is a connected open subset of G^* and $G(f_{M-A}) \subset R(G^*)$. Let G_0 denote the closure of a connected component of $R(G^*)$ containing $G(f_{M-A})$. Then G_0 is an irreducible component of G^*, and by (4.5.4)

$$G(f_{M-A}) = G_0 - G_0 \cap p^{-1}(A),$$

where $p : M \times \mathbf{P}^N(\mathbf{C}) \to M$ denotes the natural projection. Since $G_0 \cap p^{-1}(A)$ is a proper analytic subset of G_0, $\overline{G(f_{M-A})} = G_0$. Since $\mathbf{P}^N(\mathbf{C})$ is compact, $p | G_0: G_0 \to M$ is proper. *Q.E.D.*

The following theorem states that the converse of the above Theorem (4.5.3) holds at least locally on M.

(4.5.5) Theorem. *Let M be $\Delta(1)^m$ or $\mathbf{C}^m$. Then for an arbitrary meromorphic mapping $f: M \to \mathbf{P}^N(\mathbf{C})$, there are holomorphic functions $\phi^0(z), \ldots, \phi^N(z)$ satisfying*

$$I(f) = \{z \in M : \phi^0(z) = \cdots = \phi^N(z) = 0\},$$

$$(f|M - I(f))(z) = \rho(\phi^0(z), \ldots, \phi^N(z)).$$

Proof. Let $[w^0; \cdots; w^N]$ be the homogeneous coordinate of $\mathbf{P}^N(\mathbf{C})$. Put

$$X_j = \left\{ [w^0; \cdots; w^N] \in \mathbf{P}^N(\mathbf{C}); w^j = 0 \right\}.$$

Without loss of generality, we may assume that $f(M) \not\subset X_0$. Since w^j/w^0 $(1 \le j \le N)$ are meromorphic functions on $\mathbf{P}^N(\mathbf{C})$ and $f(M) \not\subset X_0 = \operatorname{supp}(w^j/w^0)_\infty$, we have meromorphic functions $f^*(w^j/w^0)$ on M by (b) of this section. We also have the pullback divisor f^*X_0 on M. It follows from Theorem (4.2.14) that there exists a holomorphic function ϕ^0 on M such that $(\phi^0) = f^*X_0$. Put

$$\phi^j = \phi^0 \cdot f^*\left(\frac{w^j}{w^0}\right), \quad 1 \le j \le N.$$

It is easy to see that ϕ^j are holomorphic on $M - I(f)$ and hence so are on M (cf. Corollary (3.3.44)). By the definition

$$f^*X_j = (\phi^j), \quad 0 \le j \le N. \tag{4.5.6}$$

We are going to show that these $\phi^0, \ldots, \phi^N$ are the required ones. Put

$$A = \{z \in M; \phi^0(z) = \cdots = \phi^N(z) = 0\}.$$

Note that

$$f(z) = \rho\left(1, \frac{w^1}{w^0}(z), \ldots, \frac{w^N}{w^0}(z)\right)$$

for $z \in \mathbf{P}^N(\mathbf{C}) - X_0$. Take an arbitrary point $x \in M - (I(f) \cup \operatorname{supp} f^*X_0)$. Noting that $f(x) \notin X_0$, we get

$$\rho(\phi^0(x), \ldots, \phi^N(x)) = \rho\left(1, f^*\left(\frac{w^1}{w^0}\right)(x), \ldots, f^*\left(\frac{w^N}{w^0}\right)(x)\right)$$

$$= \rho\left(1, \frac{w^1}{w^0}(f(x)), \ldots, \frac{w^N}{w^0}(f(x))\right) = f(x)$$

By Theorem (4.5.3) $g: x \in M - A \to \rho(\phi^0(x), \ldots, \phi^N(x))$ is meromorphic, and then f and g are mutually equivalent. It follows from Theorem (4.4.8), (v) that $I(f) \subset A$.

Suppose that $x \in M - I(f)$. Then there is some X_j such that $f(x) \notin X_j$. We deduce from (4.5.6) that $\phi^j(x) \neq 0$, so that $x \notin A$. Thus $A = I(f)$. *Q.E.D.*

The above $f(z) = [\phi^0(z); \cdots; \phi^N(z)]$ is called an **irreducible representation** of the meromorphic mapping f.

Let M be a general complex manifold and $f: M \to \mathbf{P}^1(\mathbf{C})$ a meromorphic mapping. We will see that f may be naturally identified with a meromorphic function on M. We use the same notation as in the proof of Theorem (4.5.5); i.e., w^1/w^0 is a meromorphic function on $\mathbf{P}^1(\mathbf{C})$. Suppose that $f(M) \neq [0; 1]$. Then $f^*(w^1/w^0)$ is a meromorphic function on M. Conversely, let ϕ be a meromorphic function on M such that $\phi \not\equiv 0$ and $\phi = [\xi_{M-X}]$. For an arbitrary $x \in M$ there are a neighborhood U_x of x and holomorphic functions a_x, b_x on U_x such that

(4.5.7) $a_x \xi_{M-X} = b_x$ in $U_x - X$, and $\gamma_z(a_x)$ and $\gamma_z(b_x)$ are mutually coprime for all $z \in U_x$.

Put

$$A_x = \{z \in U_x;\ a_x(z) b_x(z) = 0\}.$$

Then $A_x = U_x \cap \operatorname{supp}(\phi)$. On the other hand, the holomorphic mapping

$$f_{U_x - A_x}: U_x - A_x \to \rho(a_x(z), b_x(z)) \in \mathbf{P}^1(\mathbf{C})$$

is meromorphic with respect to U_x by Theorem (4.5.3). Now suppose that $U_x \cap U_y \neq \varnothing$. Take any $z \in U_x \cap U_y - (A_x \cup A_y) = U_x \cap U_y - \operatorname{supp}(\phi)$. Then by (4.5.7) there is a unit $u \in \mathcal{O}_{M,z}$ such that

$$\gamma_z(a_y) = u\gamma_z(a_x), \quad \gamma_z(b_y) = u\gamma_z(b_x).$$

Therefore $f_{U_x - A_x}(z) = f_{U_y - A_y}(z)$. We deduce that there is a holomorphic mapping $f_{M - \operatorname{supp}(\phi)}: M - \operatorname{supp}(\phi) \to \mathbf{P}^1(\mathbf{C})$ such that $f_{M - \operatorname{supp}(\phi)} | U_x - \operatorname{supp}(\phi) = f_{U_x - A_x}$. Of course, $f_{M - \operatorname{supp}(\phi)}$ is meromorphic with respect to M. Put $f = [f_{M - \operatorname{supp}(\phi)}]$. Then $f(M) \neq [0; 1]$, $I(f) \subset \operatorname{supp}(\phi)$ and $f^*(w^1/w^0) = \phi$.

(4.5.8) Theorem. *Let $\mathbf{L} \to M$ be a holomorphic line bundle. Let V be a vector subspace of $\Gamma(M, \mathbf{L})$ with $2 \leq \dim V < \infty$ and put $B = \{x \in M;\ s(x) = 0,\ s \in V\}$. Then the holomorphic mapping $\Phi_V: M - B \to P(V^*)$ is meromorphic with respect to M.*

Proof. Take a base $\{s_0, \dots, s_N\}$ of V and the corresponding homogeneous coordinate system $[z^0; \cdots; z^N]$ of $P(V^*)$. Then we get

$$\Phi_V: x \in M - B \to \Phi_V(x) = [s_0(x); \cdots; s_N(x)] \in \mathbf{P}^N(\mathbf{C}) = P(V^*).$$

Hence we see by Theorem (4.5.3) and Lemma (4.4.3) that Φ_V is meromorphic with respect to M. *Q.E.D.*

Notes

For general properties of analytic subsets, the reader should consult Narasimhan [72] and Gunning-Rossi [45]. While we stated the extension theorem of analytic subsets, Theorem (4.1.11), due to Remmert-Stein, the following general extension theorem due to Bishop is very useful and important:

Let $U \subset \mathbf{C}^m$ be a domain, $A \subset U$ an analytic subset and $V \subset U - V$ a purely k-dimensional analytic subset.

(i) *If the 2k-dimensional Hausdorff measure of $\bar{V} \cap A$ is 0, then $\bar{V}$ is an analytic subset of U.*

(ii) *If for any $z \in A$ there exists $r > 0$ such that*

$$\int_{(V-A)\cap B(z;r)} \alpha^k < \infty,$$

then $\bar{V}$ is an analytic subset of U.

For the proof, cf. Bishop [7] and Stolzenberg [116]. The present definition of meromorphic mappings is due to Remmert [94]. Siu [106] proved the following extension theorem for meromorphic mappings into Kähler manifolds:

Let $U \subset \mathbf{C}^m$ be a domain, $A \subset U$ an analytic subset with $\operatorname{codim} A \geq 0$, *and $f : U - A \to M$ a meromorphic mapping into a compact Kähler manifold M. Then f uniquely extends to a meromorphic mapping from U into M.*

CHAPTER V

Nevanlinna Theory

5.1 Poincaré-Lelong Formula

Let M be a paracompact complex manifold of dimension m. We here prove the so-called Poincaré-Lelong formula which is a higher dimensional residue formula relating meromorphic functions on M to closed currents of type (1, 1) on M. This formula will play an important basic role not only in the Nevanlinna theory which is the aim of the present chapter, but also in various aspects of the complex analysis.

We keep the same notation as defined in Chapter III, §2, (c).

Lemma (5.1.1). *Let X be a purely k-dimensional analytic subset of the ball $B(R)\subset \mathbf{C}^m$. Then we have*

$$0\leq \int_{B(r)\cap R(X)} \alpha^k < \infty \tag{5.1.2}$$

for any $0<r<R$.

Remark. When $k=0$, $R(X)=X$, and $\int_{X\cap B(r)}\alpha^0$ stands for the cardinality of the set $X\cap B(r)$.

Proof. Covering X by small polydiscs, we see that it is sufficient to prove (5.1.2) for some small $0<r<R$. For $J=(j_1, \ldots, j_k)\in \{m; k\}$ we put

$$p_J\colon (z^1, \ldots, z^m)\to (z^{j_1}, \ldots, z^{j_k})\in \mathbf{C}^k, \quad (dz\wedge d\bar{z})^J = p_J^*\Omega_k,$$

where Ω_k denotes the volume form $(i/2\pi)dz^1\wedge d\bar{z}^1\wedge \cdots \wedge (i/2\pi)dz^k\wedge d\bar{z}^k$ on $\mathbf{C}^k$. Then

$$(5.1.3)\qquad (dz\wedge d\bar{z})^J = \frac{i}{2\pi}dz^{j_1}\wedge d\bar{z}^{j_1}\wedge\cdots\wedge\frac{i}{2\pi}dz^{j_k}\wedge d\bar{z}^{j_k},$$

$$\alpha^k = k!\sum_{J\in\{m;\,k\}}(dz\wedge d\bar{z})^J.$$

We define complex linear spaces $L(J)\subset\mathbf{C}^m$ by

$$L(J) = \{(z^1, \ldots, z^m)\in\mathbf{C}^m;\ z^{j_1} = \cdots = z^{j_k} = 0\}.$$

After a suitable unitary transformation, we may now assume that the origin O of $\mathbf{C}^m$ is an isolated point of $X\cap L(J)$ for any $J\in\{m;k\}$. For each $J = (j_1, \ldots, j_k)\in\{m;k\}$ there are $r_\nu > 0$, $\nu = 1, 2, \ldots, m$ such that the projection

$$(5.1.4)\qquad p_J\Big|\left(\prod_{\nu=1}^{m}\Delta(r_\nu)\right)\cap X:\left(\prod_{\nu=1}^{m}\Delta(r_\nu)\right)\cap X\to\prod_{\nu=1}^{k}\Delta(r_{j_\nu})$$

is proper, surjective, and of finite fiber. Let m_J denote the covering number of (5.1.4). Then

$$(5.1.5)\qquad \int_{\left(\prod_{\nu=1}^{m}\Delta(r_\nu)\right)\cap R(X)} p_J^*\Omega_k = m_J\int_{\left(\prod_{\nu=1}^{k}\Delta(r_{j_\nu})\right)-p_J((\prod\Delta(r_\nu))\cap S(X))}\Omega_k$$

$$= m_J\int_{\prod_{\nu=1}^{k}\Delta(r_{j_\nu})}\Omega_k < \infty.$$

Take a positive $r' < r_\nu$ for all ν and for all choices of J (note that r_ν depend on J). Then we have by (5.1.5)

$$\int_{\Delta(r')^m\cap R(X)}\alpha^k = k!\sum_J\int_{\Delta(r')^m\cap R(X)}p_J^*\Omega_k < \infty.$$

Taking $r > 0$ so that $B(r)\subset\Delta(r')^m$, we obtain (5.1.2). *Q.E.D.*

Let $X\subset M$ be a purely k-dimensional analytic subset. As remarked in Example (3.2.25), we set

$$X(\phi) = T_X(\phi) = \int_{R(X)}\phi,$$

$$X(\phi) = T_X(\phi) = \sum_{x\in X}\phi(x)\ \text{ in the case of } k = 0$$

for $\phi\in D^{2k}(M)$. We are going to show that X defines a positive current on M

(5.1.6) Lemma. $|X(\phi)| < \infty$ *for any* $\phi\in D^{2k}(M)$.

Proof. Take a locally finite open covering $\{(U_\lambda, \psi_\lambda, \Delta(1)^m)\}$ of M by holomorphic local coordinate neighborhood systems. Let $\{a_\lambda\}$ be the decomposition of unity subordinated to $\{U_\lambda\}$. Then we have

$$\int_{R(X)}\phi = \sum_\lambda \int_{R(X)} a_\lambda \phi = \sum_\lambda \int_{R(U_\lambda \cap X)} a_\lambda \phi.$$

Putting $X_\lambda = \psi_\lambda(R(X) \cap U_\lambda)$, we have

$$\int_{R(X)\cap U_\lambda} a_\lambda \phi = \int_{R(X_\lambda)} (\psi_\lambda^{-1})^*(a_\lambda \phi).$$

Thus it suffices to show our lemma for $M = \Delta(1)^m$. For $\phi \in \mathcal{D}^{2k}(\Delta(1)^m)$, there is a positive constant depending only on m and k such that

$$|\iota_{R(X)}^* \phi| \le c \|\phi\|_0 \iota_{R(X)}^* \alpha^k,$$

where $\iota_{R(X)} : R(X) \to \Delta(1)^m$ denotes the immersion. Taking $0 \le r < 1$ with $\operatorname{supp} \phi \subset \Delta(r)^m$, we have

$$\left| \int_{R(X)} \phi \right| \le c \|\phi\|_0 \int_{\Delta(r)^m \cap R(X)} \alpha^k. \tag{5.1.7}$$

Therefore we have $|X(\phi)| < \infty$ by Lemma (5.1.1). *Q.E.D.*

(5.1.8) Theorem. *The mapping* $X : \phi \in \mathcal{D}^{2k}(M) \to X(\phi) \in \mathbf{C}$ *is a positive current of type* $(m-k,\ m-k)$.

Proof. Let $(U, \psi, \Delta(1)^m)$ be a holomorphic local coordinate system of M and $0 < r < 1$. By (5.1.7), there is a positive constant C such that

$$|X(\phi)| \le \left| \int_{R(X)} \psi^* \phi \right| \le C \|\phi\|_0 \int_{\Delta(r)^m \cap R(\psi(X))} \alpha^k$$

for $\phi \in \mathcal{D}^{2k}(\Delta(r)^m)$. Hence there is a positive constant C_r such that

$$|X(\psi^* \phi)| \le C_r \|\phi\|_0$$

for $\phi \in \Delta(r)^m$. Therefore $X \circ \psi^* \in \mathcal{K}'^{2m-2k}(\Delta(1)^m)$, so that $X \in \mathcal{K}'^{2m-2k}(M)$. Let $\phi^{(k,k)}$ be the (k, k)-component of $\phi \in \mathcal{D}^{2k}(M)$. Then

$$X(\phi) = \int_{R(X)} \phi^{(k,k)}.$$

Thus $X \in \mathcal{K}'^{(m-k,\, m-k)}(M)$. It is clear that the current X is real and positive (cf. Example (3.2.25)). *Q.E.D.*

Let $D = \sum_{j=1}^{\infty} \nu_j X_j$ be a divisor on M. For $\phi \in \mathcal{D}^{2m-2}(M)$ we set

$$D(\phi) = \sum_{j=1}^{\infty} \nu_j X_j(\phi).$$

Since $\{X_j\}$ is locally finite, $|D(\phi)| < \infty$ by Theorem (5.1.8). Thus we have the following corollary.

(5.1.9) Corollary. *Let D be a divisor on M. Then the mapping*

$$D : \phi \in \mathcal{D}^{2m-2}(M) \to D(\phi) \in \mathbf{C}$$

is a real current of type (1, 1)*; moreover, D is positive if D is effective.*

Let ω be a real continuous differential form of type (1, 1) on M such that $\omega \geq 0$. For a real current $T \in \mathcal{D}'^{(1,1)}(M)$ we set

$$\text{Trace}\,(\omega; T) = \frac{1}{(m-1)!} T \wedge \omega^{m-1} \in \mathcal{D}(M)'.$$

If $T \in \mathcal{K}^{(1,1)}(M)$, then $\text{Trace}(\omega; T) \in \mathcal{K}(M)'$. By Chapter III, §1, (e), Trace $(\omega; T)$ is considered to be a Radon measure.

(5.1.10) Proposition. *Let D be an effective divisor on M and $Y \subset M$ an analytic subset of dimension at most $m-2$. Then* Trace $(\omega; D)(Y) = 0$. *Especially, if* $D \,|\, U = i \sum T_{j\bar{k}} dz^j \wedge d\bar{z}^k$ *with a holomorphic local coordinate system* $(U, (z^1, \ldots, z^m))$, *then*

$$\|T_{j\bar{k}}\|(Y \cap U) = 0 \quad 1 \leq j,\, k \leq m.$$

Proof. Let $\operatorname{supp} D = \bigcup_{i=1} X_i$ be the irreducible decomposition and $D = \sum_{i=1} \nu_i X_i$ with $\nu_i \in \mathbf{Z}$. Let $A \subset M$ be any compact subset. Then

$$\text{Trace}\,(\omega; D)(A \cap Y) = \sum_{i=1} \nu_i \int_{R(X) \cap Y \cap A} \omega^{m-1}.$$

Since $R(X_i) \cap Y$ is an analytic subset of $R(X_i)$ of dimension at most $m-2$,

$$\int_{R(X) \cap Y \cap A} \omega^{m-1} = 0,$$

and hence Trace $(\omega; D)(Y) = 0$. The latter half follows from this and Corollary (3.2.23). *Q.E.D.*

Let f be a meromorphic function on M with $f \not\equiv 0$. By the definition of meromorphic function, there are holomorphic local coordinate system $\{(U_\lambda, \psi_\lambda, \Delta(1)^m)\}$ and holomorphic functions g_λ, h_λ on U_λ such that

(5.1.11) (i) $M = \bigcup_\lambda U_\lambda$ is a locally finite open covering.

(ii) $h_\lambda f = g_\lambda$ in U_λ.

(iii) $\gamma_x(g_\lambda)$ and $\gamma_x(h_\lambda)$ are coprime for all $x \in U_\lambda$.

As in Theorem (4.2.9), the following holds on all U_λ:

(5.1.12) $$(f)_0|U_\lambda = (g_\lambda), \quad (f)_\infty|U_\lambda = (h_\lambda), \quad (f)|U_\lambda = (g_\lambda) - (h_\lambda),$$
$$\log|f|^2 = \log|g_\lambda|^2 - \log|h_\lambda|^2.$$

It follows from Theorem (3.3.28) and Remark (3.3.35) that $\log|f|^2$ is locally integrable and defines a current of order 0 on M. The following formula is called the **Poincaré-Lelong formula.**

(5.1.13) Theorem. *Let $f \in \mathrm{Mer}(M)$ with $f \not\equiv 0$. Then we have*

$$dd^c[\log|f|^2] = (f).$$

We need some preparation before going into the proof. Let u be a plurisubharmonic function on $\Delta(1)^m$. By Theorem (3.3.28), $u \in L_{\mathrm{loc}}(\Delta(1)^m)$ and $dd^c[u]$ is a positive current of type (1, 1). Put

$$dd^c[u] = i\sum_{j,k} T_{j\bar{k}} dz^j \wedge d\bar{z}^k.$$

Then $T_{j\bar{k}}$ are Radon measures on $\Delta(1)^m$ by Corollary (3.2.23).

(5.1.14) Lemma. *Let $Y \subset \Delta(1)^m$ be an analytic subset of dimension at most $m-2$. Then $\|T_{j\bar{k}}\|(Y) = 0$ for all j, k.*

Proof. It follows from Corollary (3.2.23) that $T_{j\bar{k}}$ are absolutely continuous with respect to $\sum_j T_{j\bar{j}}$. Hence it suffices to show that $T_{j\bar{j}}(Y) = 0$ for all j. For instance, we consider the case of $j = 1$. Then we have

$$T_{1\bar{1}} = \frac{\partial^2[u]}{\partial z^1 \partial \bar{z}^1}.$$

Take an arbitrary $\phi \in D(\Delta(1)^m)$. Then

$$<T_{1\bar{1}}, \phi> = \int u \frac{\partial^2 \phi}{\partial z^1 \partial \bar{z}^1} \bigwedge_{j=1}^{m} \left(\frac{i}{2} dz^j \wedge d\bar{z}^j \right)$$

$$= \int_{w \in \Delta(1)^{m-1}} \left\{ \int_{z^1 \in \Delta(1)} u(z^1, w) \frac{\partial^2 \phi(z^1, w)}{\partial z^1 \partial \bar{z}^1} \frac{i}{2} dz^1 \wedge d\bar{z}^1 \right\}$$

$$\cdot \bigwedge_{j=2}^{m} \left(\frac{i}{2} dw^j \wedge d\bar{w}^j \right)$$

Continued

$$= \int_{w \in \Delta(1)^{m-1}} \left\langle \frac{\partial^2 [u(\cdot, w)]}{\partial z^1 \partial \bar{z}^1}, \phi(\cdot, w) \right\rangle \bigwedge_{j=2}^{m} \left(\frac{i}{2} dw^j \wedge d\bar{w}^j \right).$$

By Fubini's theorem, the function

$$\tau(\phi): w \in \Delta(1)^{m-1} \to \left\langle \frac{\partial^2 [u(\cdot, w)]}{\partial z^1 \partial \bar{z}^1}, \phi(\cdot, w) \right\rangle \in \mathbf{C}$$

is integrable over $\Delta(1)^{m-1}$. Furthermore, the function $u(\cdot, w): z \to u(z, w)$ is subharmonic for almost all w. Therefore

(5.1.15) if $\phi \geq 0$, then $\tau(\phi) \geq 0$ almost everywhere;

(5.1.16) $\operatorname{supp}(\tau(\phi)) \subset p(\operatorname{supp} \phi)$,

where $p: (z^1, w) \in \Delta(1)^m \to w \in \Delta(1)^{m-1}$ is the projection. Now put $\bar{Y}(r) = Y \cap \overline{\Delta(r)}^m$ for $0 < r < 1$. Take a sequence of $\phi_l \in D(\Delta((r+1)/2))$, $l = 1, 2, \ldots$, such that $\phi_l \downarrow \chi_{\bar{Y}(r)}$. Then by (5.1.15)

$$\text{(5.1.17)} \qquad \tau(\phi_1)(w) \geq \tau(\phi_2)(w) \geq \cdots \geq 0 \quad (w \in \Delta(1)^{m-1}).$$

By Lebesgue's convergence theorem we have

$$\text{(5.1.18)} \qquad \lim_{j \to \infty} \tau(\phi_l)(w) = \left\langle \frac{\partial^2 [u(\cdot, w)]}{\partial z^1 \partial \bar{z}^1}, \chi_{\bar{Y}(r)}(\cdot, w) \right\rangle,$$

$$\text{(5.1.19)} \qquad \lim_{l \to \infty} \langle T_{1\bar{1}}, \phi_j \rangle = \langle T_{1\bar{1}}, \chi_{\bar{Y}(r)} \rangle.$$

By (5.1.17) we have

$$\text{(5.1.20)} \qquad \lim_{l \to \infty} \langle T_{1\bar{1}}, \phi_l \rangle = \lim_{l \to \infty} \int_{w \in \Delta(1)^{m-1}} \tau(\phi_l) \bigwedge_{j=2}^{m} \left(\frac{i}{2} dw^j \wedge d\bar{w}^j \right)$$

$$= \int_{w \in \Delta(1)^{m-1}} \lim_{l \to \infty} \tau(\phi_l) \bigwedge_{j=2}^{m} \left(\frac{i}{2} dw^j d\bar{w}^j \right).$$

We infer from (5.1.18) and (5.1.19) that

$$\text{(5.1.21)} \qquad \langle T_{1\bar{1}}, \chi_{\bar{Y}(r)} \rangle = \int_{w \in \Delta(1)^{m-1}} \left\langle \frac{\partial^2 [u(\cdot, w)]}{\partial z^1 \partial \bar{z}^1}, \chi_{\bar{Y}(r)}(\cdot, w) \right\rangle$$

$$\cdot \bigwedge_{j=2}^{m} \left(\frac{i}{2} dw^j \wedge d\bar{w}^j \right).$$

$$= \int_{w \in \Delta(1)^{m-1}} \tau(\chi_{\bar{Y}(r)}) \bigwedge_{j=2}^{m} \left(\frac{i}{2} dw^j \wedge d\bar{w}^j \right).$$

Here $\tau(\chi_{\bar{Y}(r)})$ is defined by

$$\tau(\chi_{\bar{Y}(r)})\colon w \in \Delta(1)^{m-1} \to \left\langle \frac{\partial^2[u(\cdot, w)]}{\partial z^1 \partial \bar{z}^1}, \chi_{\bar{Y}(r)}(\cdot, w) \right\rangle,$$

which is non-negative and integrable over $\Delta(1)^{m-1}$ by (5.1.17), (5.1.18) and Fubini's theorem. It is clear by (5.1.6) that

$$\text{supp}\, \tau(\chi_{\bar{Y}(r)}) \subset p(\bar{Y}(r)). \tag{5.1.22}$$

By the assumption, the Lebesgue measure of $p(\bar{Y}(r))$ is 0. Thus we see by (5.1.21) and (5.1.22) that

$$\langle T_{1\bar{1}}, \chi_Y \rangle = \lim_{r \to 1} \langle T_{1\bar{1}}, \chi_{\bar{Y}(r)} \rangle$$

$$= \lim_{r \to 1} \int_{w \in p(\bar{Y}(r))} \tau(\chi_{\bar{Y}(r)}) \bigwedge_{j=2}^{m} \left(\frac{i}{2} dw^j \wedge d\bar{w}^j \right) = 0. \quad Q.E.D.$$

Proof of Theorem (5.1.13). We use the same notation as in (5.1.12). Let $\{a_\lambda\}$ be the partition of unity subordinated to $\{U_\lambda\}$. For an arbitrary $\phi \in D^{2(m-1)}(M)$ we obtain

$$\langle dd^c[\log |f|^2], \phi \rangle = \sum_\lambda \langle dd^c[\log |f|^2], a_\lambda \phi \rangle,$$

$$\langle (f), \phi \rangle = \sum_\lambda \langle (f), a_\lambda \phi \rangle.$$

Therefore it suffices to show

$$\langle dd^c[\log |f|^2], \phi \rangle = \langle (f), \phi \rangle \tag{5.1.23}$$

for ϕ with supp $\phi \in U_\lambda$. In this case it follows from (5.1.12) that

$$\langle dd^c[\log |f|^2], \phi \rangle = \langle dd^c[\log |g_\lambda|^2], \phi \rangle - \langle dd^c[\log |h_\lambda|^2], \phi \rangle,$$

$$\langle (f), \phi \rangle = \langle (g_\lambda), \phi \rangle - \langle (h_\lambda), \phi \rangle.$$

It is reduced to prove that

$$dd^c[\log |g_\lambda|^2] = (g_\lambda), \quad dd^c[\log |h_\lambda|^2] = (h_\lambda). \tag{5.1.24}$$

Therefore we claim that

$$dd^c[\log |f|^2] = (f) \tag{5.1.25}$$

for a holomorphic function $f \not\equiv 0$ on $\Delta(1)^m$. Put

$$(f) = \frac{i}{2} \sum T_{j\bar{k}} dz^j \wedge d\bar{z}^k.$$

Then

$$\sum_{j=1}^{m} T_{j\bar{j}} = \text{Trace}\left[\frac{i}{2}\sum_{j=1}^{m} dz^j \wedge d\bar{z}^j; (f)\right].$$

We obtain from Proposition (5.1.10)

$$\|T_{j\bar{k}}\|(S(\text{supp}(f))) = 0. \tag{5.1.26}$$

Similarly, putting

$$dd^c[\log|f|^2] = \frac{i}{2}\sum S_{j\bar{k}} dz^j \wedge d\bar{z}^k,$$

we obtain from Lemma (5.1.14) and Corollary (3.3.13)

$$\|S_{j\bar{k}}\|(S(\text{supp}(f))) = 0. \tag{5.1.27}$$

Hence it suffices to show (5.1.25) over $\Delta(1)^m - S(\text{supp}(f))$. Note that $R(\text{supp}(f))$ is a complex submanifold of $\Delta(1)^m - S(\text{supp}(f))$. Take an arbitrary point $z_0 \in R(\text{supp}(f))$. Then there is a neighborhood $U \subset \Delta(1)^m - S(\text{supp}(f))$ of x_0 with a holomorphic local coordinate $(z^1, \ldots, z^m)$ such that

$$f(z) = (z^1)^\nu g(z),$$

where $\nu \in \mathbf{Z}$, $\nu \geq 0$ and $h \neq 0$ in U. It is also sufficient to show (5.1.25) for $\phi \in D^{2(m-1)}(U)$. For such ϕ we have

$$\begin{aligned}
<dd^c[\log|f|^2, \phi> &= <dd^c[\log|z^1|^{2\nu}], \phi> \\
&= \lim_{\varepsilon\to 0}\int_{U-\{|z^1|\leq\varepsilon\}} \log|z^1|^{2\nu} dd^c\phi \\
&= \lim_{\varepsilon\to 0}\left\{\int_{U-\{|z^1|\leq\varepsilon\}}\left[d(\log|z^1|^{2\nu}\wedge d^c\phi) - d\log|z^1|^{2\nu}\wedge d^c\phi\right]\right\} \\
&= \lim_{\varepsilon\to 0}\left\{-\int_{\{|z^1|=\varepsilon\}} 2\nu(\log\varepsilon)d^c\phi + \int_{U-\{|z^1|\leq\varepsilon\}} 2\nu d^c\log|z^1|\wedge d\phi\right\} \\
&= \lim_{\varepsilon\to 0}\left\{O(\varepsilon\log\varepsilon) - 2\nu\int_{U-\{|z^1|\leq\varepsilon\}} d(d^c\log|z^1|\wedge\phi)\right\}. \\
&= \lim_{\varepsilon\to 0}\left\{-2\nu\int_{\{|z^1|=\varepsilon\}} d^c\log|z^1|\wedge\phi\right\}.
\end{aligned}$$

Putting $z = re^{i\theta}$ with the polar coordinate, we have

$$d^c\log|z^1| = \frac{1}{4\pi} r\frac{\partial}{\partial r}\log r\, d\theta = \frac{1}{4\pi}d\theta \text{ on } \{|z^1| = r\}.$$

Finally we have

$$<dd^c[\log |f|^2], \phi> = \nu \int_{\{z^1=0\}} \phi = <(f), \phi>. \quad Q.E.D.$$

(5.1.28) Corollary. *The current D defined by a divisor D on M is closed; i.e.,* $dD = 0$.

Proof. The closedness of currents is a local property. Hence we may assume that there is a meromorphic function f with $(f) = D$. Then by Theorem (5.1.13) we have

$$dD = ddd^c[\log |f|^2] = 0. \quad Q.E.D.$$

Let D be an effective divisor on a domain $U \subset \mathbf{C}^m$. Since the current D is closed and positive, the Lelong number $L(z; D)$ is defined (cf. Chapter III, §2, (c)). On the other hand, the **multiplicity** $\nu = \nu(z; D)$ of D at $z \in U$ is defined as follows. If $z \notin \operatorname{supp} D$, then we set $\nu(z; D) = 0$. Take $z \in \operatorname{supp} D$. By a translation, we may assume that $z = O$. There are a positive r and a holomorphic function f on the ball $B(r)$ such that $B(r) \subset U$ and $(f) = D|B(r)$. Then f is expanded into a power series in $B(1)$

$$f(z^1, \dots, z^m) = \sum_{\lambda \geq \nu} P_\lambda(z^1, \dots, z^m),$$

where P_λ are homogeneous polynomials of degree λ and $P_\nu \neq 0$ (cf., e.g., [48], Chapter II). It is easy to show that the number ν is independent of the choice of f and the local coordinate system around $z = O$. We set $\nu(z; D) = \nu$.

Now we are going to compute the Lelong number $L(O; D)$. It is trivial that $L(O; D) = 0$ if $O \notin \operatorname{supp} D$. Assume that $O \in \operatorname{supp} D$. It is clear by definition that $L(O; D)$ is invariant under the multiplication of positive numbers to the coordinates. Therefore we may assume that $r = 1$ and $\overline{B(1)} \subset U$. For $0 < t < r$ we define a holomorphic function f_t on $B(1)$ by $f_t(z) = f(tz)t^{-\nu}$, so that

$$f_t(z) = \sum_{\lambda \geq \nu} t^{\lambda - \nu} P_\lambda(z).$$

It follows from the definition and Theorem (5.1.13) that

$$n(O; t, D) = \frac{1}{t^{2(m-1)}} <dd^c[\log |f|^2], \chi_{B(t)} \alpha^{m-1}>.$$

Substituting $z = tw$, we have

$$n(O; t, D) = <dd^c[\log |f_t|^2], \chi_{B(1)} \alpha^{m-1}>.$$

As $t \to 0$, f_t converges uniformly on $\overline{B(1)}$ to $f_0 = P_\nu$. Hence $\log |f_t|^2$ converges to

$\log |P_\nu|^2$ in the space $L(\overline{B(1)})$ of integrable functions over $\overline{B(1)}$. Therefore $[\log |f_t|^2]$ converges to $[\log |P_\nu|^2]$ as currents, so that $dd^c[\log |f_t|^2]$ converges to $dd^c[\log |P_\nu|^2]$ as currents. Thus we obtain

$$L(O;D) = \lim_{t\to 0} n(O;t,D) = \langle dd^c[\log |P_\nu|^2], \chi_{B(1)}\alpha^{m-1}\rangle. \tag{5.1.29}$$

Take a C^∞-function $\chi_\delta(t)$ in $t \in \mathbf{R}$ so that $0 \le \chi_\delta(t) \le 1$ and

$$\chi_\delta(t) = \begin{cases} 0 & \text{for } t \ge 1, \\ 1 & \text{for } t \le 1-\delta. \end{cases}$$

It follows from (5.1.29) that

$$\begin{aligned}
L(O;D) &= \lim_{\delta\to 0} \langle dd^c[\log |P_\nu|^2], \chi_\delta(\|z\|z^2)\alpha^{m-1}\rangle \\
&= \lim_{\delta\to 0} \int_{\mathbf{C}^m} \log |P_\nu|^2 dd^c\left(\chi_\delta(\|z\|^2)\alpha^{m-1}\right) \\
&= \lim_{\delta\to 0} \int_{\mathbf{C}^m} \log |P_\nu|^2 dd^c \chi_\delta(\|z\|^2)\wedge\alpha^{m-1} \\
&= \lim_{\delta\to 0} \int_{\mathbf{C}^m} \log |P_\nu|^2 \left\{\chi_\delta'(\|z\|^2) + \frac{\|z\|^2}{m}\chi_\delta''(\|z\|^2)\right\}\alpha^m \\
&= \lim_{\delta\to 0} \int_0^\infty dt \int_{\Gamma(t)} \log |P_\nu|^2 \left\{\chi_\delta'(t^2) + \frac{t^2}{m}\chi_\delta''(t^2)\right\} 2mt^{2m-1}\eta \\
&= \lim_{\delta\to 0} \int_0^\infty 2mt^{2m-1} \left\{\chi_\delta'(t^2) + \frac{t^2}{m}\chi_\delta''(t^2)\right\} dt \int_{\Gamma(t)} \log |P_\nu|^2\eta \\
&= \lim_{\delta\to 0} \int_0^\infty 2mt^{2m-1} \left\{\chi_\delta'(t^2) + \frac{t^2}{m}\chi_\delta''(t^2)\right\} dt \int_{\Gamma(1)} (\log |P_\nu|^2 + 2\nu\log t)\eta \\
&= \lim_{\delta\to 0} \left\{\int_{\Gamma(1)} (\log |P_\nu|)\eta \int_0^\infty \left[t^{2m}\chi_\delta'(t^2)\right]' dt + 2\nu \int_0^\infty \log t \left[t^{2m}\chi_\delta'(t^2)\right]' dt\right\} \\
&= \lim_{\delta\to 0} \left\{0 - 2\nu \int_0^\infty \frac{1}{t} t^{2m}\chi_\delta'(t^2)dt\right\} = \lim_{\delta\to 0} -\nu \int_0^\infty t^{2m-2}\left[\chi_\delta(t^2)\right]' dt \\
&= \lim_{\delta\to 0} \nu \int_0^\infty (2m-2)t^{2m-3}\chi_\delta(t^2)dt = \lim_{\delta\to 0} \nu \int_0^1 (2m-2)t^{2m-3}dt = \nu.
\end{aligned}$$

Thus we have the following theorem.

(5.1.30) Theorem. *Let D be an effective divisor on a domain $U \subset \mathbf{C}^m$. Then $L(z;D) = \nu(z;D)$ for all $z \in U$.*

5.2 Characteristic Functions and the First Main Theorem

In the present section we extend Nevanlinna's famous First Main Theorem for meromorphic functions in one variable to the higher dimensional case. Let $f: \mathbf{C}^n \to M$ be a meromorphic mapping from $\mathbf{C}^n$ into a complex manifold M. Put

$$W = \mathbf{C}^n - I(f), \quad f_0 = f|W.$$

(5.2.1) Lemma. *For an arbitrary $\omega \in K^k(M)$, $f_0^* \omega \in L^k_{\mathrm{loc}}(\mathbf{C}^n)$.*

Proof. It is sufficient to prove that

$$\left| \int_W f_0^* \omega \wedge \phi \right| < \infty$$

for any $\phi \in D^{2n-k}(\mathbf{C}^n)$. Let $G(f_0) \subset W \times M$ (resp. $G(f) \subset \mathbf{C}^n \times M$) be the graph of f_0 (resp. f). Then $G(f_0)$ is a dense open subset of $G(f)$ and $G(f_0) \subset R(G(f))$. Let

$$p: \mathbf{C}^n \times M \to \mathbf{C}^n, \quad q: \mathbf{C}^n \times M \to M$$

be the natural projections. Note that

$$\int_W f_0^* \omega \wedge \phi = \int_{G(f_0)} q^* \omega \wedge p^* \phi.$$

Since $p|G(f)$ is proper, $p^{-1}(\mathrm{supp}\,\phi) \cap G(f)$ is compact. Take a non-negative function $h \in D(\mathbf{C}^n \times M)$ so that $h \equiv 1$ on $p^{-1}(\mathrm{supp}\,\phi) \cap G(f)$. Then

$$\int_{G(f_0)} q^* \omega \wedge p^* \phi = \int_{G(f_0)} h q^* \omega \wedge p^* \phi.$$

Since $G(f)$ is a purely n-dimensional analytic subset of $\mathbf{C}^n \times M$ and $hq^* \omega \wedge p^* \phi \in K^{2n}(\mathbf{C}^n \times M)$, we hence infer from Lemma (5.1.6)

$$\left| \int_W f_0^* \omega \wedge \phi \right| = \left| \int_{R(G(f))} h q^* \omega \wedge p^* \phi \right| < \infty. \quad Q.E.D.$$

We write henceforth $f^* \omega$ for $f_0^* \omega$.

For an arbitrary $T \in K'^{(1,1)}(\mathbf{C}^n)$ we set

$$n(t, T) = \frac{\langle T, \chi_{B(t)} \rangle}{t^{2n-2}}.$$

Since $n(t, T)$ is left continuous, the integral

$$N(r, T) = \int_1^r n(t, T) \frac{dt}{t}$$

makes sense and is a continuous function in r. The following lemma immediately follows from Theorem (3.2.31).

(5.2.2) Lemma. *If T is a closed positive current of type (1, 1), then $N(r, T)$ is a convex increasing function in* $\log r$.

For example, let E be an effective divisor on $\mathbf{C}^n$. Then $E = T_E$ is a closed positive current and we have

$$N(r, E) = \int_0^r n(t, E) \frac{dt}{t}, \tag{5.2.3}$$

which is called the **counting function** or **order function** of the divisor E. On the other hand, let ω be a real continuous differential form of type (1, 1) on M. Then we have by Lemma (5.2.1)

$$T_f(r, \omega) = \int_1^r n(t, [f^* \omega]) \frac{dt}{t}, \tag{5.2.4}$$

which is called the **characteristic function** or **order function** of f with respect to ω.

Let $(\mathbf{L}, H)$ be a Hermitian line bundle over M, $\| \cdot \|$ denote the norms on fibers $\mathbf{L}_x$ defined by H and ω the Chern form. Let $D \in |\mathbf{L}|$. We assume that

(5.2.5) (i) $\omega \geq 0$,

(ii) $f(\mathbf{C}^n) \not\subset \operatorname{supp} D$.

It is noted that the effective divisor $f^* D$ is defined on $\mathbf{C}^n$ (cf. Chapter IV, §5, (c)).

(5.2.6) Lemma. *Let $\sigma \in \Gamma(M, \mathbf{L})$ with $D = (\sigma)$. Then we have the following.*

(i) $\log \|\sigma \circ f_0\|^2$ *is written as the difference of two plurisubharmonic functions on $\mathbf{C}^n$; hence* $\log \|\sigma \circ f_0\|^2 \in L_{\mathrm{loc}}(\mathbf{C}^n)$ *and* $\log \|\sigma \circ f_0\|^2 \in L_{\mathrm{loc}}(\Gamma(r))$ *$(r > 0)$. (In the sequel, we write* $\log \|\sigma \circ f_0\|^2 = \log \|\sigma \circ f\|^2$.*)*

(ii) *The current $[f^* \omega]$ is closed and positive.*

(iii) $dd^c[\log \|\sigma \circ f\|^2] = f^* D - [f^* \omega]$ *as currents.*

Proof. (i) Let $(\{U_\lambda\}, \{s_\lambda\})$ be a local trivialization covering of $\mathbf{L}$. Putting $\sigma | U_\lambda = \sigma_\lambda s_\lambda$ with holomorphic functions σ_λ on U_λ, we get

$$f_0^* D \,|\, f_0(U_\lambda) = (\sigma_\lambda \circ f_0),$$

$$f_0^* \omega | f_0^{-1}(U_\lambda) = -dd^c \log \|s_\lambda \circ f_0\|^2.$$

By Theorem (5.1.13) we have

$$f_0^*D\,|\,f_0^{-1}(U_\lambda) = dd^c[\log|\sigma_\lambda\circ f_0|^2] = dd^c\left[\log\frac{\|\sigma\circ f_0\|^2}{\|s_\lambda\circ f_0\|^2}\right]$$

$$= dd^c[\log\|\sigma\circ f_0\|^2] + [f_0^*\omega]$$

as currents on $f_0^{-1}(U_\lambda)$. Therefore we obtain the following current equation on W:

(5.2.7) $$dd^c[\log\|\sigma\circ f_0\|^2] = f_0^*D - [f_0^*\omega].$$

By Theorem (4.2.14) there is a holomorphic function F with $(F) = f^*D$. Noting that $(f^*D)|W = f_0^*D$, we have by Theorem (5.1.13)

$$dd^c[\log|F|^2] = [f_0^*D] \text{ on } W.$$

Combining this with (5.2.7), we deduce

(5.2.8) $$dd^c\left[\log\frac{|F|^2}{\|\sigma\circ f_0\|^2}\right] = [f_0\omega].$$

It is clear that

(5.2.9) $$\log\frac{|F|^2}{\|\sigma\circ f_0\|^2}\in C^\infty(W).$$

By assumption (5.2.5), (i), $f_0^*\omega\geq 0$, and then (5.2.8) and (5.2.9) imply that $\log(|F|^2/\|\sigma\circ f_0\|^2)$ is C^∞ and plurisubharmonic in W (cf. Lemma (3.3.36)). Since $\operatorname{codim} I(f)\geq 2$ (cf. Theorem (4.4.8), (i)), Theorem (3.3.42) yields that $\log(|F|^2/\|\sigma\circ f_0\|^2)$ uniquely extends to a plurisubharmonic function on $\mathbf{C}^n$. Since $\log|F|^2$ is a plurisubharmonic function on $\mathbf{C}^n$, the first half of (i) is proved. The rest follows from Theorem (3.3.28) and Corollary (3.3.40).

(ii) By Lemma (5.2.1) we see that $f^*\omega\in L_{\text{loc}}^{(1,1)}(\mathbf{C}^n)$ and $[f^*\omega]$ is a positive current. It follows from $\operatorname{codim} I(f)\geq 2$, Theorem (5.1.14) and (5.2.8) that

(5.2.10) $$[f^*\omega] = dd^c\left[\log\frac{|F|^2}{\|\sigma\circ f_0\|^2}\right] \quad \text{on } \mathbf{C}^n.$$

Hence $[f^*\omega]$ is closed.

(iii) This follows from $dd^c[\log|F|^2] = f^*D$ and (5.2.10). *Q.E.D.*

Assume that M is compact. We may choose $\sigma\in\Gamma(M, \mathbf{L})$ with $(\sigma) = D$ which satisfies

(5.2.11) $$\|\sigma(x)\| < 1 \quad (x\in M).$$

By Lemma (5.2.6), (i) we may define

$$m_f(r, D) = \int_{\Gamma(r)} \log \frac{1}{\|\sigma \circ f\|} \eta \geq 0.$$

Take another $\sigma' \in \Gamma(M. \mathbf{L})$ such that $(\sigma') = D$ and $\|\sigma'(x)\| < 1$ for $x \in M$. Then there is a constant $a \in \mathbf{C}^*$ such that $\sigma' = a\sigma$. Therefore $m_f(r, D)$ is defined up to a constant term and called the **proximity function** of f with respect to $D \in |\mathbf{L}|$. Now we obtain from Lemma (5.2.6), (iii)

(5.2.12) $$T_f(r, \omega) - N(r, f^*D) = -\int_1^r \frac{\langle dd^c \log \|\sigma \circ f\|^2 \wedge \alpha^{n-1}, \chi_{B(t)} \rangle}{t^{2n-1}} dt.$$

Noting Lemma (5.2.6), (i), we apply Lemma (3.3.39) to the right-hand side of (5.2.12), so that

(5.2.13) $$T_f(r, \omega) - N(r, f^*D) = m_f(r, D) - m_f(1, D).$$

In what follows, we assume moreover that M is Kähler. Let ω' be a real closed C^∞-differential form of type (1, 1) such that $[\omega'] = c_1(\mathbf{L})$ and $\omega' \geq 0$. Since M is compact and Kähler, there is another Hermitian metric H' in $\mathbf{L}$ such that ω' is the Chern form of $(\mathbf{L}, H')$ (cf. Weil [120]). Then by the definition of Hermitian metrics, there is a positive-valued C^∞-function $b > 0$ such that $H' = bH$. It follows from (5.2.13) that

$$T_f(r, \omega) - T_f(r, \omega')$$

$$= \int_{\Gamma(r)} \log(b \circ f)\eta - \int_{\Gamma(1)} \log(b \circ f)\eta = O(1) \quad (r \to \infty).$$

In this book we denote by $O(\cdot)$ the estimation as $r \to \infty$. Under assumption (5.2.5), the **characteristic function** $T_f(r, \mathbf{L})$ of f with respect to $\mathbf{L}$ is defined by

(5.2.14) $$T_f(r, \mathbf{L}) = T_f(r, \omega),$$

where $\omega \geq 0$ and $[\omega] = c_1(\mathbf{L})$. By (5.2.13) we have the **First Main Theorem**:

(5.2.15) Theorem. *Let* $\mathbf{L}$ *be a holomorphic line bundle over a compact Kähler manifold such that* $c_1(\mathbf{L}) \geq 0$. *Let* $f : \mathbf{C}^n \to M$ *be a meromorphic mapping and* $D \in |\mathbf{L}|$ *such that* $f(\mathbf{C}^n) \not\subset \operatorname{supp} D$. *Then*

$$T_f(r, \mathbf{L}) = N(r, f^*D) + m_f(r, D) + O(1) \quad (r \geq 1).$$

Since $m_f(r, D) \geq 0$, we immediately have the following **Nevanlinna inequality**.

(5.2.16) Corollary. *Assume the same conditions as in Theorem (5.2.15). Then*

$$N(r, f^*D) \leq T_f(r, \mathbf{L}) + O(1).$$

The same conclusion as Theorem (5.2.15) holds without assumption $c_1(\mathbf{L}) \geq 0$. We do not prove it in general, but for the later use, we here give the proof only in the special case where M is a complex projective algebraic manifold. Let $\mathbf{E}_0 \to M$ be a very ample line bundle. Then there is a Hermitian metric $\|\cdot\|$ in $\mathbf{E}_0$ such that its Chern form $c_1(\mathbf{E}_0) > 0$. Since E_0 is very ample, we may take a section $\sigma_0 \in \Gamma(M, \mathbf{E}_0)$ such that supp $\sigma_0 \not\supset f(\mathbf{C}^n)$ and $\|\sigma_0\| < 1$. Let $\mathbf{E} \to M$ be an arbitrary Hermitian line bundle with Chern form ω. Let $D \in |\mathbf{E}|$ and $\sigma \in \Gamma(M, \mathbf{E})$ such that $(\sigma) = D$ and $\|\sigma\| < 1$. Let $f : \mathbf{C}^n \to M$ be a meromorphic mapping. Assume that $f(\mathbf{C}^n) \not\subset$ supp D. If $\log \|\sigma \circ f\|$ is written as the difference of two plurisubharmonic functions on $\mathbf{C}^n$ then we may define $m_f(r, D)$, and then obtain (5.2.13), provided that Lemma (5.2.6), (iii) holds. Note that Lemma (5.2.6), (iii) holds on $\mathbf{C}^n - I(f)$, and hence on the whole $\mathbf{C}^n$ if $\log \|\sigma \circ f\|$ is written as the difference of two plurisubharmonic functions on $\mathbf{C}^n$. Let $k \in \mathbf{N}$ be sufficiently large, so that $\omega + k\omega_0 > 0$. We take the natural product Hermitian metric $\|\cdot\|$ in $\mathbf{E} \otimes E_0^k$ of which Chern form is $\omega + k\omega_0$. Thus $c_1(\mathbf{E} \otimes \mathbf{E}_0^k) > 0$. Set

$$\sigma_0^k = \sigma_0 \otimes \cdots \otimes \sigma_0 \ \ (k\text{-times}) \in \Gamma(M, \mathbf{E}_0^k).$$

Then

$$\log \|\sigma\|^2 = \log \|\sigma \otimes \sigma_0^k\|^2 - k\log \|\sigma_0\|^2,$$

so that

$$\log \|\sigma \circ f\|^2 = \log \|(\sigma \otimes \sigma_0^k) \circ f\|^2 - k\log \|\sigma_0 \circ f\|^2.$$

Since $c_1(\mathbf{E} \otimes \mathbf{E}_0^k) > 0$ and $c_1(\mathbf{E}_0) > 0$, the functions $\log \|(\sigma \otimes \sigma_0^k) \circ f\|^2$ and $\log \|\sigma_0 \circ f\|^2$ are written as the differences of plurisubharmonic functions on $\mathbf{C}^n$. Therefore we have

(5.2.17) (i) $\log \|\sigma \circ f\|^2$ is written as the difference of two plurisubharmonic functions:

(ii) $dd^c[\log \|\sigma \circ f\|^2] = f^*D - [f^*\omega]$.

In the same way as in the proof of Theorem (5.2.15), we have the following.

(5.2.18) Theorem. *Let $\mathbf{E} \to M$ be an arbitrary Hermitian line bundle over a complex projective algebraic manifold M. Let $\omega \in c_1(\mathbf{E})$ be the Chern form and $D \in |\mathbf{E}|$. Let $f : \mathbf{C}^n \to M$ be a meromorphic mapping such that $f(\mathbf{C}^n) \not\subset$ supp D. Then we have the following.*

(i) (First Main Theorem)

$$T_f(r, \omega) = N(r, f^*D) + m_f(r, D) - m_f(1, D) \ \ (r > 1).$$

Especially,

$$T_f(r, \mathbf{E}) = N(r, f^*D) + m_f(r, D) + O(1) \ \ (r \geq 1).$$

(ii) (Nevanlinna inequality)

$$N(r, f^*D) \leq T_f(r, \mathbf{L}) + O(1).$$

Let $f: \mathbf{C}^n \to M$ be a meromorphic mapping into a compact complex manifold M. Let Ω be an arbitrary Hermitian metric form on M; i.e., Ω is a real positive C^∞-differential form of type (1, 1). Then by Lemma (5.2.1) we have the characteristic function $T_f(r, \Omega)$ of f with respect to Ω. Take another Hermitian metric form Ω' on M. Since M is compact, there are positive constants C and C' such that

$$C\Omega' \leq \Omega \leq C'\Omega'.$$

Thus we have

(5.2.19) $$CT_f(r, \Omega') \leq T_f(r, \Omega) \leq C'T_f(r, \Omega').$$

We define the **order** of f by

(5.2.20) $$\rho_f = \overline{\lim_{r\to\infty}} \frac{\log T_f(r, \Omega)}{\log r},$$

which is independent of the choice of Hermitian metric form Ω by (5.2.19).

(5.2.21) *Remark.* Let Ω and Ω' be Hermitian metric forms on M, which are mutually cohomologous. Then there is a C^∞-function a on M such that

$$\Omega - \Omega' = dd^c a.$$

Suppose $f: \mathbf{C}^n \to M$ to be holomorphic. It follows from Lemma (3.3.39) that

$$T_f(r, \Omega) - T_f(r, \Omega') = \int_1^r \frac{<[dd^c a \circ f \wedge \alpha^{n-1}], \chi_{B(t)}>}{t^{2n-1}} dt$$

$$= \frac{1}{2}\int_{\Gamma(r)} a \circ f\eta - \frac{1}{2}\int_{\Gamma(r)} a \circ f\eta = O(1).$$

Hence we may define the characteristic function $T_f(r, \{\Omega\})$ of f with respect to the cohomology class $\{\Omega\} \in H^{(1,1)}(M, \mathbf{R})$, up to a bounded term, by

$$T_f(r, \{\Omega\}) = T_f(r, \Omega).$$

This fact holds also for meromorphic f (cf., e.g., Shiffman [102]).

We consider the fundamental case where $M = \mathbf{P}^m(\mathbf{C})$ and $\mathbf{L} \to \mathbf{P}^m(\mathbf{C})$ is the hyperplane bundle $\mathbf{H}_0 \to \mathbf{P}^m(\mathbf{C})$ (cf. Chapter II, §1). Let $\rho: \mathbf{C}^{m+1} - \{O\} \to \mathbf{P}^m(\mathbf{C})$ be the Hopf fibering. We define hyperplanes $X_0, \ldots, X_m$ by

$$X_j = \rho\left[\{(z^0, \ldots, z^m) \in \mathbf{C}^{m+1} - \{O\}, z^j = 0\}\right], \quad 0 \leq j \leq m.$$

The space $\Gamma(\mathbf{P}^m(\mathbf{C}), \mathbf{H}_0)$ is naturally identified with $\mathbf{C}^{m+1}$. The holomorphic line bundle $\mathbf{H}_0$ carries the natural Hermitian metric such that its Chern form ω_0 satisfies

$$\rho^* \omega_0 = dd^c \log (z, z),$$

where $(z, z) = \sum_{j=0}^{m} z^j \bar{z}^j$.

Let $f : \mathbf{C}^n \to \mathbf{P}^m(\mathbf{C})$ be a meromorphic mapping. Since $\bigcap_{j=0}^{m} X_j = \varnothing$, there is a X_j, to say, X_0 such that $f(\mathbf{C}^n) \not\subset X_0$. Take meromorphic functions ψ_j on $\mathbf{P}^m(\mathbf{C})$ defined by

$$\psi_j(\rho(z^0, \ldots, z^m)) = \frac{z^j}{z^0}, \quad 1 \le j \le m.$$

Then $(\psi_j)_0 = X_j$ and $(\psi_j)_\infty = X_0$. By Theorem (4.2.14) there is a holomorphic function f_0 such that $(f_0) = f^* X_0$. The meromorphic functions $f^* \psi_j$, $1 \le j \le m$ on $\mathbf{C}^n$ are defined. Put

$$f_j = f_0 \cdot f^* \psi_j .$$

Then f_j are holomorphic functions on $\mathbf{C}^n$. Then $[f_0; \cdots ; f_m]$ is an irreducible representation of f (cf. Theorem (4.5.5)); i.e.

$$I(f) = \{x \in \mathbf{C}^n ; f_0(x) = \cdots = f_m(x) = 0\},$$

$$f(x) = \rho(f_0(x), \ldots, f_m(x)) \quad (x \in \mathbf{C}^n - I(f)).$$

Therefore we have

$$T_f(r, \mathbf{H}_0) = \int_1^r \frac{\left\langle dd^c \left[\log \left(\sum_{j=0}^{m} |f_j|^2 \right) \right] \wedge \alpha^{n-1}, \chi_{B(t)} \right\rangle}{t^{2n-1}} dt + O(1).$$

Since $\log \left[\sum_{j=0}^{m} |f_j|^2 \right]$ is a plurisubharmonic function on $\mathbf{C}^n$ and C^∞ on $\mathbf{C}^n - I(f)$ with codim $I(f) \ge 2$, we see by Lemma (5.1.14) that

$$\left[dd^c \log \left(\sum_{j=0}^{m} |f_j|^2 \right) \right] = dd^c \left[\log \left(\sum_{j=0}^{m} |f_j|^2 \right) \right].$$

It follows from Lemma (3.3.39) that

$$T_f(r, \mathbf{H}_0) = \int_{\Gamma(r)} \log \left(\sum_{j=0}^{m} |f_j|^2 \right)^{1/2} \eta + O(1).$$

Now put

$$A(z) = \log\left(\max_{0\le j\le m}|f_j(z)|\right).$$

Then

$$A(z) \le \log\left(\sum_{j=0}^{m}|f_j|^2\right)^{1/2} \le A(z) + \frac{1}{2}\log(m+1).$$

Thus we get

$$T_f(r, \mathbf{H}_0) = \int_{\Gamma(r)} A(z)\eta(z) + O(1). \tag{5.2.22}$$

The right-hand side of (5.2.22) is the characteristic function defined by H. Cartan [17].

Here we prepare several notations frequently used in the value distribution theory in one complex variable. Set

$$\log^+ t = \log\max\{t, 1\} \quad \text{for } t \ge 0.$$

Then

$$\log t = \log^+ t - \log^+\frac{1}{t}, \quad |\log t| = \log^+ t + \log^+\frac{1}{t}.$$

For $t_1, \ldots, t_l$ we have

$$\log^+\prod_{j=1}^{l} t_j \le \sum_{j=1}^{l}\log^+ t_j, \tag{5.2.23}$$

$$\log^+\sum_{j=1}^{l} t_j \le \log^+\left\{l\max_{1\le j\le l}\{t_j\}\right\} \le \sum_{j=1}^{l}\log^+ t_j + \log l.$$

In general, for a meromorphic function F on $\mathbf{C}^n$ we set

$$m(r, F) = \int_{\Gamma(r)}\log^+|F|\eta, \tag{5.2.24}$$

$$T(r, F) = N(r, (F)_\infty) + m(r, F).$$

We call $T(r, F)$ the **Nevanlinna characteristic function.** It follows from (5.2.23) that

$$T\left(r, \prod_{j=1}^{l}F_j\right) \le \sum_{j=1}^{l}T(r, F_j), \tag{5.2.25}$$

$$T\left(r, \sum_{j=1}^{l}F_j\right) \le \sum_{j=1}^{l}T(r, F_j) + \log l.$$

Now we go back to the meromorphic mapping $f : \mathbf{C}^n \to \mathbf{P}^m(\mathbf{C})$. For a simplicity, we put

$$F_j = f^*\psi_j = \frac{f_j}{f_0}, \quad 1 \le j \le m.$$

It follows from (5.2.22) that

$$\log^+|F_k| \le \log^+\left(\sum_{j=1}^{m}|F_j|^2\right)^{1/2} \le \sum_{j=1}^{m}\log^+|F_j| + \frac{1}{2}\log(m+1),$$

so that

(5.2.26)
$$m(r, F_k) \le \int_{\Gamma(r)}\log^+\left(\sum_{j=1}^{m}|F_j|^2\right)^{1/2}\eta \le \sum_{j=1}^{m} m(r, F_j) + \frac{1}{2}\log(m+1).$$

Note that

(5.2.27)
$$N(r, (F_k)_\infty) \le N(r, (f_0)) \le \sum_{j=1}^{m} N(r, (F_j)_\infty).$$

Applying Lemma (3.3.39) to $dd^c[\log|f_0|^2] = (f_0)$, and making use of (5.2.23), we have

(5.2.28)
$$\begin{aligned} T_f(r, \mathbf{H}_0) &= \int_{\Gamma(r)}\log^+\left(\sum_{j=1}^{m}|F_j|^2 + 1\right)^{1/2}\eta \\ &\quad + \int_{\Gamma(r)}\log|f_0|\eta + O(1) \\ &\le \sum_{j=1}^{m}\int_{\Gamma(r)}\log^+|F_j|\eta + \int_{\Gamma(r)}\log|f_0|\eta + O(1) \\ &= \sum_{j=1}^{m} m(r, F_j) + N(r, (f_0)) + O(1). \end{aligned}$$

Summerizing (5.2.26)~(5.2.28), we have the following theorem.

(5.2.29) Theorem. *Let $f_0, \ldots, f_m$ be holomorphic functions on $\mathbf{C}^n$ such that $[f_0; \cdots; f_m]$ is an irreducible representation of a meromorphic mapping $f : \mathbf{C}^n \to \mathbf{P}^m(\mathbf{C})$. Suppose that $f_k \not\equiv 0$ and put $F_j = f_j/f_k$, $0 \le j \le m$ (note $F_k \equiv 1$). Then*

$$T(r, F_j) + O(1) \le T_f(r, \mathbf{H}_0) \le \sum_{j=0}^{m} T(r, F_j) + O(1).$$

(5.2.30) Corollary. *Let F be a non-constant meromorphic function on $\mathbf{C}^n$. Consider F also to be a meromorphic mapping from $\mathbf{C}^n$ into $\mathbf{P}^1(\mathbf{C})$. Let $a \in \mathbf{C}$ be a constant. Then*

$$T_F(r, \mathbf{H}_0) = T(r, F) + O(1) = T\left(r, \frac{1}{F-a}\right) + O(1).$$

Proof. Take holomorphic functions f_0 and f_1 on $\mathbf{C}$ such that $F = f_1/f_0$ and $\operatorname{codim}\{x \in \mathbf{C}; f_0(x) = f_1(x) = 0\} \geq 2$ Then the meromorphic mapping $F: \mathbf{C}^n \to \mathbf{P}^1(\mathbf{C})$ is given by $F(x) = \rho(f_0(x), f_1(x))$. Since $f_0 \not\equiv 0$ and $f_1 \not\equiv 0$, Theorem (5.2.29) implies the above two equalities, provided that $a = 0$. For general a, we deduce from the case of $a = 0$ and (5.2.23) that

$$T\left(r, \frac{1}{F-a}\right) = T(r, F-a) + O(1)$$

$$= N(r, (F-a)_\infty) + m(r, F-a) + O(1)$$

$$= N(r, (F)_\infty) + m(r, F) + O(1). \quad Q.E.D.$$

We next consider the case where M is a complex torus $\mathbf{C}^m/\Lambda$, which is a compact Kähler manifold.

(5.2.31) Theorem. *Let N be a complex manifold. Then any meromorphic mapping $f: N \to \mathbf{C}^m/\Lambda$ is necessarily holomorphic.*

We need the following topological lemma for the proof.

(5.2.32) Lemma. *Let N be a simply connected complex manifold and X an analytic subset of* $\operatorname{codim} X \geq 2$. *Then $N - X$ is simply connected.*

Proof. This is clear if X has no singular point, since the real codimension of X is greater than 3. We use the induction on $\operatorname{codim} X$. Let $n = \dim N$. The case of $\operatorname{codim} X = n$ is trivial. Suppose that our assertion holds for codimension $\geq k+1 \geq 3$. Now we suppose that $\operatorname{codim} X = k \geq 2$. Since $\operatorname{codim} S(X) \geq k+1$, $N - S(X)$ is simply connected by the induction hypothesis. Since $R(X)$ is an analytic subset of $N - S(X)$ without singular point, $(N - S(X)) - R(X) = N - X$ is simply connected. *Q.E.D.*

Proof of Theorem (5.2.31). Let $\phi: \mathbf{C}^m \to \mathbf{C}^m/\Lambda$ be the universal covering and $(w^1, \ldots, w^m)$ the natural coordinate system of $\mathbf{C}^m$. For any $x \in N$ there is a simply connected neighborhood U of x. Since $\operatorname{codim} I(f) \geq 2$, $U - I(f)$ is simply connected by the above lemma. Take a lifting $g: U - I(f) \to \mathbf{C}^m$ of $f|U - I(f)$. Then g extends to a holomorphic mapping $\bar{g}: U \to \mathbf{C}^m$ by Corollary (3.3.44). Hence $f|U - I(f)$ must extends holomorphically over U; that is, $I(f) = \varnothing$. *Q.E.D.*

We fix the standard Kähler form Ω on $\mathbf{C}^m/\Lambda$ given by

$$\pi^*\Omega = \frac{i}{2\pi}\sum_{j=1}^{m} dw^j \wedge d\overline{w}^j.$$

(5.2.33) Lemma. *Let $f: \mathbf{C}^n \to \mathbf{C}^m/\Lambda$ be a non-constant holomorphic mapping. Then there are positive constants C, and r_0 such that*

$$T_f(r, \Omega) \geq Cr^2 \ \textit{for all } r \geq r_0.$$

Proof. Let $\tilde{f}: \mathbf{C}^n \to \mathbf{C}^m$ be a lifting of f and put $\tilde{f} = (f^1, \ldots, f^m)$. Then

$$T_f(r, \Omega) = \int_1^r \frac{<dd^c \sum_{j=0}^{m} |f^j|^2 \wedge \alpha^{n-1}, \chi_{B(t)}>}{t^{2n-1}} dt.$$

Since f is not constant, there is some f^j which is not constant. Therefore it is sufficient to show that if F is a non-constant holomorphic function on $\mathbf{C}^n$, there are positive constants C and r_0 such that

$$\text{(5.2.34)} \qquad \int_1^r \frac{<dd^c|F|^2 \wedge \alpha^{n-1}, \chi_{B(t)}>}{t^{2n-1}} dt \geq Cr^2$$

for all $r \geq r_0$. It follows from Lemma (3.3.20) that

$$\int_1^r \frac{<dd^c|F|^2 \wedge \alpha^{n-1}, \chi_{B(t)}>}{t^{2n-1}} dt = \frac{1}{2}\int_{\Gamma(r)} |F|^2 \eta - \frac{1}{2}\int_{\Gamma(1)} |F|^2 \eta.$$

Let $F(z) = \sum_{\lambda \geq 0} P_\lambda(z^1, \ldots, z^n)$ be the expansion with homogeneous polynomials P_λ of degree λ. Then there is some $P_\mu \neq 0$ with $\mu \geq 1$ and

$$\text{(5.2.35)} \qquad \frac{1}{2}\int_{\Gamma(r)} |F|^2 \eta = \frac{1}{2\pi}\int_0^{2\pi} d\theta \frac{1}{2}\int_{\Gamma(r)} |F(e^{i\theta}z)|^2 \eta(e^{i\theta}z)$$

$$= \frac{1}{2\pi}\int_0^{2\pi} d\theta \frac{1}{2}\int_{\Gamma(r)} |F(e^{i\theta}z)|^2 \eta(z) = \frac{1}{2}\int_{\Gamma(r)} \left[\int_0^{2\pi} |F(e^{i\theta}z)|^2 d\theta\right] \eta(z)$$

$$= \frac{1}{2}\sum_{\lambda, \nu \geq 0}\int_{\Gamma(r)} \left[\int_0^{2\pi} e^{(\lambda-\nu)i\theta} P_\lambda(z)\overline{P_\nu(z)}\, d\theta\right] \eta(z) = \frac{1}{2}\sum_{\lambda \geq 0}\int_{\Gamma(r)} |P_\lambda(z)|^2 \eta(z)$$

$$= \frac{1}{2}\sum_{\lambda \geq 0} r^{2\nu}\int_{\Gamma(1)} |P_\lambda|^2 \eta \geq r^{2\mu}\frac{1}{2}\int_{\Gamma(1)} |P_\mu|^2 \eta \geq r^2 \frac{1}{2}\int_{\Gamma(1)} |P_\mu|^2 \eta.$$

Thus (5.2.34) is proved. *Q.E.D.*

(5.2.36) Corollary. *If a holomorphic mapping* $f: \mathbf{C}^n \to \mathbf{C}^m/\Lambda$ *is not constant, then* $\rho_f \geq 2$.

5.3 Elementary Properties of Characteristic Functions

We begin with studying the relation between the Nevanlinna characteristic function $T(r, F)$ and the **maximum modulus** $M(r, F)$ for a holomorphic function F on $\mathbf{C}^n$, where

$$M(r, F) = \max\{|F(z)|;\ z \in B(r)\}.$$

Let A be an (n, n)-matrix, B an n-column vector, C an n-row vector and D a scalar such that

$$\text{(5.3.1)} \quad \text{(i)} \quad A^*A - C^*C = I_n \quad ((n, n)\text{-unit matrix}),$$

$$\text{(ii)} \quad A^*B - C^*D = O,$$

$$\text{(iii)} \quad |D|^2 - B^*B = 1.$$

Then we set

$$g = \begin{pmatrix} A & B \\ C & D \end{pmatrix}.$$

We denote by $SU(n, 1)$ the set of all those g. Then

$$SU(n, 1) = \left\{ g = \begin{pmatrix} A & B \\ C & D \end{pmatrix} \in SL(n+1);\ {}^t\bar{g} \begin{pmatrix} 1 & & & O \\ & \ddots & & \\ & & 1 & \\ O & & & -1 \end{pmatrix} g = \begin{pmatrix} 1 & & & O \\ & \ddots & & \\ & & 1 & \\ O & & & -1 \end{pmatrix} \right\}.$$

Thus $SU(n, 1)$ is a group and moreover acts on $B(1)$ by

$$\gamma_g : z = \begin{pmatrix} z^1 \\ \vdots \\ z^n \end{pmatrix} \in B(1) \to (Az + B)(Cz + D)^{-1} \in B(1)$$

for $g = \begin{pmatrix} A & B \\ C & D \end{pmatrix} \in SU(n, 1)$. In fact, by (5.3.1) we have

$$|Cz + D|^2 - \|Az + B\|^2 = 1 - \|z\|^2.$$

Hence $Cz + D \neq 0$ for $z \in B(1)$. Suppose that $Cz + D = 0$ for some $z \in \Gamma(1)$. Then $D = -Cz$ and $B = -Az$. It follows from (5.3.1), (iii) that

$$|Cz|^2 - \|Az\|^2 = 1.$$

On the other hand, (5.3.1), (i) implies that

$$\|Az\|^2 - |Cz|^2 = \|z\|^2 = 1,$$

so that we get a contradiction. Therefore it follows that

$$Cz + D \neq 0 \ \text{ for } z \in \overline{B(1)},$$

$$\|z\| < 1 \iff \|\gamma_g(z)\| < 1,$$

$$\|z\| = 1 \iff \|\gamma_g(z)\| = 1.$$

Furthermore

$$\gamma_{g'} \circ \gamma_g = \gamma_{g'g},$$

and hence the inverse $(\gamma_g)^{-1}$ exists and is given by

$$(\gamma_g)^{-1} = \gamma_{g^{-1}}.$$

Let $\gamma_g(O) = O$. Then $B = O$, $C = O$, $A^*A = I_n$ and $D = \overline{\det A}$. Hence we may write

$$\gamma_g(z) = Az, \quad {}^t A\bar{A} = I_n;$$

that is, such γ_g is a unitary transformation, of which group is denoted by $U(n)$. It is well known that $U(n)$ acts transitively on $\Gamma(r)$. Hence, for any $w \in B(1)$ there is an element $A \in U(n)$ such that $Aw = {}^t(\|w\|, 0, \ldots, 0)$ $(\|w\| < 1)$. For $0 \leq r < 1$, put

$$A = \begin{pmatrix} \frac{1}{\sqrt{1-r^2}} & & & O \\ & 1 & & \\ & & \ddots & \\ O & & & 1 \end{pmatrix}, \quad B = \begin{pmatrix} -r \\ 0 \\ \vdots \\ 0 \end{pmatrix}$$

$$C = {}^t B, \quad D = 1,$$

and define $g \in SU(n, 1)$ as above. Then

(5.3.2)
$$\gamma_g(z) = \frac{1}{1 - rz^1} \begin{pmatrix} z^1 - r \\ \sqrt{1-r^2}\, z^2 \\ \vdots \\ \sqrt{1-r^2}\, z^n \end{pmatrix}$$

and hence $\gamma_g({}^t(r, \ldots, 0)) = O$. Therefore we have

(5.3.3) $SU(n, 1)$ acts transitively on $B(1)$.

(5.3.4) Theorem. *The holomorphic automorphism group* $\mathrm{Aut}(B(1))$ *acts transitively on* $B(1)$ *and*

$$\mathrm{Aut}(B(1)) = \{\gamma_g;\, g \in SU(n, 1)\}, \quad \mathrm{Aut}_O(B(1)) = U(n),$$

where $\mathrm{Aut}_O(B(1))$ *denotes the isotropy subgroup of* $\mathrm{Aut}(B(1))$ *at* O.

Proof. Take an arbitrary $T \in \mathrm{Aut}(B(1))$. By (5.3.3) there is an element $g \in SU(n, 1)$ such that $\gamma_g(T(O)) = O$. Put

$$S = \gamma_g \circ T \in \mathrm{Aut}(B(1)).$$

We claim that S is linear (hence unitary). Put

$$H(z) = S^{-1}\left[e^{-i\theta} S(e^{i\theta} z)\right] \quad (z \in B(1)),$$

where $\theta \in \mathbf{R}$. Then $dH = I_n$. Hence H has a homogeneous polynomial expansion

$$H(z) = z + \sum_{\nu=k}^{\infty} P_\nu(z) \quad (k \geq 2),$$

where P_ν are homogeneous $\mathbf{C}^n$-valued polynomials of degree ν. Suppose that $P_k \neq 0$. Taking the l-th iteration H^l of H, we have

$$H^l(z) = z + lP_k(z) + \cdots.$$

Moreover we have

$$lP_k(z) = \frac{1}{2\pi}\int_0^{2\pi} H^l(e^{it}z)e^{-ikt}\,dt,$$

where the integral is $\mathbf{C}^n$-valued one. Thus

$$l\|P_k(z)\| \leq 1, \quad l = 1, 2, \ldots$$

This implies that $P_k(z) \equiv 0$, and so $P_\nu = 0$ for all ν. It follows that $H(z) \equiv z$, and then

$$S(e^{i\theta} z) = e^{i\theta} S(z), \quad z \in B(1).$$

Therefore S is linear (unitary) and $T = \gamma_{g^{-1}} \circ S$. Hence

$$\mathrm{Aut}(B(1)) = \left\{ \gamma_g;\, g \in SU(n, 1) \right\}.$$

The rest follows immediately. *Q.E.D.*

Let $z_0 \in B(1)$ be an arbitrarily fixed point. Take a transformation $\gamma(z_0; z) \in \mathrm{Aut}(B(1))$ with $\gamma(z_0; z_0) = O$ (Theorem (5.3.4)). Then we set

(5.3.5) $$\beta(z_0; z) = \gamma(z_0; \cdot)^* \eta(z) = dd^c \log \|\gamma(z_0; z)\|^2,$$

$$\eta(z_0; z) = \gamma(z; \cdot)^* \eta(z) = d^c \log \|\gamma(z_0; z)\|^2 \wedge \beta(z_0; z)^{n-1}.$$

It follows from Theorem (5.3.4) that $\beta(z_0; z)$ and $\eta(z_0; z)$ are independent of the choice of $\gamma(z_0; z)$. Note that

$$\beta(O; z) = \beta(z), \quad \eta(O; z) = \eta(z).$$

(5.3.6) Lemma. *Let F be a holomorphic function on a neighborhood of $\overline{B(1)}$ and $z_0 \in B(1)$. Then*

$$\log |F(z_0)| \leq \int_{\Gamma(1)} \log |F(z)| \eta(z_0; z).$$

Proof. Suppose that $z_0 = O$. Since $\log |F(z)|$ is plurisubharmonic, Corollary (3.3.40) implies

$$\log |F(z_0)| \leq \int_{\Gamma(1)} \log |F(z)| \eta(O; z). \tag{5.3.7}$$

For general z_0, put

$$G(w) = F(\gamma(z_0; \cdot)^{-1}(w)).$$

We see by Theorem (5.3.4) that $G(w)$ is holomorphic in a neighborhood of $\overline{B(1)}$. It follows from (5.3.7) that

$$\log |F(z_0)| = \log |G(O)| \leq \int_{\Gamma(1)} \log |G(w)| \eta(w).$$

Substituting $w = \gamma(z_0; z)$ to the above integral, we obtain

$$\log |F(z_0)| \leq \int_{\Gamma(1)} \log |F(z)| \eta(z_0; z). \quad Q.E.D.$$

(5.3.8) Lemma. *Let F be as in Lemma (5.3.6) and $0 < r < 1$. Then*

$$\log M(r, F) \leq \frac{1 - r^2}{(1 - r)^{2n}} \int_{\Gamma(1)} \log^+ |F(z)| \eta(z).$$

Proof. There is a point $z_0 \in \Gamma(r)$ such that $|F(z_0)| = M(r, F)$. Since η is $U(n)$-invariant, we may assume that $z_0 = {}^t(r, 0, ..., 0)$. By Lemma (5.3.6) we have

$$\log M(r, F) \leq \int_{\Gamma(1)} \log |F(z)| \eta(z_0; z) \leq \int_{\Gamma(1)} \log^+ |F(z)| \eta(z_0; z). \tag{5.3.9}$$

By (5.3.2) we may take

$$\gamma(z_0; z) = \frac{1}{1 - rz^1} \begin{pmatrix} z^1 - r \\ \sqrt{1 - r^2}\, z^2 \\ \vdots \\ \sqrt{1 - r^2}\, z^n \end{pmatrix}$$

We are going to compare $\eta(z_0; z)$ and $\eta(z)$ on $\Gamma(1)$. A direct computation yields

$$(5.3.10)\qquad d^c \log \|z\|^2 = \frac{i}{4\pi}\sum_{j=1}^{n}\left(z^j d\bar{z}^j - \bar{z}^j dz^j\right),$$

$$(5.3.11)\qquad d^c \log \|\gamma(z_0; z)\|^2 = \frac{1-r^2}{|1-rz^1|^2}\frac{i}{4\pi}\sum_{j=1}^{n}\left(z^j d\bar{z}^j - \bar{z}^j dz^j\right)$$
$$= \frac{1-r^2}{|1-rz^1|^2} d^c \log \|z\|^2.$$

In the following, $\beta(z)$ and $\beta(z_0; z)$ stand for the forms induced on $\Gamma(1)$. Note that $d\|z\|^2 = \partial\|z\|^2 + \bar{\partial}\|z\|^2 = 0$ on $\Gamma(1)$. Hence

$$\left(\sum_{j=1}^{n} \bar{z}^j dz^j\right) \wedge \left(\sum_{j=1}^{n} z^j d\bar{z}^j\right) = 0.$$

It follows that

$$\beta(z) = \frac{i}{2\pi}\sum_{j=1}^{n} dz^j \wedge d\bar{z}^j.$$

Moreover we have

$$\beta(z_0; z) = dd^c \log\left\{\frac{1}{|1-rz^1|^2}\left(|z^1 - r|^2 + (1-r^2)\sum_{j=2}^{n}|z^j|^2\right)\right\}$$
$$= \frac{i}{2\pi}\left(\frac{dz^1\wedge d\bar{z}^1 + (1-r^2)\sum_{j=2}^{n} dz^j\wedge d\bar{z}^j}{|z^1-r|^2 + (1-r^2)\sum_{j=2}^{n}|z^j|^2} - \xi\wedge\bar{\xi}\right),$$

where
$$\xi = \frac{(\bar{z}^1 - r)dz^1 + (1-r^2)\sum_{j=2}^{n}\bar{z}^j dz^j}{|z^1-r|^2 + (1-r^2)\sum_{j=2}^{n}|z^j|^2}.$$

Since $|z^1 - r|^2 + (1-r^2)\sum_{j=2}^{n}|z^j|^2 = |1-rz^1|^2$ and $\frac{i}{2\pi}\xi\wedge\bar{\xi} \geq 0$,

$$(5.3.12)\qquad \beta(z_0; z) = \frac{i}{2\pi}\frac{1}{|1-rz^1|^2}\left(dz^1\wedge d\bar{z}^1 + (1-r^2)\sum_{j=2}^{n} dz^j\wedge d\bar{z}^j\right)$$
$$\leq \frac{1}{(1-r)^2}\frac{i}{2\pi}\sum_{j=1}^{n} dz^j\wedge d\bar{z}^j = \frac{1}{(1-r)^2}\beta(z).$$

Here one notes that the real cotangent space of $\Gamma(1)$ at $z \in \Gamma(1)$ is the orthogonal sum of the subspace spanned by $d^c \log \|z\|^2$ and a complex linear subspace of complex dimension $n-1$. Therefore we infer from (5.3.11) and (5.3.12)

$$\eta(z_0; z) = d^c \log \|\gamma(z_0; z)\|^2 \wedge \beta^{n-1} \le \frac{1-r^2}{(1-r)^{2n}} \eta(z).$$

Combining this with (5.3.9), we obtain the required inequality. *Q.E.D.*

(5.3.13) Theorem. *Let F be a holomorphic function on* $\mathbf{C}^m$ *and* $0 < r < R < \infty$. *Then*

$$T(r, F) \le \log M(r, F) \le \frac{1-(r/R)^2}{(1-r/R)^{2n}} T(R, F).$$

Proof. The first inequality is clear by the definitions. The change of variables $z = Rw$ and Lemma (5.3.8) imply the second inequality. *Q.E.D.*

The following proposition gives a criterion of holomorphic functions on $\mathbf{C}^n$ to be polynomials.

(5.3.14) Proposition. *Let F be a holomorphic function on* $\mathbf{C}^n$. *Then F is polynomial of degree p if and only if*

$$T(r, F) = (p + o(1)) \log r.$$

Proof. We identify F with the meromorphic mapping $z \in \mathbf{C}^n \to [1; F(z)] \in \mathbf{P}^1(\mathbf{C})$, and denote by ω_0 the Fubini-Study Kähler form on $\mathbf{P}^1(\mathbf{C})$. Then by Corollary (5.2.30)

$$T_F(r, \omega_0) = T(r, F) + O(1).$$

It follows from Lemmas (5.2.6), (ii), (5.2.2) that $T_F(r, \omega_0)$ is a convex increasing function in $\log r$. Thus $T_F(r; \omega_0)/\log r$ is monotone increasing. Therefore it suffices to show that for any $p \in \mathbf{R}$ with $p \ge 0$, F is a polynomial of degree $\le p$ if and only if $T(r, F) \le (p + o(1)) \log r$. If F is a polynomial of degree $\le p$, then a direct computation implies that

$$T(r, F) = m(r, F) \le (p + o(1)) \log r.$$

Conversely, suppose that $T(r, F) \le (p + o(1)) \log r$. Let $0 < \varepsilon < 1$ and put

$$\delta(\varepsilon) = \frac{1-\varepsilon^2}{(1-\varepsilon)^{2n}}.$$

It follows from Theorem (5.3.13) that

$$\log M(r, F) \le \delta(\varepsilon) T\left(\frac{r}{\varepsilon}, F\right) \le \delta(\varepsilon)(p + o(1)) \log \frac{r}{\varepsilon}.$$

Therefore

$$(5.3.15)\qquad M(r,\ F) \le \varepsilon^{-\delta(\varepsilon)(p+o(1))} r^{\delta(\varepsilon)(p+o(1))}.$$

Let $F(z) = \sum_{\nu=0}^{\infty} P_\nu(z)$ be the expansion with homogeneous polynomials P_ν of degree ν. Then we have by (5.3.15) and (5.2.35)

$$\int_{\Gamma(r)} |F|^2 \eta = \sum_{\nu=0}^{\infty} r^{2\nu} \int_{\Gamma(1)} |P_\nu|^2 \eta \le \varepsilon^{-2\delta(\varepsilon)(p+o(1))} r^{2\delta(\varepsilon)(p+o(1))}.$$

Take a sufficiently small $0 < \varepsilon \ll 1$ and a sufficiently large $r \gg 1$ so that $\delta(\varepsilon)(p + o(1)) < p + 1/4$. Letting $r \to \infty$, we infer that $P_\nu = 0$ for all $\nu > p$. Hence F is polynomial of degree $\le p$. *Q.E.D.*

A divisor D on $\mathbf{C}^n$ is called an **algebraic divisor** if there is a rational function Q with $(Q) = D$. Let D be an algebraic effective divisor on $\mathbf{C}^n$. Then there is a polynomial P with $(P) = D$. Then by Proposition (5.3.14) and Corollary (5.2.30) we get

$$N(r,\ D) \le T\left(r,\ \frac{1}{P}\right) \le T(r,\ P) + O(1) \le O(\log r).$$

Stoll [110, 112] proved the converse.

(5.3.16) Lemma. *An effective divisor D on $\mathbf{C}^n$ is algebraic if and only if $N(r,\ D) = O(\log r)$.*

It requires another rather long preparation to present the self-contained proof of this lemma, but nowadays there are several ways to prove it. We do not give the proof here, but in Appendix II, depending on Weierstrass-Stoll canonical functions (see Theorem (3.25) in Appendix II).

Remark. In the case of $n = 1$, the readers easily see that

$$N(r,\ D) = O(\log r) \Longleftrightarrow n(r,\ D) = O(1) \Longleftrightarrow \operatorname{supp} D \text{ is a finite set.}$$

(5.3.17) Proposition. *A meromorphic function F on $\mathbf{C}^n$ is rational if and only if $T(r,\ F) = O(\log r)$.*

Proof. If F is rational, then there are polynomials P_0 and P_1 such that $F = P_1/P_0$. It follows from (5.2.25), Corollary (5.2.30) and Proposition (5.3.14) that

$$T(r,\ F) = T\left(r,\ \frac{P_1}{P_0}\right) = T\left(r,\ \frac{1}{P_0}\right) + T(r,\ P_1)$$

$$= T(r,\ P_0) + T(r,\ P_1) + O(1) = O(\log r).$$

Conversely, suppose that $T(r, F) = O(\log r)$. Then by definition

$$N(r, (F)_\infty) \le T(r, F) = O(\log r).$$

By Lemma (5.3.16) there is a polynomial Q with $(Q) = (F)_\infty$. Hence $G(z) = Q(z)F(z)$ is a holomorphic function on $\mathbf{C}^n$. We obtain

$$T(r, G) \le T(r, F) + T(r, Q) = O(\log r),$$

so that F is a polynomial. Therefore F is a rational function. *Q.E.D.*

A meromorphic function F on $\mathbf{C}^n$ is said to be **transcendental** if F is not rational.

(5.3.18) Corollary. *A meromorphic function F on $\mathbf{C}^n$ is transcendental if and only if*

$$\varlimsup_{r\to\infty} \frac{T(r, F)}{\log r} = +\infty.$$

Proof. It remains to show the "only if" part. Assume that F is transcendental; i.e., by Proposition (5.3.17) $T(r, F) \ne O(\log r)$. As in the proof of Proposition (5.3.14) we see that $T_F(r, \omega_0)$ $(= T(r, F) + O(1))$ is a convex increasing function in $\log r$. Hence the assumption implies that

$$\varliminf_{r\to\infty} \frac{T(r, F)}{\log r} = +\infty. \quad Q.E.D.$$

Let M be a complex projective algebraic manifold. A meromorphic mapping $f : \mathbf{C}^n \to M$ is said to be **rational** if $f^*\psi$ are rational functions for all meromorphic functions $\psi \in \mathrm{Mer}(M)$ such that $\mathrm{supp}\,(\psi)_\infty \not\supset f(\mathbf{C}^n)$; otherwise, f is said to be **transcendental**

(5.3.19) Theorem. *Let $f : \mathbf{C}^n \to M$ be a meromorphic mapping. Then the following conditions are equivalent.*

(i) *f is rational.*

(ii) *Let $\mathbf{L} \to M$ be any very ample line bundle. Let Φ: $M \to P(\Gamma(M, \mathbf{L})^*)$ be the imbedding determined by $\Gamma(M, \mathbf{L})$ and fix a homogeneous coordinate system $[u^0; \cdots; u^N]$ $(\dim \Gamma(M, \mathbf{L}) = N+1)$. Then f carries a reduced representation $f(z) = [P^0(z); \cdots; P^N(z)]$ with respect to this homogeneous coordinate system such that all $P^i(z)$ are polynomials.*

(iii) *Let Ω be an arbitrary Hermitian metric form on M. Then*

$$T_f(r, \Omega) = O(\log r).$$

Proof. (i)$\Rightarrow$(ii). Take k so that $f(\mathbf{C}^n) \not\subset \{u^k = 0\}$. Then $u^j/u^k|M \in \mathrm{Mer}(M)$ $(0 \le j \le N)$. Hence $f^*(u^j/u^k)$ are rational functions. Write

$$f^*\left(\frac{u^j}{u^k}\right) = \frac{P^j}{P^k}, \quad 0 \le j \le N, \ j \ne k,$$

where P^i, $1 \le i \le N$ are polynomials. Here we may assume that there is no common irreducible component of P^i, $0 \le i \le N$. Thus we obtain the desired reduced representation

$$f(z) = [P^0(z); \cdots; P^N(z)].$$

(ii)⟹(iii). Let $\mathbf{H}_0$ be the hyperplane bundle over $\mathbf{P}^N(\mathbf{C}) = P(\Gamma(M, \mathbf{L})^*)$. Then $\Phi^{-1}\mathbf{H}_0 = \mathbf{L}$. Let ω_0 be the Fubini-Study Kähler form on $\mathbf{P}^N(\mathbf{C})$. We consider M a complex submanifold and f a meromorphic mapping into $\mathbf{P}^N(\mathbf{C})$, of which image $f(\mathbf{C}^n)$ is contained in M. Then Proposition (5.3.17) and Theorem (5.2.29) imply

$$T_f(r, \omega_0) = T_f(r, \mathbf{H}_0) + O(1) = O(\log r).$$

It follows from (5.2.19) that

$$(5.3.20) \qquad T_f(r, \Omega) = O(\log r) \Longleftrightarrow T_f(r, \omega_0) = O(\log r).$$

(iii)⟹(i). Let $\psi \in \mathrm{Mer}(M)$ be an arbitrary meromorphic function on M such that $\mathrm{supp}\,(\psi)_\infty \not\supset f(\mathbf{C}^n)$. Let $\mathbf{E} \to M$ be the holomorphic line bundle determined by the polar divisor $(\psi)_\infty$. Then there is a holomorphic section $\sigma_0 \in \Gamma(M, \mathbf{E})$ with $(\sigma_0) = (\psi)_\infty$. We get a section $\sigma_1 = \psi\sigma_0 \in \Gamma(M, \mathbf{E})$. Take any Hermitian metric in $\mathbf{E}$ with Chern form ω. Since M is compact, there is a positive constant C such that $C\Omega - \omega > 0$ and $C\Omega + \omega > 0$. Then the assumption implies that $T_f(r, \omega) = O(\log r)$. We may assume that $\|\sigma_i\| < 1$, $i = 0, 1$. By Theorem (5.2.18)

$$(5.3.21) \qquad T_f(r, \omega) = N(r, f^*(\psi)_\infty) + m_f(r, (\psi)_\infty) + O(1) = O(\log r).$$

We infer that

$$(5.3.22) \qquad N(r, (f^*\psi)_\infty) \le N(r, f^*(\psi)_\infty),$$

$$(5.3.23) \qquad m(r, f^*\psi) = \int_{\Gamma(r)} \log^+ |f^*\psi| \eta$$

$$= \int_{\Gamma(r)} \log \frac{\|\sigma_1 \circ f\|}{\|\sigma_0 \circ f\|} \eta \le \int_{\Gamma(r)} \log^+ \frac{\|\sigma_1 \circ f\|}{\|\sigma_0 \circ f\|} \eta$$

$$\le \int_{\Gamma(r)} \log^+ \frac{1}{\|\sigma_0 \circ f\|} \eta = m_f(r, (\psi)_\infty).$$

We see by (5.3.21)~(5.3.23) that

$$T(r, f^*\psi) = O(\log r),$$

and hence by Proposition (5.3.17) that $f^*\psi$ is rational. *Q.E.D.*

(5.3.24) Corollary. *Let* $f : \mathbf{C}^n \to M$ *be as in Theorem (5.3.19) and* Ω *a Hermitian metric form. Then f is transcendental if and only if*

$$\varliminf_{r\to\infty} \frac{T_f(r, \Omega)}{\log r} = +\infty.$$

Proof. By (5.2.19) we may assume that Ω is d-closed. Then the proof is the same as in that of Corollary (5.3.18). *Q.E.D.*

5.4 Casorati-Weierstrass' Theorem

In this section we assume that M is a compact complex manifold and $\mathbf{L} \to M$ a holomorphic line bundle satisfying

(5.4.1) $\Gamma(M, \mathbf{L})$ generates $\mathbf{L}_x$ at all points $x \in M$; i.e., $e_x\colon \sigma \in \Gamma(M, \mathbf{L}) \to \sigma(x) \in \mathbf{L}_x$ are surjective for all $x \in M$.

We put

$$V = \Gamma(M, \mathbf{L}), \quad \dim V = N + 1.$$

We fix a Hermitian inner product (,) in V. We recall several facts from Chapter II, §1. For $x \in M$, we put $V(x) = \{\sigma \in V; e_x(\sigma) = 0\}$ and $V(x)^\perp = \{\alpha \in V^*; \alpha(V(x)) = 0\}$. Then we have a holomorphic mapping

$$\Phi_V\colon x \in M \to V(x)^\perp \in P(V^*).$$

Let $\mathbf{H}_{V^*} \to P(V^*)$ be the hyperplane bundle over $P(V^*)$. Then

$$\Phi_V^* \mathbf{H}_{V^*} = \mathbf{L}. \tag{5.4.2}$$

The inner product (,) naturally induces a Hermitian metric h_{V^*} in $\mathbf{H}_{V^*}$. The Hermitian metric in $\mathbf{L}$ induced from h_{V^*} through (5.4.2) is denoted by h. Let ω_{V^*} (resp. ω) denote the Chern form of $(\mathbf{H}_{V^*}, h_{V^*})$ (resp. $(\mathbf{L}, h)$) Then

$$\omega = \Phi_V^* \omega_{V^*}. \tag{5.4.3}$$

Since $\omega_{V^*} > 0$, we have

$$\omega \geq 0. \tag{5.4.4}$$

Let $\mathbf{H}_V \to P(V)$ be the hyperplane bundle over $P(V)$. The inner product (,) naturally induces a Hermitian metric h_V in $\mathbf{H}_V$. Let ω_V denote the Chern form of $\{\mathbf{H}_V, h_V\}$. Let $\rho\colon V - \{O\} \to P(V)$ be the Hopf fibering and put

$$S_x\colon \rho(\sigma) \in P(V) \to \frac{\sqrt{h(\sigma(x), \sigma(x))}}{\sqrt{(\sigma, \sigma)}} \in \mathbf{R},$$

which is a C^∞-function.

(5.4.5) Lemma. (i) $0 \le S_x \le 1$.

(ii) *The integral* $A = -\int_{D \in P(V)} \log S_x(D)(\omega_V)^N$ *is finite and independent of* $x \in M$.

Proof. Take an orthonormal base $\{\sigma_0, \ldots, \sigma_N\}$ of V. Let $(\{U_\lambda\}, \{s_\lambda\})$ be a local trivialization covering of $\mathbf{L}$. For $\tau \in V$, a holomorphic function τ_λ on U_λ is defined by $\tau | U_\lambda = \tau_\lambda s_\lambda$; in special, $\sigma_j | U_\lambda = \sigma_{j\lambda} s_\lambda$, $0 \le j \le N$. A direct computation yields

$$h(\tau(x), \tau(x)) = \frac{|\tau_\lambda(x)|^2}{\sum_{j=0}^{N} |\sigma_{j\lambda}(x)|^2}. \tag{5.4.6}$$

By the base $\{\sigma_0, \ldots, \sigma_N\}$, we identify $P(V) = \mathbf{P}^N(\mathbf{C})$. Put $V_j = \{[z^0; \cdots; z^N]; z^j \ne 0\}$ $(0 \le j \le N)$. Then

$$S_x([z^0; \cdots; z^N]) = \frac{\left| \sum_{j=0}^{N} z^j \sigma_{j\lambda}(x) \right|}{\left(\sum_{j=0}^{N} |\sigma_{j\lambda}(x)|^2 \right)^{1/2} \left(\sum_{j=0}^{N} |z^j|^2 \right)^{1/2}} \tag{5.4.7}$$

$$= \frac{\left| \sum_{j=0}^{N} z_k^j \sigma_{j\lambda}(x) \right|}{\left(\sum_{j=0}^{N} |\sigma_{j\lambda}(x)|^2 \right)^{1/2} \left(\sum_{j=0}^{N} |z_k^j|^2 \right)^{1/2}}$$

for $(x, [z^0; \cdots; z^N]) \in U_\lambda \times V_k$, where $z_k^j = z^j / z^k$ (especially, $z_k^k = 1$). Therefore (i) immediately follows from (5.4.7). The integrability of $D \in |[D]| \to \log S_x(D) \in \mathbf{R}$ follows from the local expression (5.4.7). Let $x \in U_\lambda$ and $y \in U_\mu$. Then there are a unitary matrix $g = (g_{ij})$ and a constant $a \in \mathbf{C}$ such that

$$\sigma_{j\mu}(y) = a \sum_{k=0}^{N} g_{jk} \sigma_{k\lambda}(x).$$

Defining $g : \mathbf{P}^N(\mathbf{C}) \to \mathbf{P}^N(\mathbf{C})$ by $g(\rho(z)) = \rho(gs)$, we have by (5.4.7)

$$S_y(z) = S_x(g^{-1}(z)) \quad (z \in \mathbf{P}^N(\mathbf{C})). \tag{5.4.8}$$

Since $g^* \omega_V = \omega_V$, (5.4.8) implies (iii). *Q.E.D.*

Let $f : \mathbf{C}^n \to M$ be a meromorphic mapping. Set

$$X(f)=\{D\in P(V);\ \operatorname{supp} D\supset f(\mathbf{C}^n)\},$$

which is an analytic subset of $P(V)$ and not identical to $P(V)$. It follows from (5.4.4) that for any $D\in P(V)-X(f)$, ω and D satisfy condition (5.2.5). Hence we have

(5.4.9) $$T_f(r,\omega)=N(r,f^*D)+m_f(r,D)-m_f(1,D).$$

By making use of Lemma (5.4.5) and Fubini's theorem, we get

(5.4.10) $$\int_{D\in P(V)} m_f(r,D)(\omega_V)^N$$
$$=\int_{D\in P(V)}\left\{\int_{z\in\Gamma(r)}\log\frac{1}{S_{f(z)}(D)}\eta(z)\right\}(\omega_V)^N$$
$$=\int_{z\in\Gamma(r)}\left\{\int_{D\in P(V)}-\log S_{f(z)}(D)(\omega_V)^N\right\}\eta(z)=A\int_{z\in\Gamma(r)}\eta(z)=A,$$

where $A\in\mathbf{R}$ is a constant. We next note that the mapping

$$D\in P(V)\to[f^*D]\wedge\alpha^{n-1}\in K'(\mathbf{C}^n)$$

is continuous outside $X(f)$. Since $X(f)$ is a closed subset of measure 0 with respect to $(\omega_V)^N$, the function $D\in P(V)\to N(r,f^*D)$ is a measurable function. Thus it follows from (5.4.9) that

$$T_f(r,\omega)=\int_{P(V)}T_f(r,\omega)(\omega_V)^N=\int_{P(V)}N(r,f^*D)(\omega_V)^N.$$

Noting (5.2.14), we have the following.

(5.4.11) Theorem. *Let* $f:\mathbf{C}^n\to M$ *be a meromorphic mapping. Then*

$$T_f(r,\mathbf{L})=\int_{P(V)}N(r,f^*D)(\omega_V)^N+O(1).$$

Since $T_f(r,\omega)$ is non-negative, increasing and convex function in $\log r$, we see that

(5.4.12) if $T_f(r,\omega)\not\equiv 0$, then $T_f(r,\omega)\uparrow\infty$ as $r\uparrow\infty$.

In this case we define the **defect** $\delta_f(D)$ of f for D by

(5.4.13) $$\delta_f(D)=1-\varlimsup_{r\to\infty}\frac{N(r,f^*D)}{T_f(r,\omega)}=1-\varlimsup_{r\to\infty}\frac{N(r,f^*D)}{T_f(r,\mathbf{L})}.$$

It follows from Corollary (5.2.16) that

(5.4.14) $$0\le\delta_f(D)\le 1,$$

$$f(\mathbf{C}^n) \cap \operatorname{supp} D = \varnothing \Rightarrow \delta_f(D) = 1.$$

(5.4.15) Theorem. *Let the notation be as above and suppose that $T_f(r, \mathbf{L}) \not\equiv 0$. Then*

$$\int_{D \in P(V)} \delta_f(D)(\omega_V)^N = 0.$$

Proof. By the assumption there is $R > 0$ such that $T_f(r, \omega) > 1$ for $r > R$. For $D \in P(V) - X(f)$ and $r > R$ we have by (5.4.9)

$$(5.4.16) \qquad -\frac{m_f(1, D)}{T_f(r, \omega)} \leq 1 - \frac{N(r, f^*D)}{T_f(r, \omega)}$$

$$= \frac{m_f(r, D) - m_f(1, D)}{T_f(r, \omega)} \leq \frac{m_f(r, D)}{T_f(r, \omega)}.$$

It follows from (5.4.10) that

$$(5.4.17) \qquad \int \left[1 - \frac{N(r, f^*D)}{T_f(r, \omega)}\right] (\omega_V)^N \leq \frac{A}{T_f(r, \omega)}.$$

By (5.4.16) we can use Fatou's lemma and form (5.4.17), (5.4.14) and (5.4.12) we finally obtain

$$0 \leq \int_{D \in P(V)} \delta_f(D)(\omega_V)^N \leq \varliminf_{r \to \infty} \int_{D \in P(V)} \left[1 - \frac{N(r, f^*D)}{T_f(r, \omega)}\right] (\omega_V)^N$$

$$\leq \varliminf_{r \to \infty} \frac{A}{T_f(r, \omega)} = 0. \quad Q.E.D.$$

(5.4.18) Lemma. *Assume that $\omega > 0$ and $f : \mathbf{C}^n \to M$ is a non-constant meromorphic mapping. Then $T_f(r, \omega) \not\equiv 0$.*

Proof. Put $f_0 = f | \mathbf{C}^n - I(f)$. Suppose that $T_f(r, \omega) \equiv 0$. Then by definition

$$f_0^* \omega \wedge \alpha^{n-1} \equiv 0.$$

This implies that the trace of of the semi-positive definite (1, 1)-form $f_0^* \omega$ is identically 0 in $\mathbf{C}^n - I(f)$, so that $f_0^* \omega \equiv 0$. Since $\omega > 0$, f_0 must be constant, and so is f. *Q.E.D.*

By Lemma (5.4.18), Theorem (5.4.15) and (5.4.14) we have the following corollary.

(5.4.19) Corollary. *Let M be a complex projective algebraic manifold and $f : \mathbf{C}^n \to M$ a meromorphic mapping. Let $\mathbf{L} \to M$ be an ample line bundle over M, $V = \Gamma(M, \mathbf{L})$ and* $\dim V = N + 1$. *If there is a subset $E \subset P(V)$ such that*

$$\int_E (\omega_V)^N > 0$$

and $f(\mathbf{C}^n) \cap \operatorname{supp} D = \varnothing$ *for all* $D \in E$, *then* f *is constant.*

In the case where $M = \mathbf{P}^1(\mathbf{C})$ and $\mathbf{L}$ is the hyperplane bundle, the above corollary is classical Casorati-Weierstrass's theorem. The above Theorem (5.4.15) and Corollary (5.4.19) mean that $\delta_f(D) = 0$ for almost all $D \in |\mathbf{L}|$ in the measure theoretic sense. The Second Main Theorem and defect relations dealt with in the next section are the far refined version of this fact.

5.5 The Second Main Theorem

In the present section we assume that M is an m-dimensional complex projective algebraic manifold. Let D be an analytic hypersurface of M. We assume that D has **only simple normal crossings**; i.e., all irreducible components of D have no singular point and D has only normal crossings. We consider also D a divisor on M. Let $D = \sum_{j=1}^{l} D_j$ be the decomposition into irreducible components. As in (4.2.13), we denote by $[D]$ (resp. $[D_j]$) the holomorphic line bundle determined by D (resp. D_j). Then there are holomorphic sections $\sigma \in \Gamma(M, [D])$ and $\sigma_j \in \Gamma(M, [D_j])$ such that

$$(\sigma) = D, \quad (\sigma_j) = D_j,$$

$$[D] = [D_1] \otimes \cdots \otimes [D_l],$$

$$\sigma = \sigma_1 \otimes \cdots \otimes \sigma_l.$$

For a convenience, we set in this section

$$\widehat{\mathrm{Ric}}\, \Omega = \frac{1}{2\pi} \mathrm{Ric}\, \Omega \quad (\text{cf. } (2.1.23))$$

for a volume form Ω on a complex manifold.

(5.5.1) Lemma. *Assume that* $c_1([D]) + c_1(\mathbf{K}(M)) > 0$. *Then there exist a* C^∞-*volume form* Ω *and Hermitian metrics in* $[D]$ *and* $[D_j]$ *such that the singular volume form*

$$\Psi = \frac{\Omega}{\prod_{j=1}^{l} \|\sigma_j\|^2 (\log \|\sigma_j\|^2)^2} \tag{5.5.2}$$

satisfies the following conditions on $M - D$:

(i) $\widehat{\mathrm{Ric}}\, \Psi < 0$.

(ii) $(-\widehat{\mathrm{Ric}}\, \Psi)^m \geq \Psi$.

(iii) $\int_{M-D}(-\hat{\mathrm{Ric}}\,\Psi)^m < \infty$.

Proof. (i) Since $c_1([D]) + c_1(\mathbf{K}(M)) > 0$, there are a C^∞-volume form Ω and a metric in $[D]$ with Chern form ω such that

$$\omega_0 = \omega - \hat{\mathrm{Ric}}\,\Omega > 0.$$

Then we may suitably take metrics in $[D_j]$ with Chern forms ω_j so that

$$\|\sigma\| = \|\sigma_1\| \cdots \|\sigma_l\|,$$

$$\omega = \omega_1 + \cdots + \omega_l.$$

Let $0 < \delta < 1$ be a constant which will be determined in the following arguments. By multipling a positive constants to the metrics in $[D]$ and $[D_j]$, we may assume that $\|\sigma_j\| < \delta$. By definition (5.5.2) we have

$$-\hat{\mathrm{Ric}}\,\Psi = \omega - \hat{\mathrm{Ric}}\,\Omega - \sum_{j=1}^{l} dd^c \log(\log\|\sigma_j\|^2)^2 \tag{5.5.3}$$

on $M - D$. Moreover

$$\begin{aligned} &-dd^c\log(\log\|\sigma_j\|^2)^2 \\ &= \frac{-2dd^c\log\|\sigma_j\|^2}{\log\|\sigma_j\|^2} + \frac{2d\log\|\sigma_j\|^2 \wedge d^c\log\|\sigma_j\|^2}{(\log\|\sigma_j\|^2)^2} \\ &= \frac{2\omega_j}{\log\|\sigma_j\|^2} + \frac{2dd^c\log\|\sigma_j\|^2\wedge d^c\log\|\sigma_j\|^2}{(\log\|\sigma_j\|^2)^2}. \end{aligned} \tag{5.5.4}$$

Since $\omega_j/\log\|\sigma_j\|^2$ are continuous on M, we choose a sufficiently small δ so that

$$\omega_0 + 2\sum_{j=1}^{l}\frac{\omega_j}{\log\|\sigma_j\|^2} \geq C_1\omega_0 \tag{5.5.5}$$

on M, where C_1 is a positive constant. Note that

$$d\log\|\sigma_j\|^2\wedge d^c\log\|\sigma_j\|^2 = \frac{i}{2\pi}\partial\log\|\sigma_j\|^2\wedge\overline{\partial\log\|\sigma_j\|^2} \geq 0.$$

It follows from (5.5.3)~(5.5.5) that

$$-\hat{\mathrm{Ric}}\,\Psi \geq C_2\omega_0 + 2\sum_{j=1}^{l}\frac{d\log\|\sigma_j\|^2\wedge d^c\log\|\sigma_j\|^2}{(\log\|\sigma_j\|^2)^2} > 0. \tag{5.5.6}$$

Hence (i) is shown.

(ii) Let $x \in D$ be an arbitrary point. For instance, $D_1, \ldots, D_k$ are the irreducible components of D passing through x. By the assumption there is a holomorphic local coordinate neighborhood system $(U, (z^1, \ldots, z^m))$ with $x = (0, \ldots, 0)$ such that

$$U \cap D = U \cap \left(\bigcup_{j=1}^{k} D_j \right),$$

$$U \cap D_j = \left\{ (z^1, \ldots, z^m) \in \Delta(1)^m ;\, z^j = 0 \right\}, \quad 1 \le j \le k.$$

Then there are C^∞-functions $b_j > 0$ on U such that

$$\log \|\sigma_j\|^2 = \log b_j + \log |z^j|^2. \tag{5.5.7}$$

Therefore

$$d\log \|\sigma_j\|^2 \wedge d^c \log \|\sigma_j\|^2 = \frac{i}{2\pi} \left(\frac{dz^j \wedge d\bar{z}^j}{|z^j|^2} + \rho_j \right) \quad \text{on } U, \tag{5.5.8}$$

where
$$\rho_j = \frac{\partial b_j \wedge \bar{\partial} b_j}{b_j^2} + \frac{\partial b_j \wedge \bar{\partial} d\bar{z}^j}{b_j \bar{z}^j} + \frac{\partial z^j \wedge \bar{\partial} b_j}{b_j z^j}.$$

It is noted that $|z^j|^2 \rho_j$ are C^∞ on U and vanish on D_j. Reminding $\omega_0 > 0$, we have by (5.5.6) and (5.5.8)

$$(-\hat{\text{Ric}}\, \Psi)^m \tag{5.5.9}$$

$$\ge C_3 \left(i \sum_{j=1}^{m} dz^j \wedge d\bar{z}^j + \sum_{j=1}^{k} \frac{d\log \|\sigma_j\|^2 \wedge d^c \log \|\sigma_j\|^2}{(\log \|\sigma_j\|^2)^2} \right)^m$$

$$= C_3 \left(i \sum_{j=1}^{m} dz^j \wedge d\bar{z}^j + \sum_{j=1}^{k} \frac{i}{\pi} \sum_{j=1}^{k} \frac{dz^j \wedge d\bar{z}^j + |z^j|^2 \rho_j}{|z^j|^2 (\log \|\sigma_j\|^2)^2} \right)^m$$

$$= C_4 i^m \frac{dz^1 \wedge d\bar{z}^1 \wedge \cdots \wedge dz^m \wedge d\bar{z}^m + A}{\prod_{j=1}^{k} |z^j|^2 (\log \|\sigma_j\|^2)^2},$$

where C_3 and C_4 are positive constants and A is a continuous (m, m)-form with $A(x) = 0$. Thus there are a small neighborhood $U' \subset U$ of x and a positive constant C_5 such that

$$(-\hat{\text{Ric}}\, \Psi)^m \ge C_5 \Psi \quad \text{on } U' - D.$$

Since D is compact, there are a neighborhood V of D and a positive constant C_6 such that

(5.5.10)
$$(-\hat{\mathrm{Ric}}\,\Psi)^m \geq C_6 \Psi \quad \text{on } V - D.$$

On the other hand, by (5.5.6) $-\hat{\mathrm{Ric}}\,\Psi \geq C_1 \omega_0$ on $M - V$. Choosing C_6 smaller if necessary, we obtain

$$(-\hat{\mathrm{Ric}}\,\Psi)^m \geq C_6 \Psi \quad \text{on } M - V.$$

Changing Ω to $C_6 \Omega$, we have (ii).

(iii) By (5.5.3), (5.5.4), (5.5.7) and (5.5.8) it is sufficient to confirm that

$$\int_{\{|w| \leq 1/2\}} \frac{1}{|w|^2 (\log |w|^2)^2} \frac{i}{2} dw \wedge d\bar{w} < \infty.$$

This follows from

$$\int_0^{1/2} \frac{dt}{t (\log t)^2} < \infty. \quad Q.E.D.$$

In the case of dim $M = 1$, Lemma (5.5.1), (ii) says that the Gaussian curvature of the metric determined by Ψ is less than or equal to -1. This property will play an essential role in what follows.

Now let $f : \mathbf{C}^m \to M$ be a meromorphic mapping. For a convenience, we put

$$W = \mathbf{C}^m - I(f), \quad f_0 = f|W.$$

We say that f is **non-degenerate** if rank $f = m$; i.e., df_0 has rank m at some point of W. Otherwise, f is said to be **degenerate**. In the following arguments we keep the following conditions.

(5.5.11) (i) $f : \mathbf{C}^m \to M$ is a non-degenerate meromorphic mapping.

(ii) $D \subset M$ is an analytic hypersurface with only simple normal crossings such that $c_1([D]) + c_1(\mathbf{K}(M)) > 0$.

It is noted that $f(\mathbf{C}^m)$ contains a non-empty open subset of M and hence $f(\mathbf{C}^m) \not\subset D$. Let Ω and Ψ be as in Lemma (5.5.1). Put

$$f_0^* \Omega = a(z) \alpha^m, \quad f_0^* \Psi = \xi(z) \alpha^m.$$

It follows from (5.5.2) that

(5.5.12)
$$\xi = \frac{a}{\prod_{j=1}^{l} \|\sigma_j \circ f_0\|^2 (\log \|\sigma_j \circ f_0\|^2)^2}.$$

Let $J(f_0)$ be the locally defined Jacobian of f_0 with respect to a holomorphic local coordinate system of M. Then the divisor $(J(f_0))$ is globally well defined in W. Since codim $I(f) \geq 2$, $(J(f_0))$ uniquely extends to an effective divisor R_f on $\mathbf{C}^m$ (cf. Chapter IV, §5, (c)). We also have the pullback divisor f^*D on $\mathbf{C}^m$. The form

$f_0^*(-\widehat{\mathrm{Ric}}\,\Psi)$ is defined on $\mathbf{C}^m - (I(f) \cup \operatorname{supp} f^*D)$. Since $I(f) \cup \operatorname{supp} f^*D$ is a closed subset of measure 0, we consider $f_0^*(-\widehat{\mathrm{Ric}}\,\Psi)$ a form on $\mathbf{C}^m$ with coefficients in Borel measurable functions and denote it by $f^*(-\widehat{\mathrm{Ric}}\,\Psi)$. In the same way, we consider ξ, a, etc., Borel measurable functions on $\mathbf{C}^m$.

(5.5.13) Lemma. (i) $\log \xi \in L_{\mathrm{loc}}(\mathbf{C}^m)$.

(ii) $f^*(-\widehat{\mathrm{Ric}}\,\Psi) \in L_{\mathrm{loc}}^{(1,1)}(\mathbf{C}^m)$ *and the current* $[f^*(-\widehat{\mathrm{Ric}}\,\Psi)]$ *is closed and positive.*

(iii) *As currents on* $\mathbf{C}^m$,

$$dd^c[\log \xi] = [f^*(-\widehat{\mathrm{Ric}}\,\Psi)] - f^*D + R_f. \tag{5.5.14}$$

Proof. Put

$$S = I(f) \cup S(\operatorname{supp} f^*D).$$

Then $\operatorname{codim} S \geq 2$. We first show (i), (ii) and (iii) on $\mathbf{C}^m - S$. Take an arbitrary point $z_0 \in \mathbf{C}^m - S$ and a holomorphic local coordinate neighborhood system $(U, (w^1, \ldots, w^m))$ of $f(z_0) \in M$. Put

$$\Omega | U = b(w) \bigwedge_{j=1}^{m} \left(\frac{i}{2} dw^j \wedge d\bar{w}^j \right).$$

Then $b(w)$ is C^∞ and $b(w) > 0$ in U. Take a neighborhood $V \subset \mathbf{C}^m - S$ of z_0 so that $f(V) \subset U$. Then

$$\xi(z) = \frac{b(f(z)) |J(f)(z)|^2}{\prod_{j=1}^{l} \|\sigma_j \circ f\|^2 (\log \|\sigma_j \circ f\|^2)^2} \quad \text{on } V, \tag{5.5.15}$$

where $J(f) = \det(\partial w^j \circ f / \partial z^k)$. We see by (5.2.17) that $\log \|\sigma_j \circ f(z)\|^2 \in L_{\mathrm{loc}}(\mathbf{C}^m)$. In the proof of Lemma (5.5.1) we chose $0 < \delta < 1$ such that

$$\|\sigma_j\| < \delta \quad (1 \leq j \leq l). \tag{5.5.16}$$

Let $K \subset \mathbf{C}^m$ be a compact subset with positive measure. By making use of (5.5.16) and the concavity of log, we have

$$\begin{aligned}
-\infty < \log(\log \delta^2)^2 \int_K \alpha^m &\leq \int_K \log(\log \|\sigma_j \circ f\|^2)^2 \alpha^m \\
&= 2 \int_K \log(\log \|\sigma_j \circ f\|^{-2}) \alpha^m \\
\leq 2\left(\int_K \alpha^m \right) &\log \left[\left(\int_K \alpha^m \right)^{-1} \cdot \int_K \log \|\sigma_j \circ f\|^{-2} \alpha^m \right] < \infty,
\end{aligned}$$

so that $\log(\log\|\sigma_j\circ f\|^2)^2 \in L_{\mathrm{loc}}(\mathbf{C}^m)$. Thus we get

$$\text{(5.4.17)} \qquad \log\|\sigma_j\circ f\|^2\in L_{\mathrm{loc}}(\mathbf{C}^m),\ \log(\log\|\sigma_j\circ f\|^2)^2\in L_{\mathrm{loc}}(\mathbf{C}^m),$$
$$\log\xi\in L_{\mathrm{loc}}(\mathbf{C}^m-S).$$

It follows from (5.5.15), (5.5.17) and Theorem (5.1.13) that

$$\text{(5.5.18)} \qquad dd^c[\log\xi]=[f^*\omega-f^*\hat{\mathrm{Ric}}\,\Omega]-f^*D+R_f$$
$$-\sum_j dd^c\Big[\log(\log\|\sigma_j\circ f\|^2)^2\Big].$$

Now we claim the following: For $1\le j\le l$

(5.5.19) (i) $dd^c\log(\log\|\sigma_j\circ f\|^2)\in L_{\mathrm{loc}}^{(1,1)}(\mathbf{C}^m-S)$,

(ii) $dd^c\Big[\log(\log\|\sigma_j\circ f\|^2)^2\Big]=\Big[dd^c\log(\log\|\sigma_j\circ f\|^2)^2\Big]$ on $\mathbf{C}^m-S$.

We show these in a neighborhood of $z_0\in\mathbf{C}^m-S$. It is clear that (5.5.19), (i) and (ii) hold in $\mathbf{C}^m-(S\cup\operatorname{supp}f^*D_j)$. Suppose that $z_0\in\operatorname{supp}f^*D_j$. Since $z_0\notin S$, $z_0\notin S(\operatorname{supp}f^*D_j)$. Choosing V smaller if necessary, we take a holomorphic local coordinate system $(x^1,\dots,x^m)$ in V with $z^0=(0,\dots,0)$ such that

$$\operatorname{supp}f^*D_j\cap V=\{(x^1,\dots,x^m);\,x^1=0\}.$$

Then there are an integer k and a positive valued C^∞-function B on V such that

$$\text{(5.5.20)} \qquad \|\sigma_j\circ f(x)\|^2=|x^1|^{2k}B(x)\ \text{ on } V.$$

A direct computation implies that the coefficients of $dd^c\log(\log\|\sigma_j\circ f\|^2)^2$ are written in the form of

$$\frac{A}{|x^1|^2(\log|x^1|^2)^2},\quad A\in C(V).$$

Thus (5.5.19), (i) is proved. To show (ii), we take an arbitrary element $\phi\in D^{(m-1,m-1)}(V)$. Using (5.4.20) and Stokes' theorem, we have

$$\text{(5.5.21)} \qquad \int\log(\log\|\sigma_j\circ f\|^2)^2dd^c\phi$$

$$=\lim_{\varepsilon\to0}\left\{\int_{\{|x^1|\ge\varepsilon\}}d\left[\log\left[\log(|x^1|^{2k}B)\right]^2\wedge d^c\phi\right]\right.$$

$$\left.-\int_{\{|x^1|\ge\varepsilon\}}d\log\left[\log(|x^1|^{2k}B)\right]^2\wedge d^c\phi\right\}$$

Continued

$$= \lim_{\varepsilon\to 0}\left\{-\int_{\{|x^1|=\varepsilon\}}\log\left[\log(\varepsilon^{2k}B)\right]^2 d^c\phi\right.$$

$$\left.+\int_{\{|x^1|\geq\varepsilon\}} d^c\log\left[\log(|x^1|^{2k}B)\right]^2\wedge d\phi\right\}$$

$$= \lim_{\varepsilon\to 0}\left\{O\left[\varepsilon\log(\log\varepsilon)^2\right] - \int_{\{|x^1|\geq\varepsilon\}} d\left[d^c\log\left[\log(|x^1|^{2k}B)\right]^2\wedge\phi\right]\right.$$

$$\left.+\int_{\{|x^1|\geq\varepsilon\}} dd^c\log\left[\log(|x^1|^{2k}B)\right]^2\wedge\phi\right\}$$

$$= \lim_{\varepsilon\to 0}\left\{\int_{\{|x^1|=\varepsilon\}} d^c\log\left[\log(|x^1|^{2k}B)\right]^2\wedge\phi\right.$$

$$\left.+\int_{\{|x^1|\geq\varepsilon\}} dd^c\log\left[\log(|x^1|^{2k}B)\right]^2\wedge\phi\right\}.$$

Here we put $x^1 = re^{i\theta}$. Then

$$d^c = \frac{r}{4\pi}\frac{\partial}{\partial r}d\theta + d^c_{x'} \quad \text{on } \{|x^1| = r\},$$

where $d^c_{x'}$ stands for the d^c-operator in the variables $x' = (x^2, \ldots, x^m)$. Hence we have

$$\left|\int_{\{|x^1|=\varepsilon\}} d^c\log\left[\log(|x^1|^{2k}B)\right]^2\wedge\phi\right|$$

$$= \left|\int_{\{|x^1|=\varepsilon\}} 2\frac{\frac{k}{2\pi}d\theta + d^c\log B}{2k\log\varepsilon + \log B}\right| = O\left(\frac{1}{|\log\varepsilon|}\right).$$

This estimate and (5.5.19), (i) yield

$$\int \log(\log\|\sigma_j\circ f\|^2)^2 dd^c\phi = \int dd^c\log(\log\|\sigma_j\circ f\|^2)^2\wedge\phi.$$

Hence (5.5.10), (ii) is proved. Since

$$f^*(-\hat{\mathrm{Ric}}\,\Psi) = f^*\omega - f^*\hat{\mathrm{Ric}}\,\Omega - \sum_{j=1}^{l} dd^c\log(\log\|\sigma_j\circ f\|^2)^2$$

on $\mathbf{C}^m - S$, we infer from (5.5.18) and (5.5.19) that (5.5.14) holds on $\mathbf{C}^m - S$.

We take a holomorphic function F on $\mathbf{C}^m$ so that $(F) = f^*D$. Then

$$dd^c[\log(\xi|F|^2)] = [f^*(-\hat{\mathrm{Ric}}\,\Psi)] + R_f \text{ on } \mathbf{C}^m - S.$$

The right-hand side of this equation is a positive current by Lemma (5.5.1), (i). Since codim $S \geq 2$, Theorem (3.3.42) implies that $\log(\xi|F|^2)$ uniquely extends to a plurisubharmonic function ξ_1 on $\mathbf{C}^m$. Since $\xi_2 = \log|F|^2$ is plurisubharmonic in $\mathbf{C}^m$, $\log\xi = \xi_1 - \xi_2 \in L_{\mathrm{loc}}(\mathbf{C}^m)$. Thus (i) is proved. Since (5.5.14) holds on $\mathbf{C}^m - S$, we deduce (ii) from (5.5.22), Proposition (5.1.10) and Lemma (5.1.4). Therefore we see that (5.5.14) holds on the whole $\mathbf{C}^m$. *Q.E.D.*

(5.5.23) Corollary. *Let the notation be as in Lemma (5.5.13). Then there are plurisubharmonic functions ξ_1 and ξ_2 on $\mathbf{C}^m$ such that*

$$\log\xi = \xi_1 - \xi_2.$$

(5.5.24) Lemma. *We have*

$$\xi^{1/m}\alpha^m \leq f^*(-\hat{\mathrm{Ric}}\,\Psi)\wedge\alpha^{n-1} \quad \text{on } \mathbf{C}^m;$$

especially, $\xi^{1/m} \in L_{\mathrm{loc}}(\mathbf{C}^m)$.

Proof. Put

$$f^*(-\hat{\mathrm{Ric}}\,\Psi) = \frac{i}{2\pi}\sum_{j,k=1}^{m} A_{j\bar{k}}\,dz^j \wedge d\bar{z}^k.$$

Then $A_{j\bar{k}} \in L_{\mathrm{loc}}(\mathbf{C}^m)$ by Lemma (5.5.13), (ii) and Lemma (5.5.1), (i); moreover, $(A_{j\bar{k}}(x))$ is a positive semidefinite Hermitian matrix at almost all points $x \in \mathbf{C}^m$. Hence

$$\left[\det(A_{j\bar{k}}(x))\right]^{1/m} \leq \frac{1}{m}\mathrm{Trace}\,(A_{j\bar{k}}(x)).$$

It follows from Lemma (5.5.1), (ii) that

$$\xi\alpha^m = f^*\Psi \leq (f^*(-\hat{\mathrm{Ric}}\,\Psi))^m = \det(A_{j\bar{k}})\alpha^m.$$

That is, $\xi \leq \det(A_{j\bar{k}})$. On the other hand,

$$\begin{aligned} f^*(-\hat{\mathrm{Ric}}\,\Psi)\wedge\alpha^{m-1} &= \frac{1}{m}\mathrm{Trace}\,(A_{j\bar{k}})\alpha^m \\ &\geq (\det(A_{j\bar{k}}))^{1/m}\alpha^m \geq \xi^{1/m}\alpha^m. \end{aligned}$$

Therefore $0 \leq \xi^{1/m} \leq \dfrac{1}{m}\sum_{j=1}^{m} A_{j\bar{j}}$ and then $\xi^{1/m} \in L_{\mathrm{loc}}(\mathbf{C}^m)$. *Q.E.D.*

(5.5.25) Lemma. (i) $\sum_{j=1}^{l} \log(\log\|\sigma_j \circ f\|^2)^2$ *is written as the difference of two plurisubharmonic functions on* $\mathbf{C}^m$.

(ii) $dd^c\sum_{j=1}^{l} \log(\log\|\sigma_j \circ f\|^2)^2 \in L_{\mathrm{loc}}^{(1,1)}(\mathbf{C}^m)$ *and*

$$dd^c\left[\sum_{j=1}^{l}\log(\log\|\sigma_j\circ f\|^2)^2\right]=\left[dd^c\sum_{j=1}^{l}\log(\log\|\sigma_j\circ f\|^2)^2\right]$$

on $\mathbf{C}^m$.

Proof. We use the same notation as in Lemma (5.5.13) and its proof. A holomorphic function F is taken so that $(F)=f^*D$. Put

$$\xi_2=\log|F|^2,\quad \xi_3=\log\frac{|F|^2a}{\prod_{j=1}^{l}\|\sigma_j\circ f\|^2}.$$

Then $\xi_3\in L_{\mathrm{loc}}(\mathbf{C}^m-I(f))$ and

$$\begin{aligned}dd^c[\xi_3]&=dd^c[\log a]+dd^c[\log|F|^2]-dd^c\left[\sum_{j=1}^{l}\log\|\sigma_j\circ f\|^2\right]\\&=[f^*(-\widehat{\mathrm{Ric}}\,\Omega)]+R_f+f^*D+[f^*\omega]-f^*D\\&=[f^*(-\widehat{\mathrm{Ric}}\,\Omega+\omega)]+R_f\end{aligned}$$

on $\mathbf{C}^m-I(f)$. By the assumption, $-\widehat{\mathrm{Ric}}\,\Omega+\omega>0$, and hence $dd^c[\xi_3]\geq 0$. Thus ξ_3 is a plurisubharmonic function on $\mathbf{C}^m-I(f)$ and plurisubharmonically extended over $\mathbf{C}^m$. Note that $\xi_3\in L_{\mathrm{loc}}(\mathbf{C}^m)$. Now we have

$$\sum_{j=1}^{l}\log(\log\|\sigma_j\circ f\|^2)^2=\xi_3-\xi_2-\log\xi.$$

Thus Corollary (5.5.23) implies (i). By (5.5.19) we see that (ii) is valid on $\mathbf{C}^m-S$. Since $\operatorname{codim} S\geq 2$, (i) and Theorem (5.1.14) implies that (ii) is valid on $\mathbf{C}^m$. *Q.E.D.*

Let $A(r)$ and $B(r)$ be real valued functions in $r\in[1,\infty)$. We write

$$A(r)\leq B(r)\|_E$$

if $E\subset[0,\infty)$ is a Borel subset with finite measure and if $A(r)\leq B(r)$ for $r\in[1,\infty)-E$. Now we prove the **Second Main Theorem** which is the aim of the present section:

(5.5.26) Theorem. *Let $f:\mathbf{C}^m\to M$ be a non-degenerate meromorphic mapping into a complex projective algebraic manifold M. Let $D\subset M$ be an analytic hypersurface with only simple normal crossings such that $c_1([D])+c_1(\mathbf{K}(M))>0$. Then for any $0<\varepsilon<1$*

$$\begin{aligned}&T_f(r,\mathbf{K}(M))+T_f(r,[D])\\&\leq N(r,f^*D)-N(r,R_f)+\frac{m(2m-1)\varepsilon}{2}\log r\end{aligned}$$

Continued

$$+O\left[\log^+\{T_f(r,\ \mathbf{K}(M))+T_f(r,\ [D])\}\right]+O(1)\|_{E(\varepsilon)},$$

where $O(*)$ are estimates independent of r and ε.

Proof. Take ξ defined by (5.5.12). It follows from Lemma (5.5.13), Corollary (5.5.23) and Lemma (3.3.39) that

$$\frac{1}{2}\int_{\Gamma(r)}(\log\xi)\eta-\frac{1}{2}\int_{\Gamma(1)}(\log\xi)\eta \tag{5.5.27}$$

$$=\int_1^r\frac{<[f^*(-\hat{\mathrm{Ric}}\,\Psi)]\wedge\alpha^{m-1},\ \chi_{B(t)}>}{t^{2m-1}}dt-N(r,\ f^*D)+N(r,\ R_f).$$

On the other hand,

$$f^*(-\hat{\mathrm{Ric}}\,\Psi)=f^*(-\hat{\mathrm{Ric}}\,\Omega)+f^*\omega-\sum_{j=1}^{l}dd^c\log(\log\|\sigma_j\circ f\|^2)^2$$

on $\mathbf{C}^m-(I(f)\cup\operatorname{supp}f^*D)$. By Lemma (5.2.1) and Lemma (5.5.25) we have a current equation on $\mathbf{C}^m$:

$$[f^*(-\hat{\mathrm{Ric}}\,\Psi)]=[f^*(-\hat{\mathrm{Ric}}\,\Omega)]+[f^*\omega] \tag{5.5.28}$$

$$-dd^c\left[\sum_{j=1}^{l}\log(\log\|\sigma_j\circ f\|^2)^2\right].$$

Put

$$\mu(r)=\frac{1}{2}\int_{\Gamma(r)}(\log\xi)\eta,$$

$$U(r)=\int_1^r\frac{<[f^*(-\hat{\mathrm{Ric}}\,\Psi)]\wedge\alpha^{m-1},\ \chi_{B(t)}>}{t^{2m-1}}dt.$$

Then by (5.5.27)

$$U(r)=N(r,\ f^*D)-N(r,\ R_f)+\mu(r)-\mu(1). \tag{5.5.29}$$

Now we claim that

$$T_f(r,\ \mathbf{K}(M))+T_f(r,\ [D])\le U(r)+O(1) \tag{5.5.30}$$

$$\le T_f(r,\ \mathbf{K}(M))+T_f(r,\ [D])+l\log^+(T_f(r,\ \mathbf{K}(M))$$

$$+T_f(r,\ [D]))+O(1).$$

In fact, we have by (5.5.28), Lemma (5.5.25) and Lemma (3.3.39)

$$U(r)=T_f(r,\ \mathbf{K}(M))+T_f(r,\ [D])+O(1) \tag{5.5.31}$$

$$-\int_1^r \frac{< dd^c\left[\sum_{j=1}^{l} \log(\log\|\sigma_j\circ f\|^2)^2\right]\wedge\alpha^{m-1}, \chi_{B(t)}>}{t^{2m-1}}dt$$

$$= T_f(r, \mathbf{K}(M)) + T_f(r, [D])$$

$$+\int_{\Gamma(r)}\left(\sum_{j=1}^{l}\log|\log\|\sigma_j\circ f\|^2|\right)\eta + O(1).$$

Since $0 \le \|\sigma_j\| < \delta < 1$, $\log 1/\|\sigma_j\|^2 > 0$. By making use of this and the concavity of log, we proceed to computations:

$$\int_{\Gamma(r)}\sum_{j=1}^{l}\left(\log\log\frac{1}{\|\sigma_j\circ f\|^2}\right)\eta \tag{5.5.32}$$

$$\le \sum_{j=1}^{l}\log\left(\int_{\Gamma(r)}\log\frac{1}{\|\sigma_j\circ f\|^2}\eta\right) + O(1)$$

$$\le \sum_{j=1}^{l}\log(m_f(r, D_j)) + O(1) \le \sum_{j=1}^{l}\log^+ T_f(r, [D_j]) + O(1).$$

Here we used the First Main Theorem (5.2.18). On the other hand, since $c_1(\mathbf{K}(M)) + c_1([D]) > 0$, there is a positive integer k such that

$$\omega_j \le k(-\widehat{\mathrm{Ric}}\,\Omega + \omega).$$

Here, note that ω_j are Chern forms of the Hermitian metrics in $[D_j]$ given in Lemma (5.5.1). Therefore

$$\log^+ T_f(r, [D_j]) \le \log^+\{T_f(r, \mathbf{K}(M)) + T_f(r, [D])\} + O(1). \tag{5.5.33}$$

Thus (5.5.30) is derived from (5.5.31)~(5.5.33).

We next show the following estimate for $\mu(r)$: Let $\varepsilon > 0$ be arbitrarily fixed. Then

$$\mu(r) \le \frac{m(2m-1)}{2}\varepsilon\log r + \frac{m(1+\varepsilon)^2}{2}\log U(r)\|_{E(\varepsilon)}. \tag{5.5.34}$$

(5.5.35) Lemma. *Let $h(r) > 0$ be a monotone increasing function in $r \ge 1$. Then $h(r)$ is differentiable at almost all points, and for $\delta > 0$*

$$\frac{dh(r)}{dr} \le (h(r))^{1+\delta}\|_{E(\delta)}.$$

Proof. It is a known fact in the elementary calculus that $h(r)$ is differentiable at almost all points. Put

$$E(\delta)=\left\{ r\geq 1; \frac{dh(r)}{dr} > h(r)^{1+\delta} \right\}.$$

Then

$$\int_{E(\delta)} dr \leq \int_{E(\delta)} \frac{dh(r)}{h(r)^{1+\delta}} \leq \int_1^\infty \frac{dh(r)}{h(r)^{1+\delta}}$$

$$=\left[-\frac{1}{\delta} h(r)^{-\delta} \right]_1^\infty \leq \frac{h(1)^{-\delta}}{\delta}. \quad Q.E.D.$$

We go back to $\mu(r)$. By Lemma (5.5.24), $\xi^{1/m} \in L_{\mathrm{loc}}(\mathbf{C}^m)$. It follows from the concavity of log that

$$\mu(r)=\frac{m}{2}\int_{\Gamma(r)} \log \xi^{1/m}\eta \leq \frac{m}{2}\log\left[\int_{\Gamma(r)} \xi^{1/m}\eta \right]$$

$$=\frac{m}{2}\log\left[r^{1-2m}\frac{1}{2}m\frac{d}{dr}\int_{B(r)} \xi^{1/m}\eta \right].$$

Now we put $h(r)=\int_{B(r)} \xi^{1/m}\alpha^m$, and apply Lemma (5.5.35) to obtain

$$\mu(r) \leq \frac{m}{2}\log\left\{ \frac{r^{1-2m}}{2m}\left[\int_{B(r)} \xi^{1/m}\alpha^m \right]^{1+\varepsilon} \right\} \Big\|_{E_1(\varepsilon)}.$$

Combining this with Lemma (5.5.24), we get

$$\text{(5.5.36)} \qquad \mu(r) \leq \frac{m}{2}\log\left\{ \frac{r^{1-2m}}{2m}\left[\int_{B(r)} f^*(-\hat{\mathrm{Ric}}\,\Psi)\wedge\alpha^m \right]^{1+\varepsilon} \right\} \Big\|_{E_1(\varepsilon)}$$

$$=\frac{m}{2}\log\left\{ \frac{r^{1-2m}}{2m}\left[r^{2m-1}\frac{d}{dr}\int_1^r \frac{dt}{t^{2m-1}}\int_{B(t)} f^*(-\hat{\mathrm{Ric}}\,\Psi)\wedge\alpha^{m-1} \right]^{1+\varepsilon} \right\} \Big\|_{E_1(\varepsilon)}.$$

Using Lemma (5.5.35) again, we have

$$\text{(5.5.37)} \qquad \frac{d}{dr}\int_1^r \frac{dt}{t^{2m-1}}\int_{B(t)} f^*(-\hat{\mathrm{Ric}}\,\Psi)\wedge\alpha^{m-1}$$

$$\leq \left[\int_1^r \frac{dt}{t^{2m-1}}\int_{B(t)} f^*(-\hat{\mathrm{Ric}}\,\Psi)\wedge\alpha^{m-1} \right]^{1+\varepsilon} \Big\|_{E_2(\varepsilon)}.$$

Put $E(\varepsilon)=E_1(\varepsilon)\cup E_2(\varepsilon)$. Then it follows from (5.5.36) and (5.5.37) that

$$\mu(r) \leq \frac{m}{2}\log\left\{\frac{r^{(2m-1)\varepsilon}}{2m}\left[\int_1^r \frac{<[f^*(-\hat{\mathrm{Ric}}\,\Psi)\wedge\alpha^{m-1}],\,\chi_{B(t)}>}{t^{2m-1}}dt\right]^{(1+\varepsilon)^2}\right\}\Bigg\|_{E(\varepsilon)}$$

$$\leq \frac{m(2m+1)}{2}\varepsilon\log r + \frac{m}{2}(1+\varepsilon)^2\log U(r)\|_{E(\varepsilon)}.$$

Thus (5.5.34) is proved. By (5.5.30) we have

(5.5.38) $$\log U(r) \leq \log^+\{T_f(r, \mathbf{K}(M)) + T_f(r, [D])\} + O(1).$$

Finally, the Second Main Theorem follows from (5.5.29), (5.5.30), (5.5.34) and (5.5.38). *Q.E.D.*

We continue to assume that $f: \mathbf{C}^m \to M$ and $D = \sum_{j=1}^{l} D_j$ satisfy condition (5.5.11). Moreover, we assume that

(5.5.39) $$c_1([D_1]) = \cdots = c_l([D_l]) > 0.$$

For a convenience. we put

$$\gamma = c_1([D_j]) \ (1 \leq j \leq l).$$

We fix a real positive closed form $\omega_0 \in \gamma$ of type (1, 1). Now we set

$$\left[\frac{c_1(\mathbf{K}(M)^{-1})}{\gamma}\right] = \inf\{a \in \mathbf{R};\ a\gamma + c_1(\mathbf{K}(M)) > 0\}.$$

Since f is not constant and $f(\mathbf{C}^m) \not\subset D$, we have the defects $\delta_f(D_j)$ given by

(5.5.40) $$\delta_f(D_j) = 1 - \overline{\lim_{r\to\infty}}\frac{N(r, f^*D_j)}{T_f(r, [D_j])} = 1 - \overline{\lim_{r\to\infty}}\frac{N(r, f^*D_j)}{T_f(r, \omega_0)}.$$

We also define another defect:

$$\Theta_f(D_j) = 1 - \overline{\lim_{r\to\infty}}\frac{N(r, \operatorname{supp} f^*D_j)}{T_f(r, [D_j])}.$$

It is clear that

(5.5.41) $$0 \leq \delta_f(D_j) \leq \Theta_f(D_j) \leq 1.$$

One of the most important consequence of the Second Main Theorem is the following.

(5.5.42) Corollary (Defect Relations). *Let the notation be as above. Then*

(i) $$\sum_{j=1}^{l} \delta_f(D_j) \leq \left[\frac{c_1(\mathbf{K}(M)^{-1})}{\gamma}\right] - \underline{\lim_{r\to\infty}}\frac{N(r, R_f)}{T_f(r, \omega_0)}.$$

(ii) $\sum_{j=1}^{l} \delta_f(D_j) \leq \sum_{j=1}^{l} \Theta_f(D_j) \leq \left[\dfrac{c_1(\mathbf{K}(M)^{-1})}{\gamma} \right].$

Proof. We first take an integer l_0 as follows: If $l > [c_1(\mathbf{K}(M)^{-1})/\gamma]$, then we put $l_0 = 0$; otherwise, we take l_0 so that $[l_0 D_1]$ is very ample and $l + l_0 > [c_1(\mathbf{K}(M)^{-1})/\gamma]$. Take $D_{l+1} \in |l_0 D_1|$ such that $\sum_{j=1}^{l+1} D_j$ has only simple normal crossings. We put

$$p = \begin{cases} l & \text{for } l_0 = 0, \\ l+1 & \text{for } l_0 > 0. \end{cases}$$

Since $c_1(\mathbf{K}(M)) + c_1\left(\sum_{j=1}^{p} D_j\right) = c_1(\mathbf{K}(M)) + (l + l_0)\gamma > 0$, Theorem (5.5.26) implies that

(5.5.43)
$$\begin{aligned} &T_f(r, \mathbf{K}(M)) + (l + l_0) T_f(r, \omega_0) \\ &\leq \sum_{j=1}^{p} N(r, f^* D_j) - N(r, R_f) + \frac{m(2m-1)\varepsilon}{2} \log r \\ &+ O\left(\log^+ \left\{ T_f(r, \mathbf{K}(M)) + T_f\left(r, \left[\sum_{j=1}^{p} D_j \right] \right) \right\} \right) + O(1) \|_{E(\varepsilon)}, \end{aligned}$$

Since M is compact and $\omega_0 > 0$,

(5.5.44)
$$T_f(r, \mathbf{K}(M)) + T_f\left(r, \left[\sum_{j=1}^{p} D_j \right] \right) = O(T_f(r, \omega_0)).$$

By Theorem (5.2.18), (ii) we have

(5.5.45)
$$N(r, f^* D_{l+1}) < l_0 T_f(r, \omega_0) + O(1).$$

On the other hand, it follows from Lemma (5.4.8) that

(5.4.6)
$$T_f(r, \omega_0) \uparrow \infty \quad \text{as } r \uparrow \infty.$$

Since D has only simple normal crossings,

(5.5.47)
$$f^* D - R_f \leq \operatorname{supp} f^* D.$$

To see this, it is sufficient to check it around points of $R(\operatorname{supp} f^* D)$, and the proof is easy. It follows from (5.5.43) ~ (5.5.45) and (5.5.47) that

(5.5.48)
$$\sum_{j=1}^{l} \left\{ 1 - \frac{N(r, \operatorname{supp} f^* D_j)}{T_f(r, \omega_0)} \right\}$$

$$\leq \sum_{j=1}^{l}\left\{1-\frac{N(r, f^*D_j)}{T_f(r, \omega_0)}\right\}+\frac{N(r, R_f)}{T_f(r, \omega_0)}$$

$$\leq \frac{T_f(r, \mathbf{K}(M)^{-1})}{T_f(r, \omega_0)}+\frac{m(2m-1)\varepsilon\log r}{2T_f(r, \omega_0)}$$

$$=\frac{O(\log^+ T_f(r, \omega_0))}{T_f(r, \omega_0)}+\frac{O(1)}{T_f(r, \omega_0)}\Big\|_{E(\varepsilon)}.$$

By definition, we have

$$\text{(5.5.49)}\qquad \frac{T_f(r, \mathbf{K}(M)^{-1})}{T_f(r, \omega_0)}\leq\left[\frac{c_1(\mathbf{K}(M)^{-1})}{\gamma}\right]+\kappa+\frac{O(1)}{T_f(r, \omega_0)}$$

for any $\kappa > 0$. By (5.5.46), (5.5.48) and (5.5.49) there is a positive constant C such that

$$\sum_{j=1}^{l}\Theta_f(D_j)\leq\sum_{j=1}^{l}\delta_f(D_j)+\overline{\lim_{r\to\infty}}\frac{N(rr\, R_f)}{T_f(r, \omega_0)}$$

$$\leq\left[\frac{c_1(\mathbf{K}(M)^{-1})}{\gamma}\right]+\kappa+C\varepsilon.$$

Letting $\varepsilon \to 0$ and $\kappa \to 0$, we obtain our assertions. *Q.E.D.*

(5.5.50) Corollary (Ramification Theorem). *Let the notation be as in Corollary (5.5.42). Assume, moreover, that*

$$f^*D_j \geq \nu_j \operatorname{supp} f^*D_j.$$

Then

$$\sum_{j=1}^{l}\left(1-\frac{1}{\nu_j}\right)\leq\left[\frac{c_1(\mathbf{K}(M)^{-1})}{\gamma}\right].$$

Proof. It follows from Corollaries (5.2.16) and (5.5.42) that

$$\sum_{j=1}^{l}\left(1-\frac{1}{\nu_j}\right)\leq\sum_{j=1}^{l}\left(1-\overline{\lim_{r\to\infty}}\frac{N(r, \operatorname{supp} f^*D_j)}{N(r, f^*D_j)}\right)$$

$$\leq\sum_{j=1}^{l}\left(1-\overline{\lim_{r\to\infty}}\frac{N(r, \operatorname{supp} f^*D_j)}{T_f(r, \omega_0)}\right)=\sum_{j=1}^{l}\Theta_f(D_j)$$

$$\leq\left[\frac{c_1(\mathbf{K}(M)^{-1})}{\gamma}\right].\quad Q.E.D.$$

(5.5.51) Corollary (Degeneracy Theorem). *Let D be an analytic hypersurface with only simple normal crossings such that $c_1(\mathbf{K}(M))+c_1([D])>0$ (in case $c_1(\mathbf{K}(M))>0$, D may be empty). If a meromorphic mapping $f:\mathbf{C}^m\to M$ omits D (i.e., $f(\mathbf{C}^m)\cap D=\varnothing$), then f is degenerate.*

The proof will be clear.

Among the examples described below, the first three show that the condition of D having only simple normal crossings can not be simply dropped.

(5.5.52) *Example* (Sakai-Kodaira [59]). Let $[u^0;u^1;u^2]$ be the homogeneous coordinate system of $\mathbf{P}^2(\mathbf{C})$. Let $\mathbf{H}_0\to\mathbf{P}^2(\mathbf{C})$ be the hyperplane bundle. Define $D_d\in|\mathbf{H}_0^d|$ $(d\in\mathbf{N})$ by

$$D_d=\{[u^0;u^1;u^2]\in\mathbf{P}^2(\mathbf{C});\ (u^0)^{d-1}u^2-(u^1)^d=0\}.$$

Then D_d is irreducible and has a singularity at [0; 0; 1]. Put

$$f:(z^1,z^2)\in\mathbf{C}^2\to[1;z^1;(z^1)^d+e^{z^2}]\in\mathbf{P}^2(\mathbf{C}).$$

Then f is non-degenerate everywhere. On the other hand,

$$(1)^{d-1}((z^1)^d+e^{z^2})-(z^1)^d=e^{z^2}\neq 0,$$

so that $f(\mathbf{C}^2)\cap D_d=\varnothing$. By Example (2.1.26), $\mathbf{K}(\mathbf{P}^2(\mathbf{C}))=\mathbf{H}_0^{-3}$. Hence

$$c_1([D_d])+c_1(\mathbf{K}(\mathbf{P}^2(\mathbf{C})))=(d-3)c_1(\mathbf{H}_0)>0$$

for $d\geq 4$.

(5.5.53) *Example* (Green-Shiffman [38]). We define $C_d\subset\mathbf{P}^2(\mathbf{C})$ by

$$C_d=\{[u^0;u^1;u^2]\in\mathbf{P}^2(\mathbf{C});\ (u^1)^d-u^0\{(u^1)^{d-1}+(u^2)^{d-1}\}=0\}.$$

Then $C_d\in|\mathbf{H}_0^d|$, C_d is irreducible and has a singularity at [1; 0; 0], where, locally, $d-1$ non-singular curves with distinct tangent lines are crossing. Now we put

$$f:(z^1,z^2)\in\mathbf{C}^2\to\left[\frac{1-\exp\{z^2(1+(z^1)^{d-1})\}}{1+(z^1)^{d-1}};\,1;\,z^1\right]\in\mathbf{P}^2(\mathbf{C}).$$

Then f is non-degenerate everywhere and omits C_d.

(5.5.54) *Example* (Green [38]). We put

$$D=\{[u^0;u^1;u^2]\in\mathbf{P}^2(\mathbf{C});\ u^0u^1\{(u^0+u^1)^2+u^2(u^0-u^1)\}=0\}.$$

Then $c_1([D])+c_1(\mathbf{K}(\mathbf{P}^2(\mathbf{C})))=c_1(\mathbf{H}_0)>0$. The analytic hypersurface D consists of two lines and one singular irreducible curve. Put

$$f:(z^1,z^2)\in\mathbf{C}^2\to\left[1;\,e^{z^1};\,e^{z^1}+3+\frac{e^{\phi}+4}{e^{z^1}-1}\right]\in\mathbf{P}^2(\mathbf{C}),$$

where $\phi = z^2(e^{z^1} - 1)$. Then f is non-degenerate everywhere and omits D.

(5.5.55) *Example.* Let $M \subset \mathbf{P}^2(\mathbf{C})$ be an analytic hypersurface of degree d without singularity. Let $\mathbf{L} \to M$ be the restriction of the hyperplane bundle over M. By Example (2.1.27)

$$\mathbf{K}(M) = \mathbf{L}^{d-4}.$$

In the case where $1 \le d \le 3$, it is known that M is a rational surface (cf. Shafarevich [101]); i.e., the function field $\mathrm{Mer}(M)$ is a purely transcendental extension of $\mathbf{C}$ with two variables. Hence there is a non-degenerate meromorphic (even, rational) mapping $f: \mathbf{C}^2 \to M$. If $d \ge 5$, then $c_1(\mathbf{K}(M)) > 0$ and hence any meromorphic mapping $f: \mathbf{C}^2 \to M$ is degenerate. Let $d = 4$. Put

$$M = \left\{ [u^0; u^1; u^2; u^3] \in \mathbf{P}^3(\mathbf{C});\ (u^0)^4 + (u^1)^4 + (u^2)^4 + (u^3)^4 = 0 \right\}.$$

Then it is known that M is a complex manifold called a Kummer surface; i.e., there is a 2-dimensional Abelian variety (= complex projective algebraic torus) $A = \mathbf{C}^2/\Lambda$ such that there is a proper modification $p: M \to A/\langle\iota\rangle$, where $\iota: a \in A \to -a \in A$ is the involution. Letting $\pi: \mathbf{C}^2 \to A$ be the universal covering, we have a non-degenerate meromorphic mapping

$$f = p^{-1} \circ \pi: \mathbf{C}^2 \to M.$$

It is not known if there is a non-degenerate holomorphic mapping $f: \mathbf{C}^2 \to M$ for $M \subset \mathbf{P}^3(\mathbf{C})$ of degree 4.

(5.5.56) *Example.* Let $\mathbf{H}_0 \to \mathbf{P}^m(\mathbf{C})$ be the hyperplane bundle and $D_j \in |\mathbf{H}_0^d|$ $(1 \le j \le l)$ non-singular analytic hypersurfaces such that $\sum_{j=1}^{l} D_j$ has only normal crossings. By Example (2.1.26) we see that

$$\left[\frac{c_1(\mathbf{K}(\mathbf{P}^m(\mathbf{C}))^{-1})}{c_1(\mathbf{H}_0^d)} \right] = \frac{(m+1)}{d}.$$

Hence, for instance, Corollary (5.5.42), (ii) is written as

$$\sum_{j=1}^{l} \delta_f(D_j) \le \sum_{j=1}^{l} \Theta_f(D_j) \le \frac{(m+1)}{d}.$$

In the case of $d = 1$, $\sum_{j=1}^{l} D_j$ has only normal crossings if and only if $D_1, \ldots, D_l$ are in general position.

(5.5.57) *Example.* Let $M = \mathbf{C}^m/\Lambda$ be an Abelian variety and $D \subset M$ an analytic hypersurface without singular point such that $c_1([D]) > 0$. Since $c_1(\mathbf{K}(M)) = 0$ (Example (2.1.25)), $[c_1(\mathbf{K}(M)^{-1})/c_1([D])] = 0$. Thus for any non-degenerate holomorphic mapping $f : \mathbf{C}^m \to M$

$$\delta_f(D) = \Theta_f(D) = 0.$$

In Corollary (5.5.50), all $\nu_j = 1$.

(5.5.58) *Example* (the case of $m = 1$). Let M be a compact Riemann surface of genus g. Let $D_j = a_j \in M$ $(1 \le j \le l)$ be distinct points. Then $\gamma = c_1([a_1]) = \cdots = c_1([a_l])$. Let $\omega_0 \in \gamma$ be a positive (1, 1)-form. Here one notes that

$$\left[\frac{c_1(\mathbf{K}(M)^{-1})}{\gamma}\right] = 2 - 2g = \chi(M),$$

where $\chi(M)$ denotes the Euler number of M. Let $f : \mathbf{C} \to M$ be a non-degenerate (= non-constant) holomorphic mapping. Then by Theorem (5.5.26) we have

(5.5.59) $$\{l - \chi(M)\} T_f(r, \omega_0) \le \sum_{j=1}^{l} N(r, f^* a_j)$$

$$-N(r, R_f) + \frac{\varepsilon}{2} \log r + O(\log^+ T_f(r, \omega_0)) + O(1) \|_{E(\varepsilon)}.$$

If $M = \mathbf{P}^1(\mathbf{C})$, then $g = 0$ and $\chi(M) = 2$. In this case, (5.5.59) is the same as the Second Main Theorem established by Nevanlinna. For general M, it is the same as those obtained by Ahlfors, Chern, etc. The defect relation takes the following form:

$$\sum_{j=1}^{l} \delta_f(a_j) \le \sum_{j=1}^{l} \Theta_f(a_j) \le \chi(M).$$

Notes

(a) In §3 we proved that an effective divisor D on $\mathbf{C}^n$ is algebraic if and only if $N(r, D) = O(\log r)$. More generally, a locally finite sum of analytic subsets of pure dimension k of a paracompact complex manifold M with coefficients in $\mathbf{Z}$ is called an **analytic cycle** of dimension k on M. The set of all analytic cycles of dimension k forms an Abelian group. If an analytic cycle is written as the sum of analytic subsets with positive coefficients, then it is said to be effective. An effective analytic cycle defines a closed positive current (cf. Theorem (5.1.8) and Example (3.2.25)).

An analytic subset X of $\mathbf{C}^n$ is called an **algebraic subset** if there are finitely many polynomials P_j, $1 \le j \le l$ such that

$$X = \{P_1 = \cdots = P_l = 0\}.$$

An analytic cycle on $\mathbf{C}^n$ is said to be **algebraic** if it is written as a finite sum of algebraic subsets with integer coefficients. Let E be an effective analytic cycle of dimension k on $\mathbf{C}^n$. We define the counting (order) function $N(r, E)$ by

$$N(r, E) = \int_1^r \frac{<E, \chi_{B(t)}\alpha^k>}{t^{2k-1}} dt.$$

Stoll [113] proved that

$$E \text{ is algeraic } \Longleftrightarrow N(r, E) = O(\log r).$$

(b) Let F be a holomorphic function on $\mathbf{C}^n$. To measure the growth order of F, it may be the first idea for everyone to use the maximum modulus $M(r, F)$ (cf. §3). In fact, Borel, Hadamard and many others used $M(r, F)$ to investigate the value distribution of F in the case of $n = 1$. The maximum modulus $M(r, F)$, however, fails to make sense for meromorphic functions, especially in the case of $n \geq 2$. It is Nevanlinna [73] who defined the characteristic function $T(r, F)$ for a meromorphic function F on $\mathbf{C}$ by making use of Jensen's formula. The comparison Theorem (5.3.13) between $M(r, F)$ and $T(r, F)$ was proved in Kneser [53].

(c) The First Main Theorem given in §2 is due to Griffiths and his coauthors (Carlson-Griffiths [16] and Griffiths-King [44]). They proved it for holomorphic mappings and Shiffman [102] extended it for meromorphic mappings. R. Nevanlinna [73] proved the First and Second Main Theorems and established the fundamentals of the distribution theory. One may find in Nevanlinna [74] how the function theory evolved by them. Thus it is the first important problem to establish the First Main Theorem for the value distribution in several complex variables. Other types of the First Main Theorems have been obtained by, e.g., Stoll [115], Chern [20] and Wu [123].

(d) The Second Main Theorem and defect relations are the essence of the Nevanlinna theory. R. Nevanlinna [73] established these for meromorphic functions on $\mathbf{C}$. F. Nevanlinna gave a proof to them by making use of Hermitian metrics (cf. Note in [74]). Ahlfors [1] proved them by applying the Gauss-Bonnet theorem and showed that the number "2" which appears in Picard's theorem and defect relations $\sum \delta_f(a) \leq 2$ is the Euler number $\chi(\mathbf{P}^1(\mathbf{C}))$ of the target space $\mathbf{P}^1(\mathbf{C})$ (cf. Example (5.5.57)).

The Second Main Theorem described here is due to Griffiths and his coauthors (Carlson-Griffiths [16] and Griffiths-King [44]). This is a rare case in the function theory in several complex variables where the Second Main Theorem and defect relations are satisfactorily established. They dealt with only holomorphic mappings and Shiffman [102] generalized them for meromorphic mappings. Noguchi [76] proved them for meromorphic mappings from finite covering spaces over $\mathbf{C}^m$

into complex projective algebraic manifolds. Stoll [115] dealt with meromorphic mappings from parabolic spaces.

Here we considered only those analytic hypersurfaces D which have only simple normal crossings. For the case of D with more complex singularities, see Shiffman [102] and Sakai [99].

Kodaira [58] applied the present theory to prove that compactifications of $\mathbf{C}^2$ and $\mathbf{C} \times \mathbf{C}^*$ are rational surfaces.

CHAPTER VI

Value Distribution of Holomorphic Curves

6.1 Preparation from Function Theory in One Complex Variable

Let F be a meromorphic function on $\mathbf{C}$. We defined the Nevanlinna characteristic function $T(r, F)$ in (5.2.24). While $T(r, 1/F) = T(r, F) + O(1)$ by Corollary (5.2.30), we see this more precisely. By Theorem (5.1.13)

$$dd^c\left[\log^+|F|^2 - \log^+\frac{1}{|F|^2}\right] = dd^c[\log|F|^2] = (F)_0 - (F)_\infty.$$

It follows from Lemma (3.3.20) that

(6.1.1) $$T(r, F) - T\left(r, \frac{1}{F}\right) = \frac{1}{2\pi}\int_0^{2\pi}\log|F(e^{i\theta})|\,d\theta.$$

For a constant $a \in \mathbf{C}$, $(F)_\infty = (F + a)_\infty$ and hence by (5.2.23)

(6.1.2) $$|T(r, F+a) - T(r, F)| \le \log^+|a| + \log 2.$$

For finitely many meromorphic functions $F_1, \ldots, F_l$ we recall (5.2.25):

(6.1.3) $$T\left(r, \prod_{j=1}^{l} F_j\right) \le \sum_{j=1}^{l} T(r, F_j),$$

$$T\left(r, \sum_{j=1}^{l} F_j\right) \le \sum_{j=1}^{l} T(r, F_j) + \log l,$$

From now on, we will frequently use the above (6.1.1)~(6.1.3) without explicitly mentioning. The following is easily obtained from the Cauchy integral formula.

(6.1.4) Lemma. $\frac{1}{2\pi}\int_0^{2\pi}\log|e^{i\theta} - a|\,d\theta = \log^+|a|$ *for* $a \in \mathbf{C}$.

(6.1.5) Lemma. *Let F be a meromorphic function on* $\mathbf{C}$ *and* $A_0, \ldots, A_l$ *holomorphic functions on* $\mathbf{C}$ *such that* $A_0 \not\equiv 0$ *and*

$$A_0F^l + A_1F^{l-1} + \cdots + A_{l-1}F + A_l \equiv 0.$$

Then

$$T(r, F) \le \sum_{j=1}^{l} T(r, A_j) - \frac{1}{2\pi}\int_0^{2\pi} \log |A_0(e^{i\theta})|\, d\theta + \log(l+1).$$

Proof. Take $z \in \mathbf{C}$ with $A_0(z) \neq 0$ and define a polynomial in t by

$$B(z;t) = A_0t^l + A_1t^{l-1} + \cdots + A_{l-1}t + A_l.$$

Let $F(z), F_2(z), \ldots, F_l(z)$ be the roots of the equation $B(z;t) = 0$ with counting multiplicities. Then

$$B(z;t) = A_0(z)(t - F(z))(t - F_2(z)) \cdots (t - F_l(z)).$$

Therefore, using Lemma (6.1.4), we have

$$\frac{1}{2\pi}\int_0^{2\pi} \log |B(z;e^{i\theta})|\, d\theta = \log |A_0(z)| + \frac{1}{2\pi}\int_0^{2\pi} \log |e^{i\theta} - F(z)|\, d\theta$$

$$+ \sum_{j=2}^{l} \frac{1}{2\pi}\int_0^{2\pi} \log |e^{i\theta} - F_j(z)|\, d\theta$$

$$= \log |A_0(z)| + \log^+ |F(z)| + \sum_{j=2}^{l} \log^+ |F_j(z)|.$$

We obtain

$$\frac{1}{2\pi}\int_0^{2\pi} \log |B(z;e^{i\theta})|\, d\theta \ge \log |A_0(z)| + \log^+ |F(z)|. \tag{6.1.6}$$

On the other hand,

$$\log |B(z;e^{i\theta})| = \log |A_0e^{il\theta} + \cdots + A_l(z)|$$

$$\le \log^+ \left(\sum_{j=0}^{l} |A_j(z)| \right) + \log(l+1) \le \sum_{j=0}^{l} \log^+ |A_j| + \log(l+1).$$

Hence

$$\frac{1}{2\pi}\int_0^{2\pi} \log |B(z;e^{i\theta})|\, d\theta \le \sum_{j=0}^{l} \log^+ |A_j| + \log(l+1). \tag{6.1.7}$$

It follows from (6.1.6) and (6.1.7) that

(6.1.8) $$m(r, F) + \frac{1}{2\pi}\int_0^{2\pi} \log|A_0(re^{i\theta})|\,d\theta \le \sum_{j=0}^{l} T(r, A_j) + \log(l+1).$$

We see by Theorem (5.1.13) and Lemma (3.3.20) that

(6.1.9) $$\frac{1}{2\pi}\int_0^{2\pi} \log|A_0(re^{i\theta})|\,d\theta = N(r, (A_0)) + \frac{1}{2\pi}\int_0^{2\pi} \log|A_0(e^{i\theta})|\,d\theta.$$

Since $N(r, (F)_\infty) \le N(r, (A_0))$, we have the desired estimate by (6.1.8) and (6.1.9). *Q.E.D.*

Let F be a meromorphic function. Considering F a holomorphic mapping from $\mathbf{C}$ into $\mathbf{P}^1(\mathbf{C})$, we deduce from Corollary (5.2.16) and Theorem (5.2.29) that

$$N(r, (F-a)_0) \le T(r, F) + O(1) \quad (a \in \mathbf{P}^1(\mathbf{C})),$$

where $(F-a)_0 = (F)_\infty$ for $a = \infty = [0; 1]$. We investigate this in more details.

(6.1.10) Lemma. *There is a positive constant C depending only on F such that*

$$N(r, (F-a)_0) \le T(r, F) + C \quad \textit{for all } a \in \mathbf{P}^1(\mathbf{C}).$$

Proof. In the case of $a = \infty$, the inequality holds with $C = 0$. Suppose that $a \in \mathbf{C}$. Let $0 < \delta < 1$. Then we infer from Theorem (5.1.13) and Lemma (3.3.20) that

$$T(r, F-a) + \int_\delta^1 \frac{n(t, (F-a)_\infty)}{t}\,dt$$
$$= T\left(r, \frac{1}{F-a}\right) + \int_\delta^1 \frac{n(t, (F-a)_0)}{t}\,dt + \frac{1}{2\pi}\int_0^{2\pi} \log|F(\delta e^{i\theta})|\,d\theta.$$

By $(F-a)_\infty = (F)_\infty$ and (6.1.2)

(6.1.11) $$N(r, (F-a)_0) \le T\left(r, \frac{1}{F-a}\right)$$
$$= T(r, F) + \log^+|a| + \log 2 + \int_\delta^1 \frac{n(t, (F)_\infty)}{t}\,dt$$
$$- \int_\delta^1 \frac{n(t, (F-a)_0)}{t}\,dt + \frac{1}{2\pi}\int_0^{2\pi} \log|F(\delta e^{i\theta}) - a|^{-1}\,d\theta.$$

From now on, we denote by $C_1, C_2, \ldots$ constants independent of r and a. Assume first that $|a| \le 1$. If 0 is a pole of F, then $|F(\delta e^{i\theta})| \ge 2$ $(0 \le \theta \le 2\pi)$ for a sufficiently small δ, so that $\log|F(\delta e^{i\theta}) - a|^{-1} \le 0$. Thus our assertion follows from (6.1.11). We next suppose that F is holomorphic at 0. Then

$$N(r,\ (F-a)_0) \le T(r,\ F) + \log 2 + \int_\delta^1 \frac{n(t,\ (F)_\infty)}{t} dt \tag{6.1.12}$$

$$-\int_\delta^1 \frac{n(t,\ (F-a)_0)}{t} dt + \frac{1}{2\pi}\int_0^{2\pi} \log|F(\delta e^{i\theta}) - a|^{-1} d\theta.$$

Put

$$S(a) = \overline{\lim_{\delta\to 0}} \left\{ \frac{1}{2\pi}\int_0^{2\pi} \log|F(\delta e^{i\theta}) - a|^{-1} d\theta - \int_\delta^1 \frac{n(t,\ (F-a)_0)}{t} dt \right\}.$$

We claim that $S(a)$ is bounded from above on $\{|a| \le 1\}$. We put in a neighborhood of 0

$$F(z) - F(0) = z^\nu G(z) \quad (G(0) \ne 0).$$

If $a = F(0)$, then $n(t,\ (F-a)_0) \ge \nu$, so that

$$S(a) \le \log|G(0)|. \tag{6.1.13}$$

If $a \ne F(0)$, then

$$S(a) = \log|F(0) - a|^{-1} - \int_0^1 \frac{n(t,\ (F-a)_0)}{t} dt. \tag{6.1.14}$$

If $|F(0)| > 1$, (6.1.14) implies that $S(a)$ is bounded from above on $\{|a| \le 1\}$. We suppose that $|F(0)| \le 1$. By (6.1.13) and (6.1.14) it suffices to show that $S(a)$ is bounded from above as $a \to F(0)$. Taking a small neighborhood U of 0, we may obtain a holomorphic function $\Phi(z) = z(G(z))^{1/\nu}$ on U such that $\Phi\colon U \to \Delta(\varepsilon)$ $(\varepsilon > 0)$ is biholomorphic. Put

$$\psi\colon w \in \Delta(\varepsilon) \to w^\nu \in \Delta(\varepsilon^\nu).$$

Then $F(z) - F(0) = \psi \circ \Phi(z)$. Let $\lambda = |a - F(0)| < \varepsilon^\nu$. Then the set $\{z \in U;\ F(z) = a\}$ consists of ν distincts points $z_1, \ldots, z_\nu$ such that

$$|z_\mu| = \lambda^{\frac{1}{\nu}} |G(z_\mu)|^{-\frac{1}{\nu}} = \lambda^{\frac{1}{\nu}}(1 + o(1))|G(0)|^{-\frac{1}{\nu}},\quad 1 \le \mu \le \nu,$$

where $o(1) \to 0$ as $\lambda \to 0$. Therefore

$$\int_0^1 \frac{n(t,\ (F-a)_0)}{t} dt = \int_{\lambda^{1/\nu}(1+o(1))|G(0)|^{-1/\nu}}^1 \frac{n(t,\ (F-a)_0)}{t} dt$$

$$\ge \int_{\lambda^{1/\nu}(1+o(1))|G(0)|^{-1/\nu}}^1 \frac{\nu dt}{t} = -\log\lambda + \log|G(0)| - \nu\log(1 + o(1)).$$

Combining this with (6.1.14), we see that $S(a) \le C_2$ for $|a| \le 1$. Thus (6.1.12) implies that

$$(6.1.15) \qquad N(r, (F-a)_0) \le T(r, F) + C_3,$$

provided that $|a| \le 1$. By making use of $1/F$, the case of $|a| > 1$ is reduced to the above case. *Q.E.D.*

We prove Nevanlinna's lemma on logarithmic derivative. We give a proof due to F. Selberg [100].

(6.1.16) Lemma. *Let F be a meromorphic function on* $\mathbf{C}$ *and* $0 < \delta < 1$. *Then*

$$m\left(r, \frac{F'}{F}\right) \le 3\log^+ T(r, F) + \delta \log r + O(1) \|_{E(\delta)}.$$

Proof. Identify $\mathbf{C}$ with $\mathbf{P}^1(\mathbf{C}) - \{[0; 1]\}$ by $w \in \mathbf{C} \to [1; w] \in \mathbf{P}^1(\mathbf{C})$. Take two holomorphic functions f_0 and f_1 without common zero such that $F = f_1/f_0$. Then the holomorphic mapping $f : \mathbf{C} \to \mathbf{P}^1(\mathbf{C})$ identified with F is given by

$$f(z) = [f_0(z); f_1(z)] \in \mathbf{P}^1(\mathbf{C}).$$

Define a singular volume form Ω on $\mathbf{P}^1(\mathbf{C})$ by

$$\Omega | \mathbf{C} = \frac{1}{\{1 + (\log |w|)^2\} |w|^2} \cdot \frac{i}{2\pi} dw \wedge d\bar{w}.$$

It is easy to see that

$$\int_{\mathbf{P}^1(\mathbf{C})} \Omega = C_0 < \infty.$$

Since
$$f^* \Omega = \frac{|F'|^2}{\{1 + (\log |F|)^2\} |F|^2} \frac{i}{2\pi} dz \wedge d\bar{z},$$

$$(6.1.17) \qquad \int_{\Delta(r)} \frac{|F'|^2}{\{1 + (\log |F|)^2\} |F|^2} \frac{i}{2\pi} dz \wedge d\bar{z}$$
$$= \int_{\Delta(r)} f^* \Omega = \int_{w \in \mathbf{P}^1(\mathbf{C})} n(r; (F - w)_0) \Omega.$$

It follows from Fubini's theorem, Lemma (6.1.10) and (6.1.17) that

$$(6.1.18) \qquad \int_1^r \frac{dt}{t} \int_{\Delta(t)} f^* \Omega = \int_{w \in \mathbf{P}^1(\mathbf{C})} N(r, (F - w)_0) \Omega$$
$$\le C_0 (T(r, F) + C).$$

Using the concavity of log, we compute $m(r, F'/F)$:

$$m\left(r, \frac{F'}{F}\right)$$

$$= \frac{1}{4\pi}\int_{\{|z|=r\}} \log^+\left\{(1+(\log|F|)^2)\frac{|F'|^2}{(1+(\log|F|)^2)|F|^2}\right\} d\theta$$

$$\le \frac{1}{2}\log 2 + \frac{1}{2\pi}\int_{\{|z|=r\}} \log(1+|\log|F||)d\theta$$

$$+ \frac{1}{4\pi}\int_{\{|z|=r\}} \log\left\{1+\frac{|F'|^2}{(1+(\log|F|)^2)|F|^2}\right\} d\theta$$

$$\le \frac{1}{2}\log 2 + \log\left[\frac{1}{2\pi}\int_{\{|z|=r\}}\left\{1+\log^+|F|+\log^+\frac{1}{|F|}\right\} d\theta\right]$$

$$+ \frac{1}{2}\log\left[\frac{1}{2\pi}\int_{\{|z|=r\}}\left\{1+\frac{|F'|^2}{(1+(\log|F|)^2)|F|^2}\right\} d\theta\right]$$

$$\le \log^+ T(r, F) + O(1) + \frac{1}{2}\log^+\left\{\frac{1}{r}\frac{d}{dr}\int_{\Delta(r)} f^*\Omega\right\}.$$

Using Lemma (5.5.35), we continue the computation:

$$m\left(r, \frac{F'}{F}\right) \le \log^+ T(r, F) + \frac{1}{2}\log^+\left\{\frac{1}{r}\left[\int_{\Delta(r)} f^*\Omega\right]^{1+\delta}\right\} + O(1) \Big\|_{E_1(\delta)}$$

$$\le \log^+ T(r, F) + \frac{1}{2}\log^+\left\{r^\delta\left[\frac{d}{dr}\int_1^r \frac{dt}{t}\int_{\Delta}(t),)f^*\Omega\right]^{1+\delta}\right\} + O(1) \Big\|_{E_1(\delta)}$$

$$\le \log^+ T(r, F) + \frac{\delta}{2}\log r$$

$$+ \frac{1}{2}\log^+\left\{\int_1^r \frac{dt}{t}\int_{\Delta}(t),)f^*\Omega\right\}^{(1+\delta)^2} + O(1) \Big\|_{E_2(\delta)}.$$

Now it finally follows from (6.1.8) that

$$m\left(r, \frac{F'}{F}\right) \le \log^+ T(r, F) + \frac{\delta}{2}\log r$$

$$+ 2\log^+ T(r, F) + O(1) \|_{E_2(\delta)}$$

$$\leq 3\log^+ T(r, F) + \delta \log r + O(1) \|_{E_2(\delta)} . \quad Q.E.D.$$

(6.1.19) Corollary. *Let $F^{(k)}$ denote the k-th derivative of F ($k \in \mathbf{Z}^+$) and $0 < \delta < 1$. Then the following hold.*

(i) $T(r, F^{(k)}) \leq (k+1)T(r, F) + O(\log^+ T(r, F) + \delta \log r) \|_{E(\delta)}$, $k \geq 0$;

(ii) $m\left(r, \dfrac{F^{(k)}}{F}\right) \leq O(\log^+ T(r, F) + \delta \log r) \|_{E(\delta)}$, $k \geq 1$.

Proof. We use the induction on k. The assertion (i) for $k = 0$ is trivial. Lemma (6.1.16) is (ii) with $k = 1$. Assume that (i) holds for $k-1$ and (ii) holds for k ($k \geq 1$). Then it follows from the induction hypothesis that

$$m(r, F^{(k)}) = m\left(r, F \cdot \frac{F^{(k)}}{F}\right)$$

$$\leq m(r, F) + m\left(r, \frac{F^{(k)}}{F}\right)$$

$$= m(r, F) + O(\log^+ T(r, F) + \delta \log^+ r) \|_{E(\delta)} .$$

Since $N(r, (F^{(k)})_\infty) \leq (k+1)N(r, (F)_\infty)$, we have

$$T(r, F^{(k)}) \leq (k+1)N(r, (F)_\infty) + m(r, F)$$

$$+ O(\log^+ T(r, F) + \delta \log^+ r) \|_{E(\delta)}$$

$$\leq (k+1)T(r, F) + O(\log^+ T(r, F) + \delta \log^+ r) \|_{E(\delta)} .$$

Hence (i) is proved for k. We show (ii) for $k+1$. It follows from the induction hypothesis, Lemma (6.1.16) and (i) proved above for k that

$$m\left(r, \frac{F^{(k+1)}}{F}\right) = m\left(r, \frac{F^{(k+1)}}{F^{(k)}} \cdot \frac{F^{(k)}}{F}\right)$$

$$\leq m\left(r, \frac{F^{(k+1)}}{F^{(k)}}\right) + m\left(r, \frac{F^{(k)}}{F}\right)$$

$$\leq O(\log^+ T(r, F^{(k)}) + \delta \log r) + O(\log^+ T(r, F) + \delta \log r) \|_{E_1(\delta)}$$

$$\leq O(\log^+ T(r, F) + \delta \log r) \|_{E_2(\delta)} .$$

Thus (ii) is proved for $k+1$. *Q.E.D.*

(6.1.20) Lemma (Borel). *Let $F_1, \ldots, F_N$ ($N \geq 2$) be holomorphic functions without* *zero such that*

(6.1.21) $$F_1 + \cdots + F_N = 1.$$

Then $F_1, \ldots, F_N$ are linearly dependent over **C**.

Remark. The identity (6.1.21) is called the **Borel identity**.

Proof. Assume that $F_1, \ldots, F_N$ are linearly independent. Then there is a non-constant F_j and moreover the Wronskian

$$W = \begin{vmatrix} F_1 & \cdots & F_N \\ F_1^{(1)} & \cdots & F_N^{(1)} \\ \vdots & \vdots & \vdots \\ F_1^{(N-1)} & \cdots & F_N^{(N-1)} \end{vmatrix} \not\equiv 0.$$

Let W_j be the $(1, j)$-minor of W. Then it follows from (6.1.21) that

(6.1.22) $$F_j = \frac{W_j}{F_1 \cdots \hat{F}_j \cdots F_N} \left(\frac{W}{F_1 \cdots F_N} \right)^{-1} = \frac{\Delta_j}{\Delta},$$

where $$\Delta = \begin{vmatrix} 1 & \cdots & 1 \\ F_1^{(1)}/F_1 & \cdots & F_N^{(1)}/F_N \\ \vdots & \vdots & \vdots \\ F_1^{(N-1)}/F_1 & \cdots & F_N^{(N-1)}/F_N \end{vmatrix}$$

and Δ_j are its $(1, j)$-minors. Put

$$T(r) = \max_{1 \le j \le N} T(r, F_j) \; (= \max_{1 \le j \le N} m(r, F_j)).$$

By (6.1.22)

$$\begin{aligned} m(r, F_j) &\le m(r, \Delta_j) + T(r, \Delta) + O(1) \\ &= m(r, \Delta_j) + m(r, \Delta) + O(1). \end{aligned}$$

Applying Corollary (6.1.19) to the elements of Δ and Δ_j, we have

$$m(r, F_j) \le O(\log^+ T(r) + \delta \log r) \|_{E(\delta)}.$$

Hence

(6.1.23) $$T(r) \le O(\log^+ T(r) + \delta \log r) \|_{E(\delta)}.$$

Since some F_j is not constant, $T(r) \to \infty$ $(r \to \infty)$. Moreover, $T(r)$ is a monotone increasing function in $\log r$, up to $O(1)$-term, so that there is a constant C such that

$$\overline{\lim_{r \to \infty}} \frac{\log r}{T(r)} = C < \infty.$$

It follows from (6.1.23) that $1 \le \delta C$. Letting $\delta \to 0$, we have a contradiction. *Q.E.D.*

(6.1.24) *Remark.* Let a_j, $j = 1, 2, \ldots, N$, be meromorphic functions such that $T(r, a_j) = o(T(r))\|_E$. In the above lemma, we obtain the same conclusion by assuming $\sum_{j=1}^{N} a_j F_j = 1$ instead of (6.1.21). The same holds in Corollary below.

(6.1.25) Corollary (Borel). *Let $F_0, \ldots, F_N$ be holomorphic functions on* **C** *without zero such that*

(6.1.26)
$$F_0 + \cdots + F_N = 0.$$

Then there is a decomposition of indices

$$\{0, 1, , \ldots, N\} = I_1 \cup \cdots \cup I_l$$

such that

(i) *every I_k contains at least two indices,*

(ii) *for i, $j \in I_k$, F_i/F_j is constant,*

(iii) *for $i \in I_k$ and $j \in I_h$ with $h \ne k$, F_i/F_j is not constant,*

(iv) *for every I_k, $\sum_{j \in I_k} F_j = 0$.*

Proof. We use the induction on N. The case of $N = 1$ is trivial. Assume that our assertion holds up to $N - 1$. Now we introduce an equivalence relation $\sim$ in $\{0, 1, \ldots, N\}$ by

$$i \sim j \iff \frac{F_i}{F_j} \text{ is constant.}$$

Then we classify $\{0, 1, \ldots, N\}$ by this equivalence relation:

$$\{0, 1, \ldots, N\}/\!\sim\; = \{I_1, \ldots, I_l\}.$$

We show that (i)~(iv) hold with respect to these $I_1, \ldots, I_l$. The construction immediately implies (ii) and (iii). Now, for the proof of (ii) we may assume that an index contained in I_k is, to say, 0. Put $G_j = -F_j/F_0$ $(1 \le j \le N)$. Then they are not constant, free from 0 and satisfy the Borel identity

$$G_1 + \cdots + G_N = 1.$$

By Lemma (6.1.20), $G_1, \ldots, G_N$ are linearly dependent over **C**. Therefore there is a non-trivial linear relation

(6.1.27)
$$c_1 F_1 + \cdots + c_N F_N = 0.$$

Without loss of generality, we may assume that $c_N = 1$. It follows from (6.1.26)

and (6.1.27) that

$$F_0 + (1 - c_1)F_1 + \cdots + (1 - c_{N-1})F_{N-1} = 0.$$

Then there is one $1 \le i \le N-1$ such that $1 - c_i \ne 0$. By the induction hypothesis, there is an index $1 \le j \le N-1$ such that $1 - c_j \ne 0$ and $F_0/(1-c_j)F_j$ is constant. Thus $\{0, j\} \subset I_k$. To show (iv), we choose an index i_k from each I_k and put

$$\sum_{i \in I_k} F_i = b_k F_{i_k},$$

where b_k is a constant. Then (6.1.26) is written as

(6.1.28)
$$\sum_{k=1}^{l} b_k F_{i_k} = 0.$$

Applying (i) to (6.1.28), we infer that all $b_k = 0$; that is, $\sum_{i \in I_k} F_i = 0$. *Q.E.D.*

Let M be a compact complex manifold of dimension m and h a Hermitian metric on M with Hermitian metric form Ω. Let $(z^1, \ldots, z^m)$ be a holomorphic local coordinate system of M and write

$$h = \sum h_{i\bar{j}} dz^i d\bar{z}^j.$$

Then

$$\Omega = \frac{i}{2} \sum h_{i\bar{j}} dz^i \wedge d\bar{z}^j.$$

In general, we sometimes call a holomorphic mapping from a domain of $\mathbf{C}$ into M a **holomorphic curve** into (or in) M. Let $f : \mathbf{C} \to M$ be a holomorphic curve into M. Then the characteristic function $T_f(r, \Omega)$ is defined by (5.2.4). Let ω be a holomorphic 1-form on M and put

$$f^* \omega = \zeta(z) dz.$$

We show that an estimate similar to Lemma (6.1.16) holds for $m(r, \zeta)$.

(6.1.29) Lemma. *Let the notation be as above and* $0 < \delta < 1$. *Then*

$$m(r, \zeta) \le 2 \log^+ T_f(r, \Omega) + \delta \log r \,\|_{E(\delta)}.$$

Proof. Since M is compact, there is a positive constant C such that

(6.1.30)
$$|\omega(v)| \le C \sqrt{h(v, \bar{v})}$$

for all holomorphic tangent vectors $v \in \mathbf{T}(M)$. Put $f^*\Omega = s(z) \frac{i}{2} dz \wedge d\bar{z}$. Then $s(z) = h(f_*(\partial/\partial z), f_*(\partial/\partial \bar{z}))$. Since $\zeta(z) = \omega(f_*(\partial/\partial z))$, it follows from (6.1.30) that

$$|\zeta(z)| \le C\sqrt{s(z)}.$$

By making use of this, the concavity of log and Lemma (5.5.35), we estimate $m(r, \zeta)$:

$$m(r, \zeta) = \frac{1}{2\pi}\int_0^{2\pi} \log^+|\zeta(re^{i\theta})|\,d\theta = \frac{1}{4\pi}\int_0^{2\pi}\log^+|\zeta(re^{i\theta})|^2 d\theta$$

$$\le \frac{1}{4\pi}\int_0^{2\pi}\log(1+|\zeta(re^{i\theta})|^2)d\theta \le \frac{1}{2}\log\left\{1+\frac{1}{2\pi}\int_0^{2\pi}|\zeta(re^{i\theta})|^2 d\theta\right\}$$

$$\le \frac{1}{2}\log\left\{1+\frac{C}{2\pi}\int_0^{2\pi} s(re^{i\theta})d\theta\right\}$$

$$= \frac{1}{2}\log\left\{1+\frac{C}{2\pi r}\frac{d}{dr}\int_{\Delta(r)} s(z)\frac{i}{2}dz\wedge d\bar{z}\right\}$$

$$\le \frac{1}{2}\log\left\{1+\frac{C}{2\pi r}\left[\int_{\Delta(r)} f^*\Omega\right]^{(1+\delta)}\right\}\Big\|_{E_1(\delta)}$$

$$= \frac{1}{2}\log\left\{1+\frac{Cr^{\delta}}{2\pi}\left[\frac{d}{dr}\int_1^r \frac{dt}{t}\int_{\Delta(r)} f^*\Omega\right]^{(1+\delta)}\right\}\Big\|_{E_1(\delta)}$$

$$\le \frac{1}{2}\log\left\{1+\frac{Cr^{\delta}}{2\pi}(T_f(r, \Omega))^{(1+\delta)^2}\right\}\Big\|_{E_2(\delta)}$$

$$\le 2\log^+ T_f(r, \Omega) + \delta\log r + O(1)\|_{E(\delta)}. \quad Q.E.D.$$

6.2 Elementary Facts on Algebraic Varieties

Here we explain elementary facts on algebraic varieties which we will use in this chapter, giving proofs as far as possible. Here we do not consider general algebraic varieties and deal with only quasi-projective algebraic varieties explained below. We also explain the relation between analytic and algebraic subsets.

Let $\mathbf{P}^m(\mathbf{C})$ be the complex m-dimensional projective space and $\rho: \mathbf{C}^{m+1} \to \{O\} \to \mathbf{P}^m(\mathbf{C})$ the Hopf fibering. Let $(z^0, \ldots, z^m)$ be the natural complex coordinate system of $\mathbf{C}^{m+1}$ and $\mathbf{C}[z^0, \ldots, z^m]$ denote the polynomial ring. A subset $X \subset \mathbf{P}^m(\mathbf{C})$ is said to be **algebraic** if there are finitely many homogeneous polynomials $P_1, \ldots, P_k \in \mathbf{C}[z^0, \ldots, z^m]$ such that

$$X = \left\{[z^0; \cdots; z^m] \in \mathbf{P}^m(\mathbf{C}); P_j(z^0, \ldots, z^m) = 0,\ 1 \le j \le k\right\}$$

(cf. Chapter V, Notes, (a)). Then X is, of course, an analytic subset of $\mathbf{P}^m(\mathbf{C})$. The converse holds, too.

(6.2.1) Theorem (Chow). *An analytic subset X of $\mathbf{P}^m(\mathbf{C})$ is algebraic.*

Proof. Note that $\rho^{-1}(X)$ is an analytic subset of $\mathbf{C}^{m+1} - \{O\}$ and

$$z \in \rho^{-1}(X) \iff tz \in \rho^{-1}(X) \quad (t \in \mathbf{C}^*). \tag{6.2.2}$$

Hence any irreducible component of $\rho^{-1}(X)$ is of positive dimension. By Theorem (4.1.13), the closure $Y = \overline{\rho^{-1}(X)}$ is an analytic subset of $\mathbf{C}^{m+1}$. Let $F_1, \ldots, F_k$ be the defining functions of Y at O. Then F_i are holomorphic in a neighborhood U of O and

$$Y \cap U = \{F_1 = \cdots = F_k = 0\}.$$

Taking a sufficiently small $\varepsilon > 0$, we see by (6.2.2) that

$$z \in Y \cap B(\varepsilon),\ t \in \Delta(1) \Rightarrow F_j(tz) = 0,\ \ 1 \le j \le k. \tag{6.2.3}$$

We put

$$F_j(z) = \sum_{i=1}^{\infty} P_{ji}(z),$$

where P_{ji} are homogeneous polynomials of degree i. Noting that $F_j(tz) = \sum_{i=1}^{\infty} t^i P_{ji}(z)$, we infer from (6.2.3) that

$$Y \cap B(\varepsilon) = \{z \in B(\varepsilon);\ P_{ji}(z) = 0,\ 1 \le j \le k,\ i = 1, 2, \ldots\}.$$

Since $\mathbf{C}[z^0, \ldots, z^m]$ is Noetherian, there is a number $i_0 \in \mathbf{Z}^+$ such that

$$Y \cap B(\varepsilon) = \{z \in B(\varepsilon);\ P_{ji}(z) = 0,\ 1 \le j \le k,\ 1 \le i = i_0\}.$$

Therefore wee see by (6.2.2) that

$$Y = \{z \in \mathbf{C}^{m+1};\ P_{ji}(z) = 0,\ 1 \le j \le k,\ 1 \le i = i_0\},$$

so that X is algebraic. *Q.E.D.*

The dimension of an algebraic subset $X \subset \mathbf{P}^m(\mathbf{C})$ is defined by that of X as an analytic subset. As an application of Theorem (6.2.1) we have the following.

(6.2.4) Theorem (Siegel). *Let M be a compact complex manifold of dimension m. Then the transcendental degree of the meromorphic function field* Mer(M) *is at most m and finitely generated over* $\mathbf{C}$.

Proof. Take arbitrarily $m+1$ meromorphic functions $f_1, \ldots, f_{m+1} \in \mathrm{Mer}(M)$. Put $W = M - \bigcup_{i=1}^{m+1} \mathrm{supp}\,(f_i)_\infty$. Then the holomorphic mapping

$$f : x \in W \to [1; f_1(x); \cdots; f_{m+1}] \in \mathbf{P}^{m+1}(\mathbf{C})$$

defines a meromorphic mapping $f : M \to \mathbf{P}^{m+1}(\mathbf{C})$. Hence the graph $G(f) \subset M \times \mathbf{P}^{m+1}(\mathbf{C})$ is an analytic subset. Let $q : M \times \mathbf{P}^{m+1}(\mathbf{C}) \to \mathbf{P}^{m+1}(\mathbf{C})$ be the natural projection. Then it follows from Theorem (4.3.3) that the image $q(G(f)) = f(M)$ is an analytic subset of $\mathbf{P}^{m+1}(\mathbf{C})$. Since $\dim f(M) \le m < m+1$, $f(M) \ne \mathbf{P}^{m+1}(\mathbf{C})$. By Theorem (6.2.1) there is a non-zero polynomial $P(w^1, \ldots, w^{m+1})$ such that

$$P(f_1, \ldots, f_{m+1}) = 0.$$

Therefore the transcendental degree of Mer(M) is at most m. We fix a transcendental base $(f_1, \ldots, f_l)$ of Mer(M) ($l \le m$). For an arbitrary $g \in$ Mer(M), there is an irreducible algebraic relation

$$g^d + R_1(f_1, \ldots, f_l) g^{d-1} + \cdots + R_d(f_1, \ldots, f_l) = 0, \tag{6.2.5}$$

where R_j are rational functions in $f_1, \ldots, f_l$. As above, we consider the meromorphic mapping

$$f : x \in M \to [1;\ f_1(x); \cdots; f_l(x)] \in \mathbf{P}^l(\mathbf{C}).$$

Then $f(M) = \mathbf{P}^l(\mathbf{C})$. Put

$$Z = \bigcup_{i=1}^{l} \operatorname{supp}\,(R_i(f_1, \ldots, f_l))_\infty \quad (\ne M).$$

Then wee see by (6.2.5) that g is holomorphic on $M - Z$ and takes a constant on each fiber $f^{-1}(y)$ with $y \notin \mathbf{P}^l(\mathbf{C}) - f(Z)$. Let d_0 denote the maximum number of connected components of fibers $f^{-1}(y)$ for $y \in \mathbf{P}^l(\mathbf{C}) - f(Z)$. Then we deduce that $d \le d_0$. Since d_0 is independent of g, Mer(M) must be finitely generated. *Q.E.D.*

The notions of irreducibility and irreducible components for algebraic subsets of $\mathbf{P}^m(\mathbf{C})$ are defined as in the case of analytic subsets. Let $X \subset \mathbf{P}^m(\mathbf{C})$ be an irreducible algebraic subset. Then X is called a **complex projective algebraic variety**. If X has no singular point as an analytic subset, X is called a **non-singular complex projective algebraic variety**. Let X be a complex projective algebraic variety and $Z \underset{\ne}{\subset} X$ an algebraic subset. Then $Y = X - Z$ is called a **complex quasi-projective algebraic variety**. Then $\bar{Y} = X$. A subset A of Y is said to be algebraic if there is an algebraic subset $B \subset X$ ($\subset \mathbf{P}^m(\mathbf{C})$) such that $A = Y \cap B$. Especially, $\mathbf{P}^m(\mathbf{C}) - \{z^0 = 0\} = \mathbf{C}^m$ is called the **complex affine space** and an irreducible algebraic subset of $\mathbf{C}^m$ is called a **complex affine algebraic variety**. Let Y be a complex quasi-projective algebraic variety. Then we may define a new topology by taking algebraic subsets as "closed subsets", which is called the **Zariski topology**. Let $A \subset Y$ be an arbitrary subset. Then the closure of A with respect to the Zariski topology is the smallest algebraic subset of Y containing A. Henceforth we use the terminologies such as Zariski open, Zariski closed, Zariski closure, etc. with respect to the Zariski topology, while terminologies with respect to the ordinary topology as an analytic subset are used in the same way as before.

Let P and Q be homogeneous polynomials and $Q \neq 0$. Then the ratio P/Q defines a meromorphic function on $\mathbf{P}^m(\mathbf{C})$, which is called a **rational function**. Let $Y \subset \mathbf{P}^m(\mathbf{C})$ be a complex quasi-projective algebraic variety. A meromorphic function on Y is called a **rational function** if it is written as a rational function P/Q with $Y \not\subset \{Q=0\}$ restricted over Y. All rational functions on Y form a field named the **rational function field** of Y and denoted by $\mathbf{C}(Y)$.

(6.2.6) Theorem. *Let* $X \subset \mathbf{P}^m(\mathbf{C})$ *be a non-singular complex projective algebraic variety. Then the rational function field* $\mathbf{C}(X)$ *coincides with the meromorphic function field* $\mathrm{Mer}(X)$.

For the proof, see Remmert [94].

We consider the product manifold $\mathbf{P}^m(\mathbf{C}) \times \mathbf{P}^n(\mathbf{C})$. Let $[z^0; \cdots; z^m]$ (resp. $[z^0; \cdots; z^n]$) be the homogeneous coordinate system of $\mathbf{P}^m(\mathbf{C})$ (resp. $\mathbf{P}^n(\mathbf{C})$). Then the holomorphic mapping

$$\iota\colon ([z^i], [w^j]) \in \mathbf{P}^m(\mathbf{C}) \times \mathbf{P}^n(\mathbf{C}) \to [z^i w^j] \in \mathbf{P}^{(m+1)(n+1)-1}(\mathbf{C})$$

is a holomorphic imbedding. Through this ι, $\mathbf{P}^m(\mathbf{C}) \times \mathbf{P}^n(\mathbf{C})$ gives rise to a non-singular complex projective algebraic variety. Let $Y_1 \subset \mathbf{P}^m(\mathbf{C})$ and $Y_2 \subset \mathbf{P}^n(\mathbf{C})$ be complex quasi-projective algebraic varieties. Then $\iota(Y_1 \times Y_2)$ is a complex quasi-projective algebraic variety. Since $Y_1 \times Y_2$ is biholomorphic to $\iota(Y_1 \times Y_2)$, we may consider $Y_1 \times Y_2$ a complex quasi-projective algebraic variety.

Let Y_1 and Y_2 be as above. A **rational mapping** $f: Y_1 \to Y_2$ is defined as follows:

(6.2.7) (i) There is a non-empty Zariski open subset $W \subset Y_1$ and f is a holomorphic mapping from W into Y_2.

(ii) For a rational function F on Y_2 which is holomorphic at least one point of $f(W)$, f^*F is a rational function on Y_1.

To be precise, we have to introduce an equivalence relation among them and to call the equivalence class a rational function as in the definition of meromorphic mappings (Chapter IV, §4). We note that the closure of the graph of a rational mapping $f: W \to Y_2$ in $Y_1 \times Y_2$ is a complex quasi-projective algebraic variety and that the inverse image $f^{-1}(A)$ of an algebraic subset $A \subset Y_2$ is defined as an algebraic subset, while the image of an algebraic subset is not necessarily an algebraic subset. If a rational mapping $f: Y_1 \to Y_2$ is holomorphic everywhere on Y_1, then f is said to be **regular rational mapping**. If there are quasi-projective algebraic varieties X

and Y in $\mathbf{P}^m(\mathbf{C})$ such that $Y \subset X$, then the inclusion mapping $\iota_Y: Y \to X$ is a regular rational mapping. The natural projections $p_i: Y_1 \times Y_2 \to Y_i$, $i = 1, 2$, are regular rational mappings. Let $X_1 \subset \mathbf{P}^m(\mathbf{C})$ and $X_2 \subset \mathbf{P}^n(\mathbf{C})$ be complex projective algebraic varieties. Then a rational mapping from X_1 into X_2 is a meromorphic mapping and the converse is also true by Theorem (6.2.6):

(6.2.8) Theorem. *Let X_1 and X_2 be as above and $f: X_1 \to X_2$ a meromorphic mapping. Then f is rational.*

Let $f: Y_1 \to Y_2$ be a rational mapping which is holomorphic on a non-empty Zariski open subset $W \subset Y_1$. Then f is said to be **dominant** if $f(W)$ is Zariski dense in Y_2. In this case, f^*F is defined for an arbitrary $F \in \mathbf{C}(Y_2)$ and $f^*: \mathbf{C}(Y_2) \to \mathbf{C}(Y_1)$ is an injective homomorphism, and hence $\mathbf{C}(Y_2)$ is considered a subfield of $\mathbf{C}(Y_1)$. Assume that $\dim Y_1 = \dim Y_2$. Let d_f be the maximum number of the elements of $f^{-1}(y)$ for $y \in Y_2$ such that $f^{-1}(y)$ are discrete.

(6.2.9) Theorem. *Let Y_1 and Y_2 be complex quasi-projective algebraic varieties of the same dimension and $f: Y_1 \to Y_2$ a dominant rational mapping. Then $\mathbf{C}(Y_2)$ is a finite algebraic extension of $\mathbf{C}(Y_1)$ with extension degree $[\mathbf{C}(Y_1); \mathbf{C}(Y_2)] = d_f$.*

Proof. Take a point $y_0 \in Y_2$ so that $f^{-1}(y_0)$ consists of d_f points. Then there is a rational function $F \in \mathbf{C}(Y_1)$ such that F takes distinct values at the points of $f^{-1}(y_0)$. As in the proof of Theorem (6.2.4), we infer that $d_f = [\mathbf{C}(Y_1); \mathbf{C}(Y_2)]$. *Q.E.D.*

Let $Y \subset \mathbf{P}^m(\mathbf{C})$ be a complex quasi-projective algebraic variety and put $X = \bar{Y}$. Then X is a complex projective algebraic variety and coincides with the Zariski closure of Y in $\mathbf{P}^m(\mathbf{C})$. Let $S(X)$ (resp. $S(Y)$) denote the set of singular points of X (resp. Y) as an analytic subset. Then $S(X)$ is an algebraic subset of $\mathbf{P}^m(\mathbf{C})$ by Theorem (6.2.1). Thus $S(Y)$ is an algebraic subset of Y and $R(Y) = Y - S(Y)$ is a complex quasi-projective algebraic variety. Put $n = \dim Y$. By definition we see that

(6.2.10) for an arbitrary $x \in R(X)$, there are rational functions $u^1, \ldots, u^n$ on Y such that u^i are holomorphic in a neighborhood of x and $(u^1, \ldots, u^n)$ is a holomorphic local coordinate system around x; moreover, those points $y \in R(Y)$ around which $(u^1, \ldots, u^n)$ gives a holomorphic local coordinate system form a Zariski open subset of Y.

Let W be a non-empty Zariski open subset of $R(Y)$ and ω a holomorphic k-form on W. For an arbitrary point x, we take $u = (u^1, \ldots, u^n)$ as in (6.2.10). Then there is a Zariski open neighborhood $W' \subset W$ of x on which

$$\omega = \sum_{J \in \{n;\, k\}} \omega_J du^J.$$

If all ω_J are rational functions on Y, we call ω a **rational** k**-forms** on Y. To be precise, we have to take a suitable equivalence class for the definition of rational k-forms as in the case of meromorphic functions. Especially, if $Y = R(Y)$ and $W = Y$, then ω is called a **regular rational k-form** on Y.

(6.2.11) Theorem. *Let $X \subset \mathbf{P}^m(\mathbf{C})$ be a non-singular complex projective algebraic variety. Then all holomorphic k-forms are rational.*

Proof. Take an arbitrary $x_0 \in X$ and $u = (u^1, ..., u^n)$ as in (6.2.10) which gives a holomorphic local coordinate system around x_0 $(n = \dim X)$. It follows from Theorem (6.2.1) that $Z_1 = \bigcup_{i=1}^{n} \operatorname{supp}(U^i)_\infty$ and the closure of the set $\{x \in X - Z_1;\, du^1 \wedge \cdots \wedge du^n = 0\}$ in X are algebraic subsets different to X. Then $u = (u^1, ..., u^n)$ gives rise to a holomorphic local coordinate system around every point of $X - (Z_1 \cup Z_2)$. We write

$$\omega = \sum_{J \in \{n;\, k\}} \omega_J du^J.$$

Then ω_J are holomorphic functions on $X - (Z_1 \cup Z_2)$. Moreover, taking rational functions around the points of $Z_1 \cup Z_2$ similar to the above u^i, we see that ω_J are meromorphic functions on Y. By Theorem (6.2.6), $\omega_J \in \mathbf{C}(X)$. *Q.E.D.*

In the same way as above, we see the following.

(6.2.12) Proposition. *Let $Y \subset \mathbf{P}^m(\mathbf{C})$ be a complex quasi-projective algebraic variety of dimension n and $1 \leq k \leq n$. Put $l = \binom{n}{k}$. Let $\omega^1, ..., \omega^l$ be rational k-forms on Y such that ω^i are holomorphic on a Zariski open subset $W \subset R(Y)$ and $\omega^i(x_0)$, $1 \leq i \leq l$ generate $\bigwedge^k \mathbf{T}^*(W)_{x_0}$ at one point $x_0 \in W$. Then the set W' of points $x \in W$ at which $\omega^i(x)$, $1 \leq i \leq l$ generate $\bigwedge^k \mathbf{T}^*(W)_x$ is a Zariski open subset of W, and an arbitrary rational k-form ω on Y is uniquely written as*

$$\omega = \sum_{i=1}^{l} F_i \omega^i,$$

where F_i are rational functions on Y. Q.E.D.

The next theorem follows from (6.2.10) and Theorem (6.2.11).

(6.2.13) Theorem. *Let Y_1 and Y_2 be complex quasi-projective algebraic varieties and $f: Y_1 \to Y_2$ a rational mapping which is holomorphic on a Zariski open subset W_1 of $R(Y_1)$. Let ω be a rational k-form on Y_2 which is holomorphic on a Zariski open subset $W_2 \subset R(Y_2)$. Assume that $f(W_1) \cap W_2 \neq \emptyset$. Then $(f|W_1)^{-1}(W_2)$ is a non-empty Zariski open subset of $R(Y_1)$ and $f^*\omega$ gives rise to a rational k-form; especially, if $Y_i = R(Y_i) = W_i$ $(i = 1, 2)$, then $f^*\omega$ is a regular rational k-form.*

Let $\mathbf{C}^m/\Lambda$ be a complex torus of dimension m.

(6.2.14) Theorem. *The transcendental degree of* Mer$(\mathbf{C}^m/\Lambda)$ *is* m *if and only if* $\mathbf{C}^m/\Lambda$ *is a complex projective algebraic variety.*

For the proof, see Weil [120]. A complex torus is called an **Abelian variety** if it is a complex projective algebraic variety. An Abelian variety A has the natural group structure with zero 0. If an irreducible algebraic subset $N \subset A$ contains 0 and is an Abelian variety as itself, then N is called an **Abelian subvariety.** In this case it is known that the group structure of N is compatible with that of A.

(6.2.15) Theorem. *Let A be an Abelian variety and B a subset of A which is a subgroup of A as an abstract group. Then the Zariski closure of B is an Abelian subvariety of A.*

The proof is the same as in that of Proposition (1.7.5).

6.3 Jet Bundles and Subvarieties of Abelian Varieties

In general, let X be an n-dimensional complex manifold and $x \in X$. Let U be a neighborhood of $0 \in \mathbf{C}$ and $f_U\colon U \to X$ a holomorphic mapping such that $f_U(0) = x$. We denote by $H(\mathbf{C}, X)_x$ the set of all those holomorphic mappings from neighborhoods of 0 into X. Let $f_U, g_V \in H(\mathbf{C}, X)_x$. Then we define an equivalence relation $f_U \sim g_V$ as follows: There is a neighborhood $W \subset U \cap V$ of 0 such that $f_U | W = g_V | W$. We set

$$J(X)_x = H(\mathbf{C}, X)_x / \sim.$$

An element f of $J(X)_x$ is called a **germ of holomorphic mappings** from **C** into X and we write

$$f : (\mathbf{C}, 0) \to (X, x).$$

The germ f is represented by a holomorphic mapping $f_U\colon U \to X$ with $f_U(0) = x$. If no confusion occurs, we also write f for f_U. Let $(z^1, \ldots, z^n)$ be a holomorphic local coordinate system around x and $k \in \mathbf{N}$. Let $f,\ g \in J(X)_x$ and put $f = (f^1, \ldots, f^n)$ and $g = (g^1, \ldots, g^n)$. Then we write $f \overset{k}{\sim} g$ if

$$\frac{d^j}{dx^j} f^i(0) = \frac{d^j}{dx^j} g^i(0), \quad 0 \le j \le k, \quad 1 \le i \le n.$$

Then $\overset{k}{\sim}$ is an equivalence relation independent of the choice of a holomorphic local coordinate system around x. We denote by $j_k(f)$ the equivalence class of f and set

$$J_k(X)_x = J(X)_x / \overset{k}{\sim} = \{ j_k(f);\ f \in J(X)_x \}.$$

The space $J_k(X)_x$ is naturally biholomorphic to $\mathbf{C}^{nk}$. Set

$$J_k(X) = \bigcup_{x \in X} J_k(X)_x, \quad p : J_k(X) \to X,$$

where $p(J_k(X)_x) = x$. Then $J_k(X)$ naturally carries a structure of a complex manifold such that p is holomorphic and the triple $(J_k(X), p, X)$ forms a holomorphic fiber bundle (not vector bundle for $k \geq 2$) over X. The holomorphic fiber bundle $J_k(X)$ is called the k-**jet bundle** over X. In special, $J_1(X) = \mathbf{T}(X)$. Let Y be another complex manifold and $\alpha: X \to Y$ a holomorphic mapping. Then α naturally induces a holomorphic bundle morphism

$$\alpha_* : j_k(f) \in J_k(X) \to j_k(\alpha \circ f) \in J_k(Y).$$

Let $U \subset \mathbf{C}$ be a domain and $f : U \to X$ a holomorphic mapping. At an arbitrary point $z \in U$, f defines a holomorphic mapping $g_z(w) = f(z+w)$ from a neighborhood of 0 into X with $g_z(0) = x$. Set

$$J_k(f) \colon z \in U \to j_k(g_z) \in J_k(X).$$

Then $J_k(f)$ is a holomorphic curve from U into $J_k(X)$ and the following diagram is commutative:

$$\begin{array}{ccc} U & \xrightarrow{J_k(f)} & J_k(X) \\ \| & & \downarrow p \\ U & \xrightarrow{f} & X \end{array}$$

We call $J_k(f)$ the **lifting of order** k of f.

(6.3.1) Lemma. *Let $\omega^1, \ldots, \omega^n$ be closed holomorphic 1-forms on X such that $\omega^1 \wedge \cdots \wedge \omega^n \neq 0$ at any point of X. Then there is a trivialization*

$$\tilde{\omega} : J_k(X) \xrightarrow{\cong} X \times (\mathbf{C}^n)^k,$$

$$\tilde{\omega}(J_k(f)(z)) = \left(f(z), (\zeta^i(z)), \left(\frac{d}{dz}\zeta^i(z) \right), \ldots, \left(\frac{d^{k-1}}{dz^{k-1}}\zeta^i(z) \right) \right),$$

where $f^*\omega^i = \zeta^i(z)dz$ $(1 \leq i \leq n)$.

Proof. The integrals $\int^x \omega^i$ form a holomorphic local coordinate system around every point of X. Hence our assertion immediately follows. *Q.E.D.*

We henceforth deal with the case where $\mathbf{C}^m/\Lambda$ is an Abelian variety; Y is an irreducible algebraic subset of $\mathbf{C}^m/\Lambda$ and $X = R(Y)$. Then X is a complex quasi-projective algebraic variety. Let $\iota_X \colon X \to \mathbf{C}^m/\Lambda$ be the natural inclusion mapping. Then ι_X is a regular rational mapping. At every point of $\mathbf{C}^m/\Lambda$ the holomorphic tangent space of $\mathbf{C}^m/\Lambda$ is identified with the space of holomorphic vector fields on $\mathbf{C}^m/\Lambda$. Fix a base $\{\omega^1, \ldots, \omega^m\}$ of the space of holomorphic 1-forms on $\mathbf{C}^m/\Lambda$. Then $d\omega^i = 0$ and $\omega^1 \wedge \cdots \wedge \omega^m \neq 0$ at any point of $\mathbf{C}^m/\Lambda$. By Lemma (6.3.1) we have a trivialization

$$\tilde{\omega} : J_k(\mathbf{C}^m/\Lambda) \overset{\cong}{\rightarrow} (\mathbf{C}^m/\Lambda) \times (\mathbf{C}^m)^k$$

Let $q : (\mathbf{C}^m/\Lambda) \times (\mathbf{C}^m)^k \rightarrow (\mathbf{C}^m)^k$ be the natural projection. Then we put

(6.3.2) $$I_k = q \circ \tilde{\omega} \circ (\iota_X)_* : J_k(X) \rightarrow (\mathbf{C}^m)^k.$$

Here we remark that if X is, in general, a complex quasi-projective algebraic variety, then so is $J_k(X)$. (For readers who know abstract algebraic geometry, it will be clear that if X is an algebraic variety, then so is $J_k(X)$.) In the following argument, it is sufficient for us to know that $J_k(X)$ is an algebraic variety. Because of the present special circumstance, however, we consider $J_k(X)$ as follows. We know that $(\mathbf{C}^m/\Lambda) \times (\mathbf{C}^m)^k$ is a complex quasi-projective algebraic variety. Identifying $J_k(\mathbf{C}^m/\Lambda)$ with $(\mathbf{C}^m/\Lambda) \times (\mathbf{C}^m)^k$, we consider $J_k(\mathbf{C}^m/\Lambda)$ a complex quasi-projective algebraic variety. Then $J_k(X)$ is naturally a complex quasi-projective variety contained in $J_k(\mathbf{C}^m/\Lambda)$ and $(\iota_X)_* : J_k(X) \rightarrow J_k(\mathbf{C}^m/\Lambda)$ is a regular rational mapping. Hence I_k is also a regular rational mapping. Let Ker I_{k_*} denote the kernel of the differential I_{k*} of I_k. Take an arbitrary point $x \in X$. Then there is a simply connected neighborhood $U \subset X$ of x such that for suitable indices $1 \leq i_1 < \cdots < i_n \leq m$

$$\iota_X^*(\omega^{i_1} \wedge \cdots \wedge \omega^{i_n}) \neq 0 \text{ on } U.$$

For a simplicity, we suppose that $i_1 = 1, \ldots, i_n = n$. By Lemma (6.3.1), there is a trivialization of $J_k(X)|U$ with respect to $\iota_X^*\omega^1|U, \ldots, \iota_X^*\omega|U$:

(6.3.3) $$J_k(X)|U = J_k(U) \cong U \times (\mathbf{C}^n)^k.$$

Let

(6.3.4) $$\left(\begin{pmatrix} Z^{1(1)} \\ \vdots \\ Z^{n(1)} \end{pmatrix}, \ldots, \begin{pmatrix} Z^{1(k)} \\ \vdots \\ Z^{n(k)} \end{pmatrix}\right)$$

be the natural coordinate system of $(\mathbf{C}^n)^k$. Taking U smaller if necessary, we have a simply connected neighborhood U' of x in $\mathbf{C}^m/\Lambda$ satisfying the conditions:

(6.3.5) (i) $U' \cap X = U$:

(ii) Put $x^i(y) = \int_x^y \omega^i$, $1 \leq i \leq m$ in U'. Then they form a holomorphic local coordinate system in U', and there are holomorphic functions $F^j(x^1, ..., x^n)$, $n+1 \le j \le m$ such that

$$U' \cap X = \{(x^1, ..., x^m) \in U';$$
$$x^j = F^j(x^1, ..., x^n),\ n+1 \le j \le m\}.$$

Therefore $(x^1, ..., x^n)$ restricted over U is a holomorphic local coordinate system in U and

$$\iota_X^*\omega^j = d(x^j|U) = \sum_{i=1}^{n} \frac{\partial F^j}{\partial x^i} dx^i,\quad n+1 \le j \le m,\ \text{ on } U. \tag{6.3.6}$$

Since $\iota_X^*(\omega^1 \wedge \cdots \wedge \omega^n) \neq 0$ on U, by Proposition (6.2.12) there are rational functions P_i^j on X which are holomorphic on a Zariski open subset of X containing U and satisfy

$$(-1)^{n-i}\iota_X^*(\omega^1 \wedge \cdots \wedge \hat{\omega^i} \wedge \cdots \wedge \omega^n \wedge \omega^j) = P_i^j \iota_X^*(\omega^1 \wedge \cdots \wedge \omega^n)$$

for $1 \le i \le n$ and $n+1 \le j \le m$. Combining this with (6.3.6), we obtain

$$P_i^j|U = \frac{\partial F^j}{\partial x^i}. \tag{6.3.7}$$

It follows from (6.3.2)~(6.3.4) and (6.3.6) that

$$I_k\colon U \times (\mathbf{C}^n)^k \to (\mathbf{C}^m)^k \tag{6.3.8}$$

$$\left(z, \begin{pmatrix} Z^{1(1)} \\ \vdots \\ Z^{n(1)} \end{pmatrix}, ..., \begin{pmatrix} Z^{1(k)} \\ \vdots \\ Z^{n(k)} \end{pmatrix}\right) \to \left(\begin{pmatrix} Z^{1(1)} \\ \vdots \\ Z^{n(1)} \\ I^{n+1(1)} \\ \vdots \\ I^{m(1)} \end{pmatrix}, ..., \begin{pmatrix} Z^{1(k)} \\ \vdots \\ Z^{n(k)} \\ I^{n+1(k)} \\ \vdots \\ I^{m(k)} \end{pmatrix}\right).$$

Here $I^{j(l)}$ $(n+1 \le j \le m,\ 1 \le l \le k)$ are polynomials in $Z^{i(h)}$, $1 \le i \le n$, $1 \le h \le l$ and partial derivatives of order $\le l-1$ of P_i^j with respect to $x^1, ..., x^n$, which are rational functions on X for the same reason as (6.3.7). Now we have the direct sum decomposition

$$\mathbf{T}(U \times (\mathbf{C}^n)^k) = \mathbf{T}(U) \oplus \mathbf{T}((\mathbf{C}^n)^k).$$

We see by (6.3.8) that

$$\operatorname{Ker} I_{k*} \subset \mathbf{T}(U). \tag{6.3.9}$$

(6.3.10) Lemma (M. Green). *Let $f: \Delta(1) \to X$ be a holomorphic curve such that the Zariski closure of $f(\Delta(1))$ in $\mathbf{C}^m/\Lambda$ is Y. Assume that* Ker $(I_{k*J_k(f)(0)}) \neq \{O\}$ *for* *all* $k \in \mathbf{N}$. *Then the dimension of the algebraic subset* $G = \{a \in \mathbf{C}^m/\Lambda;\ Y + a = Y\}$ *is positive.*

Remark. Since G is a subgroup of $\mathbf{C}^m/\Lambda$, *the identity component* G^0 *of G is an Abelian subvariety of* $\mathbf{C}^m/\Lambda$.

Proof. Put $x = f(0)$ and

$$V_k = \operatorname{Ker} I_{k*J_k(f)(0)} \subset \mathbf{T}(U)_x.$$

We infer from (6.3.8) that $V_{k+1} \subset V_k$. The assumption implies that $\bigcap_{k=1}^{\infty} V_k \neq \{O\}$. Take $v \in \bigcap_{k=1}^{\infty} V_k - \{O\} \subset \mathbf{T}(X)_x - \{O\}$. Let $(x^1, \ldots, x^m)$ be the holomorphic local coordinate system around x in $\mathbf{C}^m/\Lambda$ given in (6.3.5). Using $(x^1, \ldots, x^n)$ as a holomorphic local coordinate system around x in X, we put

$$f(z) = (f^1(z), \ldots, f^n(z)),$$

$$v = \sum_{i=1}^{n} a^i \left(\frac{\partial}{\partial x^i}\right)_x \qquad (a^i \in \mathbf{C}).$$

Define a holomorphic vector field v on U by

$$v = \sum_{i=1}^{n} a^i \left(\frac{\partial}{\partial x^i}\right)_y, \qquad y \in U.$$

Since $I_{k*}(v) = O$ for all $k \in \mathbf{N}$, we deduce from (6.3.5), (iii) that

$$\left.\frac{d^k}{dz^k}(vF^j)(f^1(z), \ldots, f^n(z))\right|_{z=0} = 0, \quad n+1 \leq j \leq m,\ k \geq 1.$$

Hence $(vF^j)(f(z))$ are constant in z. On the other hand, it follows from (6.3.5), (iii) that

$$(\iota_X)_* v(y) = \sum_{i=1}^{n} a^i \left(\frac{\partial}{\partial x^i}\right)_y + \sum_{j=n+1}^{m} \left(\sum_{i=1}^{n} a^i \frac{\partial F^j}{\partial x^i}(y)\right) \left(\frac{\partial}{\partial x^j}\right)_y$$

$$= \sum_{i=1}^{n} a^i \left(\frac{\partial}{\partial x^i}\right)_y + \sum_{j=n+1}^{m} (vF^j)(y) \left(\frac{\partial}{\partial x^j}\right)_y.$$

Since $(vF^j)(f(z))$ are constant, we have

(6.3.11) $$(\iota_X)_* v(f(z)) = \sum_{i=1}^{n} a^i \left(\frac{\partial}{\partial x^i}\right)_{f(z)} + \sum_{j=n+1}^{m} (vF^j)(f(z)) \left(\frac{\partial}{\partial x^j}\right)_{f(z)}$$

Continued

$$= \sum_{i=1}^{n} a^i \left(\frac{\partial}{\partial x^i} \right)_{f(z)} + \sum_{j=n+1}^{m} (vF^j)(x) \left(\frac{\partial}{\partial x^j} \right)_{f(z)}$$

around $z = 0$. Let $\tilde{v}$ denote the holomorphic vector field on $\mathbf{C}^m/\Lambda$ determined by $(\iota_X)_* v(x) \in \mathbf{T}(\mathbf{C}^m/\Lambda)_x$. Then by (6.3.11)

(6.3.12) $$\tilde{v}(x^j - F^j)(f^1(z), \ldots, f^n(z)) \equiv 0, \quad n+1 \le j \le m.$$

Thus $\tilde{v}$ is tangent to X at all points of $f(\Delta(1))$. Since $\{w \in X; \tilde{v}_w \in \mathbf{T}(X)_w\}$ is an algebraic subset and $f(\Delta(1))$ is Zariski dense in Y, $\tilde{v}$ is tangent to X at all points of X and hence tangent to Y at all points of Y. Put

$$\tilde{v} = \sum_{i=1}^{m} a^i \frac{\partial}{\partial x^i} \quad \text{on } \mathbf{C}^m/\Lambda,$$

$$\exp \tilde{v}(t) = \lambda(ta^1, \ldots, ta^m), \quad t \in \mathbf{C}.$$

Then Y is invariant with respect to the translations by $\exp \tilde{v}(t)$ for all $t \in \mathbf{C}$. Hence $\dim G \ge 1$. *Q.E.D.*

6.4 Bloch's Conjecture

Let M be a non-singular complex projective algebraic variety and $q(M)$ the dimension of the vector space of holomorphic 1-forms on M; i.e.,

$$q(M) = \dim \Gamma(M, \mathbf{T}^*(M)),$$

and $q(M)$ is called the **irregularity** of M. A a holomorphic mapping $f: N \to M$ from a complex manifold N into M is said to be **algebraically degenerate** if the image $f(N)$ is not Zariski dense in M.

(6.4.1) Theorem (Bloch's conjecture). *Let $f: \mathbf{C} \to M$ be a holomorphic curve into M. Assume that $q(M) > \dim M$. Then f is algebraically degenerate.*

Proof. Put $q = q(M)$ and take a base $\{\omega^1, \ldots, \omega^q\}$ of $\Gamma(M, \mathbf{T}^*(M))$. Since M is a Kähler manifold, $d\omega^i = 0$ (cf. Weil [120]). Let $\{\gamma_1, \ldots, \gamma_p\}$ be a base of the free part of the first homology group $H_1(M, \mathbf{Z})$ of M. Then the following are known:

(6.4.2) (i) $p = 2q$;

(ii) the vectors $v_j = \left(\int_{\gamma_j} \omega^1, \ldots, \int_{\gamma_j} \omega^q \right) \in \mathbf{C}^q$ $(1 \le j \le 2q)$ are linearly independent over $\mathbf{R}$. The torus $A(M) = \mathbf{C}^q / \sum_{j=1}^{2q} \mathbf{Z} \cdot v_j$ is an Abelian variety.

Fix a point $x_0 \in M$ and consider a holomorphic mapping

$$\alpha: x \in M \to \left(\int_{x_0}^{x} \omega^1, \ldots, \int_{x_0}^{x} \omega^q \right) \left(\mathrm{mod} \sum_{j=1}^{2q} \mathbf{Z} \cdot v_j \right) \in A(M).$$

The Abelian variety $A(M)$ is called the **Albanese variety** of M and α called the **Albanese mapping**. By Theorem (6.2.8) α is a regular rational mapping. We infer the following from the construction:

(6.4.3) (i) Let $h: M \to T$ be a holomorphic mapping from M into another complex torus T. Then there are a holomorphic group homomorphism $\beta: A(M) \to T$ and $a \in T$ such that $h(x) = \beta(\alpha(x)) + a$.

(ii) $\alpha(M)$ generates $A(M)$; i.e., there is some $l \in \mathbf{N}$ such that the holomorphic mapping

$$(x_1, \ldots, x_l) \in \prod^{l} M \to \sum_{i=1}^{l} \alpha(x_1) \in A(M)$$

is surjective.

Now we put $Y = \alpha(M)$. Since $q > \dim M \geq \dim Y$, $Y \neq A(M)$. Let G^0 be the identity component of the group $\{a \in A(M);\, Y + a = Y\}$. Take the quotient $T_1 = A(M)/G^0$ and let $\lambda: A(M) \to T_1$ be the natural mapping. Then T_1 is also an Abelian variety. We infer from (6.4.3), (ii) that $Y_1 = \lambda(Y) \neq T_1$ is an irreducible algebraic subset of positive dimension. Put $f_1 = \lambda \circ \alpha \circ f$. It suffices to prove that f_1 is algebraically degenerate. We assume the contrary. Then

(6.4.4) (i) The Zariski closure of $f_1(\mathbf{C})$ coincides with Y_1.

(ii) $\{a \in T_1;\, Y_1 + a = Y_1\}$ is a finite group.

Put $X = R(Y_1)$ and take an open disk $\Delta \subset \mathbf{C}$ so that $f(\Delta) \subset X$. We may assume that $\Delta = \Delta(1)$. We use the holomorphic mapping defined by (6.3.2):

$$I_k: J_k(X) \to (\mathbf{C}^{q_1})^k \quad (q = \dim T_1).$$

We see by (6.4.4) and Lemma (6.3.10) that the rank of $I_{k*J_k(f)(0)}$ is maximal for a sufficiently large k. Let Z be the Zariski closure of $J_k(f)(\Delta(1))$ in $J_k(X)$. Then we have the following commutative diagram:

$$\begin{array}{ccccc} Z & \subset & J_k(X) & \xrightarrow{I_k} & (\mathbf{C}^{q_1})^k \\ J_k(f)\uparrow & & \downarrow p & & \\ \Delta(1) & \underset{f_1}{\rightarrow} & X & & \end{array}$$

Here $p: J_k(X) \to X$ is the bundle projection. Take an imbedding $T_1 \subset \mathbf{P}^N(\mathbf{C})$ and consider f_1 to be a holomorphic mapping into $\mathbf{P}^N(\mathbf{C})$ of which image is contained in T_1. Let Ω be the Fubini-Study Kähler form on $\mathbf{P}^N(\mathbf{C})$ and put $T(r) = T_{f_1}(r, \Omega)$. Take a homogeneous coordinate system $[u^0; \cdots; u^N]$ of $\mathbf{P}^N(\mathbf{C})$ so that $Y_1 \not\subset \{u^0 = 0\}$, and put

$$w^i = \frac{u^i}{u^0}, \quad 1 \le i \le N.$$

It follows from Theorem (5.2.29) that

$$T(r, f_1^* w^i) \le T(r) + O(1) \le \sum_{i=1}^{N} T(r, f_1^* w^i) + O(1).$$

Now put

$$U(r) = \max_{1 \le i \le N} T(r, f_1^* w^i).$$

Then

$$\text{(6.4.5)} \qquad U(r) \le T(r) + O(1) \le NU(r) + O(1).$$

Let V be the Zariski closure of $I_k \circ J_k(f)(\Delta)$ in $(\mathbf{C}^{q_1})^k$. Since I_{k*} has maximal rank at $J_k(f)(0)$, $\dim Z = \dim V$ and $I_k | Z: Z \to V$ is dominant. Consider w^i to be rational functions on Z through $p | Z: Z \to X$. By Theorem (6.2.9) there are algebraic relations

$$\text{(6.4.6)} \qquad (A_{i0} \circ I_k)(w^i)^{d_i} + (A_{i1} \circ I_k)(w^i)^{d_i - 1} + \cdots + (A_{id_i} \circ I_k) = 0,$$

where A_{ij} are polynomials on $(\mathbf{C}^{q_1})^k$ and $A_{i0} \circ I_k \not\equiv 0$ on Z. Let $\{\eta^1, \ldots, \eta^{q_1}\}$ be a base of the space of holomorphic 1-forms on T_1 and put

$$f_1^* \eta^\nu = \zeta^\nu(z) dz, \quad 1 \le \nu \le q_1.$$

Then $A_{ij} \circ I_k \circ J_k(f)$ are polynomials in $\zeta^\nu, \zeta^{\nu(1)}, \ldots, \zeta^{\nu(k-1)}$, which are denoted by $A_{ij}(\zeta^{\nu(\mu)})$. It follows from (6.4.4), (i) and (6.4.6) that

$$A_{i0}(\zeta^{\nu(\mu)}) \not\equiv 0, \quad 1 \le i \le N,$$

$$A_{i0}(\zeta^{\nu(\mu)})(f_1^* w^i)^{d_i} + \cdots + A_{id_i}(\zeta^{\nu(\mu)}) = 0.$$

Then Lemma (6.2.5) yields

$$\text{(6.4.7)} \qquad U(r) \le O\left(\sum_{\nu, \mu} T(r, \zeta^{\nu(\mu)}) \right) + O(1).$$

We infer from Lemma (6.1.29) and Corollary (6.1.19) that

$$\text{(6.4.8)} \qquad T(r, \zeta^{\nu(\mu)}) = O(\log^+(rT(r))) \|_E .$$

It follows from (6.4.5), (6.4.7) and (6.4.8) that

$$\text{(6.4.9)} \qquad T(r) \le O(\log r + \log^+ T(r)) \|_E .$$

Since $f_1: \mathbf{C} \to Y_1 \subset T_1$ is non-constant, we obtain from Lemma (5.2.33)

(6.4.10) $$T(r) \geq Cr^2 \quad \text{for } r \geq r_0,$$

where $C > 0$ and $r_0 \geq 1$. Hence (6.4.9) and (6.4.10) imply a contradiction, $0 < C \leq 0$. *Q.E.D.*

(6.4.11) Corollary. *Let $f: \mathbf{C}^n \to A$ be a holomorphic mapping into an Abelian variety A. Then the Zariski closure of $f(\mathbf{C}^n)$ is a translation of an Abelian subvariety of A.*

Proof. We take real numbers $\theta_1, \ldots, \theta_n$ which are linearly independent over $\mathbf{Q}$, and put

$$\phi(z) = (e^{2\pi i\theta_1 z}, \ldots, e^{2\pi i\theta_n z}), \quad z \in \mathbf{C}.$$

Then $\{\phi(z);\, z \in \mathbf{R}\}$ is dense in $\{(z^i) \in \mathbf{C}^n;\, |z^i| = 1\}$. Hence there is no holomorphic function $F \not\equiv 0$ on $\mathbf{C}^n$ such that $F \circ \phi(z) \equiv 0$. Put

$$g = f \circ \phi : z \in \mathbf{C} \to f(e^{2\pi i\theta_1 z}, \ldots, e^{2\pi i\theta_n z}) \in A.$$

Then the Zariski closure Y of the image $f(\mathbf{C}^n)$ coincides with that of g. Suppose that Y is not an Abelian variety. We may assume that $0 \in Y$. Let G be a subgroup of A generated by all elements of Y. By Theorem (6.2.15) the Zariski closure A' of G is an Abelian subvariety of A and $Y \underset{\neq}{\subset} A'$. Then, in the same way as in the proof of Theorem (6.4.1) we see that the Zariski closure of $g(\mathbf{C})$ is strictly smaller then Y. This is a contradiction. *Q.E.D.*

(6.4.12) Corollary. *Let S be a compact Riemann surface of genus $g \geq 2$. Then any holomorphic mapping form $\mathbf{C}$ into S is constant.*

To the above assertion, Corollary (6.4.12) we have given four kinds of proofs in Chapter I, Chapter II, Chapter V and the present Chapter VI.

(6.4.13) *Example* (Ueno). Let E_j, $1 \leq j \leq 4$ be four non-singular elliptic curves of degree 3 in $\mathbf{P}^2(\mathbf{C})$ and put

$$A = E_1 \times \cdots \times E_4 \subset (\mathbf{P}^2(\mathbf{C}))^4.$$

Then A is an Abelian variety. Let $[u_j^0; \cdots; u_j^2]$ be homogeneous coordinate systems of $\mathbf{P}^2(\mathbf{C}) \supset E_j$. Let $H_j(u_1^0, u_1^1, u_1^2, \ldots, u_3^2)$, $0 \leq j \leq 2$, be polynomials which are homogeneous of degree l_j in each $(u_j^0, \ldots, u_j^2)$ and put

$$D = \left\{ (\ldots, [u_j^0; u_j^1; u_j^2], \ldots) \in (\mathbf{P}^2(\mathbf{C}))^4;\ \sum_{j=0}^{2} H_j \cdot (u_4^j)^p = 0 \right\},$$

$$M = A \cap D.$$

For generic H_j, M is non-singular and $q(M) = 4$ (Lefshetz' theorem (cf. [67])). Put

$$\{H_j = 0;\, 0 \leq j \leq 2\} \cap (E_1 \times E_2 \times E_3) = \{p_1, \ldots, p_s\}.$$

Then $\bigcup_{i=1}^{s}\{p_i\}\times E_4\subset M$ and hence there is a non-constant holomorphic curve $f:\mathbf{C}\to M$.

Notes

R. Nevanlinna's lemma on logarithmic derivatives (Lemma (6.1.16)) was the most crucial step in the proof of his Second Main Theorem. Borel [12] gave an elementary proof of Picard's theorem by making use of a lemma on the Borel identity (Lemma (6.1.20)), which does not depend on the uniformization theorem. We here gave the proof due to R. Nevanlinna [74]. H. Cartan [17] studied the value distribution of holomorphic curves $f=[f^0;\cdots;f^m]$: $\mathbf{C}\to\mathbf{P}^m(\mathbf{C})$ with respect to hyperplanes in general position, where f^i, $0\le i\le m$ are linearly independent over $\mathbf{C}$. Using Nevanlinna's lemma on logarithmic derivatives, he established the Second Main Theorem and defect relations

$$\sum_{j=1}^{l}\delta_f(D_j)\le m+1$$

for hyperplanes D_j in general position. The problem of the value distribution of holomorphic curves in $\mathbf{P}^m(\mathbf{C})$ was afterward reformulated in terms of Plücker coordinates by H. Weyl-J. Weyl [122], and then Ahlfors [2] proved the Second Main Theorem and defect relations for f as well as the associated curves $f^{(k)}$ in the Grassmann spaces. It seems, however, that they had not known the work of H. Cartan. The method of H. Weyl-J. Weyl and Ahlfors have been succeeded by Chern [21], Cowen-Griffiths [23], etc. By making use of H. Cartan's method, Fujimoto [29] proved the Second Main Theorem and defect relations for the associated curves $f^{(k)}$.

Stoll [109, 111] took the first task to extend the above theory of holomorphic curves in $\mathbf{P}^m(\mathbf{C})$ to the case of meromorphic mappings f from a complex manifold N into $\mathbf{P}^m(\mathbf{C})$, dealing with a rather large class of complex manifolds for N. He proved the First, Second Main Theorems and defect relations for f. Vitter [119] generalized Nevanlinna's lemma on logarithmic derivatives to the case of several complex variables, and proved the Second Main Theorem and defect relations for meromorphic mappings from $\mathbf{C}^n$ into $\mathbf{P}^m(\mathbf{C})$ and for hyperplanes in general position by using Cartan's method.

Chern-Osserman [22] applied the theory of holomorphic curves to study the Gauss mappings of minimal surfaces in the Euclidean spaces. Fujimoto [30] finally succeeded in proving the so-called Gauss mapping conjecture which states that the Gauss mapping of a complete minimal surface in $\mathbf{R}^3$ can omit at most 4 points of the Riemann sphere $\mathbf{P}^1(\mathbf{C})$.

Theorem (6.4.1) was called Bloch's conjecture. Bloch [9] stated the theorem with an incomplete, sketchy proof. Ochiai [88] filled the gaps substantially and proved it in several important cases. The final step for the general case is Lemma (6.3.10) which is due to M. Green. Kawamata [51] proved the same result by a different method. Green-Griffiths [41] gave another proof of Bloch's conjecture by using certain metrics. Bloch [9] also states without proof that

any holomorphic curve from $\mathbf{C}$ *into an analytic hypersurface of degree* 5 *of* $\mathbf{P}^3(\mathbf{C})$ *is algebraically degenerate.*

This is still an open problem.

Theorem (6.4.1) seems to have more importance in the theory of holomorphic curves in a general algebraic variety. Let M be a non-singular complex projective algebraic variety and $D \subset M$ an analytic hypersurface. Instead of holomorphic 1-forms on M, we consider meromorphic 1-forms with logarithmic singularities on D; the space of all such forms is denoted here by $\Gamma(M, \mathbf{T}^*(M, \log D))$. Let $f: \mathbf{C} \to M$ be an algebraically non-degenerate holomorphic curve. Then, if $\Gamma(M, \mathbf{T}^*(M, \log D))$ is ample in a certain sense, Noguchi [77, 80] proved the following inequality of the Second Main Theorem type:

$$KT_f(r) \leq N(r, f^*D) + O(\log^+ T_f(r) + \delta \log r)\|_{E(\delta)},$$

where K is a positive constant independent of f. As a corollary, we see that

if $\dim \Gamma(M, T^*(M, \log D)) > \dim M$, *then any holomorphic curve from* $\mathbf{C}$ *into* $M - D$ *is algebraically degenerate.*

He applied the method to extend the big Picard theorem in higher dimensional case and to prove similar inequalities for more general cases (cf. Noguchi [82, 84]). In the course of the proof, he essentially uses the notion of logarithmic jet spaces (cf. [85]). It is noted that the above inequality implies Lemma (6.1.20) on the Borel identity. If K is determined in an explicit way, then we might obtain defect relations. In this regard, there is a conjecture due to Griffiths. Let $D = \sum_{i=1}^{l} D_i$ with $c_1([D_i]) = \gamma_0 \in H^2(M, \mathbf{R})$. Then

$$\sum_{i=1}^{l} \delta_f(D_i) \leq \left[\frac{c_1(\mathbf{K}(M)^{-1})}{\gamma_0} \right].$$

Lately, Siu [107] proved the conjecture, assuming the existence of special meromorphic connections.

Let M be an Abelian variety. The following is called the Lang conjecture:

any holomorphic curve $f: \mathbf{C} \to A - D$ *is algebraically degenerate.*

In the case where f is a 1-parameter subgroup, Ax [4] proved this. By taking the quotient by the identity component of the subgroup $\{a \in A; a + D = D\}$, we easily reduce the above Lang conjecture to the case where $c_1([D]) > 0$. Then it is equivalent to

If $c_1([D]) > 0$, *then any holomorphic curve* $f : \mathbf{C} \to A - D$ *is constant.*

In view of the Nevanlinna theory we conjecture the following:

Let $c_1([D]) > 0$ *and* $f : \mathbf{C} \to A$ *an algebraically non-degenerate holomorphic curve. Then*

$$T_f(r, c_1([D])) \leq N(r, f^*D) + O(\log^+ T_f(r, c_1([D])) + \delta \log r) \|_{E(\delta)}$$

(cf. Noguchi [79]). Remark that this implies the Lang conjecture.

APPENDIX I

Canonical Bundles of Complex Submanifolds of $P^m(C)$

1. Holomorphic Vector bundles

We need the general notion of holomorphic vector bundles. We call a triple $(\mathbf{E}, \pi, M)$ a **holomorphic vector bundle of rank** k if the following conditions are satisfied:

(1.1) (i) $\mathbf{E}$ is an $(m+k)$-dimensional complex manifold.

(ii) $\pi: \mathbf{E} \to M$ is a surjective holomorphic mapping.

(iii) For every $x \in M$, the fiber $\mathbf{E}_x = \pi^{-1}(x)$ is a complex vector space of complex dimension k.

(iv) For every $x \in M$, there exist an open neighborhood U of x and a biholomorphic mapping $\Phi: \mathbf{E}|U = \pi^{-1}(U) \to U \times \mathbf{C}^k$ such that $p \circ \Phi = \pi$ and $q \circ \Phi|_{\pi^{-1}(y)}: \pi^{-1}(y) \to \mathbf{C}^k$ are linear isomorphisms for all $y \in U$, where $p: U \times \mathbf{C} \to U$ and $q: U \times \mathbf{C} \to \mathbf{C}^k$ denote the natural projections.

The rank of $\mathbf{E}$ is denoted by rank $\mathbf{E}$. The mapping $\Phi: \mathbf{E}|U \to U \times \mathbf{C}^k$ is called a **local trivialization** of $\mathbf{E}$ over U, we consider $\mathbf{C}^k$ the column vector space of dimension k. Then we have k holomorphic sections on U

$$s_i(x) = \Phi^{-1}(x, {}^t(0, \ldots, \overset{i\text{-}th}{1}, \ldots, 0)), \quad 1 \le i \le k,$$

such that $\{s_i(x)\}_{i=1}^k$ is a base of $\mathbf{E}_x$ at every $x \in U$. Such $s = (s_1, \ldots, s_k)$ is called a **holomorphic local frame** of $\mathbf{E}$ over U. Conversely, if there is a holomorphic local frame over an open subset $U \subset M$, there is a local trivialization of $\mathbf{E}$ over U. Let $\{U_\lambda\}$ be an open covering of M such that there are local trivializations $\Phi_\lambda: \mathbf{E}|U \to U \times \mathbf{C}^k$. The pair $(\{U_\lambda\}, \{\Phi_\lambda\})$ is called a **local trivialization covering** of $\mathbf{E}$. Let $(x_\lambda, \xi_\lambda) \in U_\lambda \times \mathbf{C}^k$ and Let $(x_\mu, \xi_\mu) \in U_\mu \times \mathbf{C}^k$. Then they correspond to the same point of $\mathbf{E}$ if and only if

(1.2)
$$x_\lambda = x_\mu = x \in U_\lambda \cap U_\mu,$$
$$\xi_\lambda = T_{\lambda\mu}(x)\xi_\mu,$$

where $T_{\lambda\mu}: U_\lambda \cap U_\mu \to GL(k, \mathbf{C})$ are holomorphic. The family $\{T_{\lambda\mu}\}$ is called the **system of holomorphic transitions** subordinated to the local trivialization covering. A complex submanifold $\mathbf{F} \subset \mathbf{E}$ is called a **holomorphic vector subbundle** if $\mathbf{F}$ is itself a holomorphic vector bundle of rank h $(0 \le h \le k)$ over M of which fiber structure is compatible with that of $\mathbf{E}$. Then we naturally have the quotient bundle $\mathbf{E}/\mathbf{F}$ which is a holomorphic vector bundle of rank $k - h$. We write the above facts as follows:

$$0 \to \mathbf{F} \to \mathbf{E} \to \mathbf{E}/\mathbf{F} \to 0.$$

For two holomorphic vector bundles $\mathbf{E}_1$ and $\mathbf{E}_2$ over M, the tensor product $\mathbf{E}_1 \otimes \mathbf{E}_2$ and the direct sum $\mathbf{E}_1 \oplus \mathbf{E}_2$ are naturally defined as holomorphic vector bundles over M. Furthermore, the exterior power bundle $\overset{l}{\wedge}\mathbf{E}$ $(1 \le l \le k)$ is defined. In special, $\overset{k}{\wedge}\mathbf{E}$ is called the **determinant bundle** of $\mathbf{E}$ and denoted by $\det \mathbf{E}$. In terms of (1.2), $\det \mathbf{E}$ is a holomorphic line bundle such that it is trivial over U_λ and the system of holomorphic transition functions is given by $\{\det T_{\lambda\mu}\}$. For instance, we have

(1.3)
$$\det \mathbf{T}(M) = \mathbf{K}(M)^{-1}, \quad \det \mathbf{T}^*(M) = \mathbf{K}(M).$$

(1.4) Lemma. (i) *Let* $\mathbf{E}_i$ *(*$i = 1, 2$*) be a holomorphic vector bundle of rank* k_i *over* M. *Then*

$$\det(\mathbf{E}_1 \otimes \mathbf{E}_2) = (\det \mathbf{E}_1)^{k_2} \otimes (\det \mathbf{E}_2)^{k_1}.$$

(ii) *Let* $\mathbf{E}$ *be a holomorphic vector bundle over* M *and* $\mathbf{F}$ *a holomorphic subbundle of* $\mathbf{E}$. *Then*

$$\det \mathbf{E} = (\det \mathbf{F}) \otimes (\det \mathbf{E}/\mathbf{F}).$$

Proof. (i) This immediately follows from the tensor algebra.

(ii) Let Let $x \in M$ be an arbitrary point. Then there is a holomorphic local frame $s = (s_1, \ldots, s_h)$ $(h = \operatorname{rank} \mathbf{F})$ on a neighborhood U of x. Taking U smaller if necessary, we may extend s to a holomorphic local frame $\tilde{s} = (s_1, \ldots, s_h, \ldots, s_k)$ $(k = \operatorname{rank} \mathbf{E})$ on U. Then $\hat{s} = (s_{h+1}, \ldots, s_k)$ induces a holomorphic local frame of $\mathbf{E}/\mathbf{F}$ over U. Taking a local trivialization covering as above, we easily see that the transition $T_{\lambda\mu}$ are of the type

$$\begin{bmatrix} A_{\lambda\mu} & B_{\lambda\mu} \\ O & C_{\lambda\mu} \end{bmatrix}$$

such that $\{A_{\lambda\mu}\}$ (resp. $\{C_{\lambda\mu}\}$) is a system of holomorphic transitions for $\mathbf{F}$ (resp. $\mathbf{E}/\mathbf{F}$). Therefore

$$\det(T_{\lambda\mu}) = \det(A_{\lambda\mu})\cdot\det(C_{\lambda\mu}),$$

so that $\det \mathbf{E} = (\det \mathbf{F}) \otimes (\det \mathbf{E}/\mathbf{F})$. *Q.E.D.*

2. Non-Singular Complete Intersections of $\mathbf{P}^m(\mathbf{C})$ and the Canonical Bundles

In general, let D be an analytic hypersurface of a complex manifold M. Let $F = 0$ be a defining equation of D in a neighborhood of a point $x \in D$ (cf. Chapter IV, §2). It follows from Theorem (4.2.3) and the implicit function theorem that

$$D \cap U \text{ is non-singular} \Longleftrightarrow dF \neq O \text{ on } D \cap U.$$

Let $[z^0; \cdots; z^m]$ be a homogeneous coordinate system of $\mathbf{P}^m(\mathbf{C})$. Then we set $U_i = \{z^i \neq 0\}$ and $z_i^j = z^j/z^i$, $j \neq i$. Then (z_i^j) are the affine coordinate of U_i. Let P be a homogeneous polynomial in $(z^0, \ldots, z^m)$ of degree d without multiple factor and D an analytic hypersurface defined by

$$D = \{[z^0; \cdots; z^m]; P(z^0, \ldots, z^m) = 0\}.$$

Put

$$P_i(z_i^j) = P(z_i^0, \ldots, \overset{i\text{-}th}{1}, \ldots, z^m) \text{ on } U_i.$$

Then for $i \neq k$

(2.1) $$P_i(x) = (z_i^k)^d P_k(x).$$

Assume that D is non-singular. Then it follows that

(2.2) $$dP_i \neq O \quad \text{on } D \cap U_i.$$

Let $\iota_D \colon D \to \mathbf{P}^m(\mathbf{C})$ be the natural inclusion mapping. Then we obtain from (2.1)

(2.3) $$(\iota_D | U_i)^* dP_i = (z_i^k)^d (\iota_D | U_k)^* dP_k \text{ on } U_i \cap U_k \cap D.$$

Let $\mathbf{H}_0$ be the hyperplane bundle over $\mathbf{P}^m(\mathbf{C})$. Then $\{(\iota_D | U_i)^* dP_i\}$ defines a global holomorphic section σ_D of the holomorphic vector bundle $\mathbf{H}_0^d \otimes \mathbf{T}^*(\mathbf{P}^m(\mathbf{C}))|D$, which does not vanish anywhere in D. We identify the trivial line bundle $\mathbf{1}_D$ with a holomorphic vector subbundle of $\mathbf{H}_0^d \otimes \mathbf{T}^*(\mathbf{P}^m(\mathbf{C}))|D$ through

$$(x, a) \in D \times \mathbf{C} = \mathbf{1}_D \to a\sigma_D(x) \in \mathbf{H}_0^d \otimes \mathbf{T}^*(\mathbf{P}^m(\mathbf{C}))|D.$$

Then the quotient bundle $\mathbf{H}_0^d \otimes \mathbf{T}^*(\mathbf{P}^m(\mathbf{C}))|D/\mathbf{1}_D$ is isomorphic to $(\mathbf{H}_0|D)^d \otimes \mathbf{T}^*(D)$. Therefore we have the following.

(2.4) Lemma. *Let the notation be as above. Then*

$$0 \to \mathbf{1}_D \to \mathbf{H}_0^d \otimes \mathbf{T}^*(\mathbf{P}^m(\mathbf{C}))|D \to (\mathbf{H}_0|D)^d \mathbf{T}^*(D) \to 0.$$

(2.5) Lemma. *Let D be a non-singular hypersurface of degree d of $\mathbf{P}^m(\mathbf{C})$. Then*

$$\det \mathbf{T}^*(D) = (\mathbf{H}_0|D)^{d-m-1}.$$

Proof. This follows from Lemma (2.4), Lemma (1.4), (1.3) and Example (2.1.26). *Q.E.D.*

Let D_ν, $\nu = 1, \ldots, l$ be non-singular analytic hypersurfaces of degree d_ν of $\mathbf{P}^m(\mathbf{C})$. Let P_ν be homogeneous polynomials without multiple factor such that $D_\nu = \{P_\nu = 0\}$. Assume that $\sum D_\nu$ has only simple normal crossings. Put $M = \bigcap_{\nu=1}^{l} D_i$. Then M is a complex submanifold of dimension $m - l$ and called a **non-singular complete intersection of degree** $d = \sum d_\nu$.

(2.6) Theorem. *Let $M \subset \mathbf{P}^m(\mathbf{C})$ be a non-singular complete intersection of degree d. Then*

$$\mathbf{K}(M) = (\mathbf{H}_0|M)^{d-m-1}.$$

Proof. Let D_ν, $1 \le \nu \le l$ be non-singular analytic hypersurfaces of degree d_ν such that $M = \bigcap_{\nu=1}^{l} D_i$ as above. By Lemma (2.5)

$$\det \mathbf{T}^*(D_1) = (\mathbf{H}_0|D_1)^{d_1-m-1}. \tag{2.7}$$

Applying the same arguments as above to the non-singular analytic hypersurface $D_1 \cap D_2$ of D_1 and making use of (2.7), we have

$$\det \mathbf{T}^*(D_1 \cap D_2) = (\mathbf{H}_0|D_1 \cap D_2)^{d_1+d_2-m-1}.$$

Inductively, we get

$$\det \mathbf{T}^*(D_1 \cap \cdots \cap D_l) = (\mathbf{H}_0|D_1 \cap \cdots \cap D_l)^{d_1+\cdots+d_l-m-1},$$

so that $\det \mathbf{T}^*(M) = (\mathbf{H}_0|M)^{d-m-1}$. *Q.E.D.*

APPENDIX II

Weierstrass-Stoll Canonical Functions

1. Review of Potential Theory on $\mathbf{R}^m$

We recall several fundamental facts from the real potential theory on the euclidean space $\mathbf{R}^m$ ($m \geq 2$). Cf. [32], Chapter 2 for this section. Let $x = (x^1, ..., x^m)$ be the standard coordinate system of $\mathbf{R}^m$ and set

$$\|x\| = \left\{ \sum_{i=1}^{m} |x^i|^2 \right\}^{1/2}, \quad B(r) = \{ \|x\| < r \}.$$

The **Laplacian** Δ is defined by

$$\Delta = \frac{\partial^2}{\partial(x^1)^2} + \cdots + \frac{\partial^2}{\partial(x^m)^2}.$$

Let U be a domain of $\mathbf{R}^m$. Then the Laplacian Δ operates on the distribution space $D(U)'$. We use the same notation and terminologies as in Chapter III, §1. A locally integrable function $u: U \to [-\infty, \infty)$ is called a **subharmonic function** if the following are satisfied:

(1.1) (i) u is upper semicontinuous.

(ii) $\Delta[u]$ is a positive Radon measure.

Remark. For the definition of subharmonic functions on U, it is equivalent to take the same definition as Definition (3.3.1), where (iii) has to be replaced with spherical mean integrals.

Subharmonic functions satisfy properties similar to those of subharmonic functions on $\mathbf{C} = \mathbf{R}^2$. Let $u: U \to [-\infty, \infty)$ be a subharmonic function on U. Then

$$u(a) \leq \frac{1}{r^m |B(1)|} \int_{a+B(r)} u\, dx^1 \cdots dx^m,$$

provided that $a + B(r) \subset U$, where $|B(1)| = \int_{B(1)} dx^1 \cdots dx^m$. The smoothing u_ε defined as in Chapter III, §1 is a C^∞ subharmonic function on U_ε and satisfies

$$(1.2) \qquad u_\varepsilon \downarrow u \quad \text{as } \varepsilon \downarrow 0.$$

Moreover, if $-u$ is also subharmonic, then u is called a **harmonic** function on U. A harmonic function u on U are C^∞ by (1.2), and $\Delta u \equiv 0$. Let $S(r) \subset \mathbf{R}^m$ denote the sphere of radius $r > 0$ with center origin and dS_r the rotation invariant measure on $S(r)$ induced from the Euclidean metric, normalized as

$$\int_{S(r)} dS_r = 1.$$

Let v be an integrable function on $S(r)$ with respect to dS_r. Then the integral

$$u(x) = \int_{S(r)} v(y) \frac{r^2 - \|x\|^2}{\|y - x\|^m} r^{m-2} dS_r(y), \quad \|x\| < r$$

is called the **Poisson integral**. The function u is harmonic in the ball $B(r) = \{\|x\| < r\}$ with boundary value v. For a harmonic function u in a neighborhood of $\overline{B(r)}$ we have

$$(1.3) \qquad u(x) = \int_{S(r)} u(y) \frac{r^2 - \|x\|^2}{\|y - x\|^m} r^{m-2} dS_r(y), \quad \|x\| < r$$

(1.4) Lemma. *Let u be a harmonic function on $\mathbf{R}^m$. Assume that there are a positive increasing sequence $r_\nu \uparrow \infty$ $(\nu \uparrow \infty)$ and constants $C > 0$, $d \geq 0$ such that*

$$\sup\{u(x);\, x \in B(r_\nu)\} \leq C r_\nu^d, \quad \nu = 1, 2, \ldots$$

Then u is a polynomial of degree $\leq d$ in $x^1, \ldots, x^m$.

Proof. Taking the complexification $(z^j) = (x^j + iy^j)$ of (x^j), we put

$$(1.5) \qquad \hat{u}_r(z) = \int_{S(r)} u(w) \frac{r^2 - \sum_{j=1}^m (z^j)^2}{\left(\sum_{j=1}^m (w^j - z^j)^2\right)^{\frac{m}{2}}} r^{m-2} dS_r(w)$$

for $\|z\| < r/2$. Then u_r is holomorphic in $\{\|z\| < r/2\}$ and $\hat{u}_r | \{y^1 = \cdots = y^m = 0\} = u$. Hence $\hat{u}_r = \hat{u}_{r'}$ for $r < r'$, so that they define a holomorphic function $\hat{u}$ on $\mathbf{C}^m$ such that $\hat{u} | \{y^1 = \cdots = u^m = 0\} = u$. In (1.5) we put $r = r_\nu$ and $\|z\| < r_\nu/4$: Then there is a constant $C_1 > 0$ such that

$$|\hat{u}(z)| \leq C_1 \left(\frac{r_\nu}{4}\right)^d, \quad \nu = 1, 2 \ldots$$

Since $\hat{u}$ is a holomorphic function on $\mathbf{C}^m$, $\hat{u}$ is a polynomial of degree $\leq d$ in $z^1, \ldots, z^m$, and hence so is u in $x^1, \ldots, x^m$. *Q.E.D.*

We define the potential kernel function $P(a, x)$ on $\mathbf{R}^m$ by

$$(1.6) \qquad P(a, x) = \frac{1}{\|a - x\|^{m-2}} \quad \text{for } n \geq 3,$$

$$P(a, x) = -\log \|a - x\| \quad \text{for } n = 2.$$

In the case of $n \geq 3$ (resp. $n = 2$) $P(a, x)$ is called the **Newton kernel function** (resp. **logarithmic kernel function**). The potential kernel function $P(a, x)$ with a fixed is harmonic in $\mathbf{R}^m - \{a\}$ and $-P(a, x)$ is subharmonic in $\mathbf{R}^m$. It is a classical fact that

$$\Delta[P(a, x)] = -\frac{1}{(m-2)|S(1)|}\delta_a,$$

where $|S(1)|$ denotes the area of the unit sphere $S(1) \subset \mathbf{R}^m$ with respect to the Euclidean metric and δ_a the Dirac measure at a. This leads to the following:

(1.7) Proposition. *Let σ be a positive Radon measure on $B(R)$ and $0 < r < R$ and set*

$$(1.8) \qquad U_r(x, \sigma) = -\frac{1}{(m-2)|S(1)|}\int_{y \in B(r)} P(y, x) d\sigma(y).$$

Then $U_r(x, \sigma)$ is subharmonic in $\mathbf{R}^m$ and satisfies

$$\Delta[U_r(\cdot, \sigma)] = \sigma \quad \textit{in } B(r).$$

Here we call the above integral $U(x, \sigma)$ the **potential** of the Radon measure σ.

2. Local Potentials of Positive Closed (1, 1)-Currents and Modified Kernel Function

Let $z^j = x^{2j-1} + ix^{2j}$, $1 \leq j \leq m$, be the complex coordinates of $\mathbf{C}^m$ with real variables x^k, $1 \leq k \leq 2m$. By this we identify $\mathbf{C}^m = \mathbf{R}^{2m}$. Note that

$$\Delta = \sum_{j=1}^{m}\left(\frac{\partial^2}{(\partial x^{2j-1})^2} + \frac{\partial^2}{(\partial x^{2j})^2}\right) = 4\sum_{j=1}^{m}\left(\frac{\partial^2}{\partial z^j \partial \bar{z}^j}\right).$$

We use the same notation as in Chapter III, §3. For a locally integrable function u on a domain of $\mathbf{C}^m$ we have

$$(2.1) \qquad dd^c[u] \wedge \alpha^{m-1} = \frac{1}{4m}\Delta[u] \cdot \alpha^m.$$

Hence, if u is plurisubharmonic, then u is subharmonic. Let

$$T = \sum_{j,k} \frac{i}{2\pi} T_{j\bar{k}} dz^j \wedge d\bar{z}^k$$

be a positive current of type (1, 1) on $B(R)$. Define the **potential** of T by

$$U_r(z, T) = -\frac{1}{m-1}\int_{y\in B(r)} P(y, x) T(y) \wedge \alpha^{m-1}(y), \quad 0 < r < R. \tag{2.2}$$

Then $U_r(z, T)$ is subharmonic in $\mathbf{C}^m$ and satisfies

$$\Delta[U_r(z, T)] = 4\sum_{j=1}^{m} T_{j\bar{j}} \text{ on } B(r). \tag{2.3}$$

(2.4) Lemma. *Let T be a closed positive current of type (1, 1) on $B(R)$ and $0 < r < R$. Then there is a subharmonic function u on $B(r)$ such that*

$$dd^c[u] = T \quad \text{on } B(r).$$

Proof. Suppose first that T is C^∞ but not necessarily positive. Let $r < r' < R$. By Poincaré's lemma (Lemma (3.2.30)) there is a real 1-form v on $B(r')$ such that $dv = T$ on $B(r')$. Decompose $v = v' + v''$, where v' (resp. v'') is a 1-form of type (1, 0) (resp. (0, 1)). Since v is real and T is of type (1, 1), $v' = \overline{v''}$ and $\partial v' = \bar{\partial} v'' = 0$. By Dolbeault's lemma (cf. Hörmander [48], Chapter II, §3), there is a C^∞-function u'' on $B(r)$ with $\bar{\partial} u'' = v''$. Putting $u = 2\pi i(u'' - \overline{u''})$, we have

$$dd^c u = T.$$

In the general case, we take $U_{r'}(z, T)$ defined by (2.2) and put

$$S = T - dd^c U_{r'}(\cdot, T).$$

Put $S = \sum_{i,j} \frac{i}{2\pi} S_{i\bar{j}} dz^i \wedge d\bar{z}^j$. Then (2.1) and (2.3) imply that

$$\sum_{j=1}^{m} S_{j\bar{j}} = 0 \text{ in } B(r'). \tag{2.5}$$

Moreover, we put

$$S_{i\bar{j}k\bar{l}} = \frac{\partial^2}{\partial z^k \partial \bar{z}^l} S_{i\bar{j}}.$$

The d-closedness of S implies that $S_{i\bar{j}k\bar{l}}$ is invariant under index exchanges of i and j, and of k and l. Using (2.5), we have

$$\Delta S_{i\bar{j}} = 4\sum_{k=1}^{m} S_{i\bar{j}k\bar{k}} = 4\left(\sum_{k=1}^{m} S_{k\bar{k}}\right)_{i\bar{j}} = 0 \text{ in } B(r').$$

Therefore $S_{i\bar{j}}$ are harmonic in the sense of distribution and hence C^∞. Then the result of the first half implies our assertion. *Q.E.D.*

We are going to define a modified kernel function derived from $P(a, x)$. The potential kernel function $P(a, x)$ for $m \geq 2$ is quite different to that for $m = 1$. Since our main concern is in the case of $m \geq 2$, we restrict ourselves to the case of $m \geq 2$. Now we expand $P(a, z)$:

$$P(a, z) = \frac{1}{\|a - z\|^{2m-2}}$$

$$= P_0(a, z) + P_1(a, z) + \cdots + P_\lambda(a, z) + \cdots,$$

where $P_0(a, z) = \dfrac{1}{\|a\|^{2m-2}}$ and $P_\lambda(a, z)$ are harmonic homogeneous polynomials of degree λ in z^j and $\bar{z}^j$, $j = 1, 2, \ldots, m$. For $q \in \mathbf{Z}^+$ we set

$$e(q; a, z) = -P(a, z) + P_0(a, z) + \cdots + P_q(a, z) \tag{2.6}$$

$$= -P_{q+1}(a, z) - P_{q+2}(a, z) - \cdots.$$

This $e(q; a, z)$ is the modified kernel function, which will be used to construct Weierstrass-Stoll canonical functions in the next section. Put $t = \|z\|/\|a\| < 1$ and let θ be the angle formed by the vectors a and z. Then

$$P(a, z) = \|a\|^{2-2m}(1 - 2t\cos\theta + t^2)^{1-m} \tag{2.7}$$

$$= \|a\|^{2-2m}\left(1 + B_1(\cos\theta)t + \cdots + B_\lambda(\cos\theta)t^\lambda + \cdots\right).$$

On the other hand, we have

$$P(a, z) \leq \|a\|^{2-2m}(1 - t)^{2-2m} = \|a\|^{2-2m}\sum_{\lambda=0}^{\infty} b_\lambda t^\lambda.$$

Here note that $P_\lambda(a, z) = \|a\|^{2-2m}B_\lambda(\cos\theta)t^\lambda$. Hence

$$|B_\lambda(\cos\theta)| \leq b_\lambda, \tag{2.8}$$

$$e(q; a, z) = -\|a\|^{2-2m}\sum_{\lambda=q+1}^{\infty} B_\lambda(\cos\theta)t^\lambda. \tag{2.9}$$

Now we consider general $t = \|z\|/\|a\|$. We fix $0 < \tau < 1$. Suppose first that $t \leq \tau$. Then it follows from (2.8) and (2.9) that

$$|e(q; a, z)| \leq \|a\|^{2-2m}t^{q+1}\sum_{\lambda=q+1}^{\infty} b_\lambda \tau^{\lambda-q-1} \tag{2.10}$$

$$= C_1(q, \tau)\|a\|^{2-2m}t^{q+1},$$

where $C_1(q, \tau) = \sum_{\lambda=q+1}^{\infty} b_\lambda \tau^{\lambda-q-1}$. We next suppose that $t \geq \tau$. It follows from (2.6)~(2.7) that

$$\begin{aligned} e(q;a,z) &\le \|a\|^{2-2m}(1+B_1(\cos\theta)t+\cdots+B_q(\cos\theta)t^q) \\ &\le \|a\|^{2-2m}(1+b_1t+\cdots+b_qt^q) \\ &\le \|a\|^{2-2m}t^q(t^{-q}+b_1t^{-q+1}+\cdots+b_q) \\ &\le \|a\|^{2-2m}t^q(\tau^{-q}+b_1\tau^{-q+1}+\cdots+b_q). \end{aligned}$$

Putting $C_2(q,\tau)=(\tau^{-q}+b_1\tau^{-q+1}+\cdots+b_q)$, we have

(2.11) $$e(q;a,z)\le C_2(q,\tau)\|a\|^{2-2m}t^q.$$

Thus (2.11) and (2.10) imply the following.

(2.12) Lemma. *Let the notation be as above. Then*

$$e(q;a,z)\le C(q,\tau)\|a\|^{2-2m}\frac{\|z\|^{q+1}}{\|a\|^q(\|z\|+\|a\|)}$$

for any $z, a\in\mathbf{C}^m$ with $a\ne O$, and

$$|e(q;a,z)|\le C(q,\tau)\|a\|^{2-2m}\frac{\|z\|^{q+1}}{\|a\|^q(\|z\|+\|a\|)}$$

for $\|z\|\le\tau\|a\|$, where $C(q,\tau)=\max\{(1+\tau)C_1(q,\tau),\ (1+1/\tau)C_2(q,\tau)\}$.

3. Weierstrass-Stoll Canonical Functions

Let $T=\sum_{j,k}\frac{i}{2\pi}dz^j\wedge d\bar{z}^k$ be a closed positive current of type (1, 1) on $\mathbf{C}^m$. Consider the equation

(3.1) $$dd^c[U]=T.$$

If U is a solution of (3.1), then

$$\Delta[U]=4\sum_{j=1}^{m}T_{j\bar{j}}.$$

The potential $U_r(z,T)$ defined by (2.2) is subharmonic in $\mathbf{C}^m$ and satisfies

$$\Delta[U_r(z,T)]=4\sum_{j=1}^{m}T_{j\bar{j}}\ \text{in}\ B(r).$$

Hence, if $U_r(z,T)$ converges as $r\to\infty$, then the limit is a candidate for a solution of (3.1). In general, it is not convergent. Therefore we will use the modified kernel function $e(q;a,z)$ defined by (2.6), of which difference to $-P(a,z)$ is only a harmonic polynomial of degree q.

We first investigate the condition for the convergence. Let $q\in\mathbf{Z}^+$. Then

$$\int_1^r t^{-q} dn(t, T) = \Big[t^{-q} n(t, T) \Big]_1^r + q \int_1^r \frac{n(t, T)}{t^{q+1}} dt.$$

This implies the following lemma.

(3.2) Lemma. *Let $r_\nu \uparrow \infty$, $\nu = 1, 2, \ldots$ be a positive increasing sequence. Then*

$$\lim_{\nu \to \infty} \int_1^{r_\nu} t^{-q} dn(t, T) < \infty \tag{3.3}$$

if and only if

$$\lim_{\nu \to \infty} \frac{n(r_\nu, T)}{r_\nu^q} < \infty \text{ and } \lim_{\nu \to \infty} \int_1^{r_\nu} \frac{n(t, T)}{t^{q+1}} dt < \infty.$$

We define the **order** ρ_T of T by

$$\rho_T = \overline{\lim_{r \to \infty}} \frac{\log N(r, T)}{\log r} = \overline{\lim_{r \to \infty}} \frac{\log n(r, T)}{\log r}. \tag{3.4}$$

Assume that there is a positive increasing sequence $r_\nu \uparrow \infty$ such that

$$\lim_{\nu \to \infty} \int_1^{r_\nu} t^{-q} dn(t, T) < \infty.$$

We may use any $r_n \uparrow \infty$ if $\overline{\lim}_{r \to \infty} \int_1^r t^{-q} dn(t, T) < \infty$. For instance, we may put $q = 0$ if $n(r, T) = O(1)$; if $\rho_T < \infty$, then we may put

$$q = [\rho_T] + 1, \tag{3.5}$$

where $[\cdot]$ stands for the Gauss' symbol. We define the **canonical potential** $U(q; z, T)$ of T due to Lelong [63] by

$$U(q; z, T) = \lim_{\nu \to \infty} \frac{1}{m-1} \int_{a \in B(r_\nu)} e(q; a, z) T \wedge \alpha^{m-1}. \tag{3.6}$$

We check the convergence. For simplicity, we assume that $O \notin \operatorname{supp} T$. Let $\|z\| = r < r_\nu$. It follows from Lemma (2.12) that

$$\begin{aligned} &\frac{1}{m-1} \int_{a \in B(r_\nu)} e(q; a, z) T \wedge \alpha^{m-1} \qquad (3.7) \\ &\leq \frac{1}{m-1} \int_{a \in B(r_\nu)} C(q, \tau) \|a\|^{2-2m} \frac{r^{q+1}}{\|a\|^q (r + \|a\|)} T \wedge \alpha^{m-1} \\ &= \frac{C(q, \tau)}{m-1} \int_0^{r_\nu} t^{2-2m-q} \frac{r^{q+1}}{r+t} d(T \wedge \alpha^{m-1}(B(t))) \end{aligned}$$

Continued

$$= \frac{C(q, \tau)}{m-1} \left\{ \left[\frac{r^{q+1}}{r+t} t^{-q} n(t, T) \right]_0^{r_\nu} \right.$$

$$\left. + r^{q+1} \int_0^{r_\nu} \left[(q+2m-2) \frac{t^{1-2m-q}}{r+t} + \frac{t^{2-2m-q}}{(r+t)^2} \right] T \wedge \alpha^{m-1}(B(t)) dt \right\}$$

$$= \frac{C(q, \tau)}{m-1} \left\{ \frac{r^{q+1}}{r+r_\nu} \frac{n(r_\nu, T)}{r_\nu^q} \right.$$

$$\left. + r^{q+1} \int_0^{r_\nu} \frac{(q+2m-1)t + (q+2m-2)r}{(r+t)^2 t^{q+1}} n(t, T) dt \right\}.$$

By Lemma (3.2)

$$\text{(3.8)} \qquad \frac{r^{q+1}}{r+r_\nu} \frac{n(r_\nu, T)}{r_\nu^q} \to 0 \quad (\nu \to \infty).$$

Moreover we have

$$\text{(3.9)} \qquad \frac{C(q, \tau)}{m-1} r^{q+1} \int_0^{r_\nu} \frac{(q+2m-1)t + (q+2m-2)r}{(r+t)^2 t^{q+1}} n(t, T) dt$$

$$\leq C'(q, \tau) r^{q+1} \int_0^{r_\nu} \frac{n(t, T)}{(r+t) t^{q+1}} dt,$$

where $C'(q, \tau) = C(q, \tau)(q+2m-1)/(m-1)$. Lemma (3.2) implies that the last integral in (3.9) converges as $\nu \to \infty$. Without loss of generality, we may assume that $T \wedge \alpha^{m-1}(S(r_\nu)) = 0$, $\nu = 1, 2, \ldots$ (cf. the proof of Theorem (3.2.31)). Making use of Lemma (2.12) in the same way as above, we infer that the sequence

$$\int_{a \in B(r_\nu)} |q(q; a, z)| T \wedge \alpha^{m-1}, \quad \nu = 1, 2, \ldots$$

is a Cauchy sequence which is uniform for z belonging to a fixed compact subset. Therefore the right hand of (3.6) converges uniformly on compact subsets and $U(q; z, T)$ is a subharmonic function on $\mathbf{C}^m$. Note that

$$\text{(3.10)} \qquad \int_0^{r_\nu} \frac{n(t, T)}{(r+t) t^{q+1}} dt = \int_0^{r} \frac{n(t, T)}{(r+t) t^{q+1}} dt + \int_r^{r_\nu} \frac{n(t, T)}{(r+t) t^{q+1}} dt$$

$$\leq \frac{1}{r} \int_0^{r} \frac{n(t, T)}{t^{q+1}} dt + \int_r^{r_\nu} \frac{n(t, T)}{t^{q+2}} dt.$$

Put

$$C(q) = \inf\{C'(q, \tau); 0 < \tau < 1\}.$$

Then (3.6)~(3.10) implies the following.

(3.11) Lemma. *The canonical potential $U(q; z, T)$ is a subharmonic function on $\mathbf{C}^m$ and satisfies*

$$\Delta[U(q; z, T)] = 4\sum_{j=1}^{m} T_{j\bar{j}},$$

$$U(q; z, T) \le C(q) r^q \left\{ \int_0^r \frac{n(t, T)}{t^{q+1}} dt + r \lim_{\nu\to\infty} \int_r^{r_\nu} \frac{n(t, T)}{t^{q+2}} dt \right\}. \tag{3.12}$$

Put

$$M(r) = \sup\{U(q, z, T); z \in B(r)\}.$$

(3.13) Corollary. $\frac{1}{r^{2m}} \int_{z\in B(r)} |U(q, z, T)| \alpha^m \le 2M(r).$

Proof. Put $U^{\pm}(q; z, T) = \max\{0, \pm U((q; z, T)\}$. Then

$$U(q; z, T) = U^+(q; z, T) - U^-(q; z, T).$$

Since $U(q; z, T)$ is subharmonic and $U(q; O, T) = 0$,

$$r^{-2m} \int_{B(r)} \{U^+(q; z, T) - U^-(q; z, T)\} \alpha^m \ge 0,$$

so that

$$r^{-2m} \int_{B(r)} U^-(q; z, T) \alpha^m \le r^{-2m} \int_{B(r)} U^+(q; z, T) \alpha^m \le M(r).$$

Since $|U(q; z, T)| = U^+(q; z, T) + U^-(q; z, T)$, we have the desired estimate. *Q.E.D.*

(3.14) Theorem. *Let T be a closed positive current of type* (1, 1) *on* $\mathbf{C}^m$ *such that* $O \notin \operatorname{supp} T$. *Assume that there is a sequence* $r_\nu \uparrow \infty$ $(\nu \uparrow \infty)$ *such that*

$$\lim_{\nu\to\infty} \int_1^{r_\nu} t^{-q} dn(t, T) < \infty.$$

Then the canonical potential $U(q; z, T)$ satisfies the estimate (3.12) *and a solution of* (3.1):

$$dd^c[U(q; z, T)] = T.$$

Proof. There remains to show that $dd^c[U(q;z,T)]=T$. We may assume that $B(r_1)\cap \operatorname{supp} T=\varnothing$. By Lemma (2.4) there are plurisubharmonic functions u_ν on $B(r_\nu)$ such that $u_1\equiv 0$ and $dd^c[u_\nu]=T|B(r_\nu)$. Put

$$H_\nu(z)=U(q,z,T)-u_\nu(z),\quad z\in B(r_\nu).$$

Then $\Delta[H_\nu]=0$, so that H_ν are C^∞ harmonic functions on $B(r_\nu)$. Put

$$H_{\nu i\bar{j}}=\frac{\partial^2 H_\nu}{\partial z^i\partial \bar{z}^j}.$$

Then

$$H_{(\nu+1)i\bar{j}}-H_{\nu i\bar{j}}=0\quad \text{on } B(r_\nu).$$

Therefore $H_{\nu i\bar{j}}$ define harmonic functions $H_{i\bar{j}}$ on $\mathbf{C}^m$. It follows from (2.6) that

(3.15) all partial derivatives of order $\le q-2$ of $H_{i\bar{j}}$ vanish at O.

Let χ and χ_ε be the convolution kernels defined in Chapter III, §1, (c). Then

$$\chi_\varepsilon(z)=\frac{1}{\varepsilon^{2m}}\chi\left(\frac{z}{\varepsilon}\right).$$

Put

$$C=\max\left\{\frac{\partial^2\chi}{\partial z^i\partial \bar{z}^j}(z);\ z\in\mathbf{C}^m,\ 1\le i,\ j\le m\right\}.$$

Then

$$\left|\frac{\partial^2\chi_\varepsilon}{\partial z^i\partial \bar{z}^j}(z)\right|\le C\varepsilon^{-2m-2}. \tag{3.16}$$

Since $H_{i\bar{j}}$ is harmonic, $H_{i\bar{j}\varepsilon}=H_{i\bar{j}}*\chi_\varepsilon=H_{i\bar{j}}$. By definition

$$H_{i\bar{j}}=\frac{\partial^2[U(q;z,T)]}{\partial z^i\partial \bar{z}^j}*\chi_\varepsilon-\frac{\partial^2[u_\nu]}{\partial z^i\partial \bar{z}^j}*\chi_\varepsilon.$$

Put

$$S_1=\frac{\partial^2[U(q;z,T)]}{\partial z^i\partial \bar{z}^j}*\chi_\varepsilon,$$

$$S_2=\frac{\partial^2[u_\nu]}{\partial z^i\partial \bar{z}^j}*\chi_\varepsilon=T_{i\bar{j}}*\chi_\varepsilon.$$

It follows from (3.16) that

$$|S_1(z)|\le C\varepsilon^{-2m-2}\int_{w\in B(\varepsilon)}|U(q;z+w,T)|\alpha^m(w).$$

Put $r = \|z\|$ and $\varepsilon = r$. Then by Corollary (3.13)

$$(3.17)\qquad |S_1(z)| \le Cr^{-2m-2}\int_{w\in B(2r)}|U(q;z,T)|\alpha^m$$
$$\le Cr^{-2m-2}(2r)^{2m}2M(2r) \le 2^{2m+1}Cr^{-2}M(2r).$$

Since $\|T_{i\bar{j}}\| \le \frac{1}{\sqrt{2}}\sum T_{i\bar{i}}$ as measures,

$$(3.18)\qquad |S_2(z)| = \left|\int_{w\in B(z;\varepsilon)} T_{i\bar{j}}\chi_\varepsilon(w-z)\alpha^m(w)\right|$$
$$\le \int_{w\in B(z;\varepsilon)}\|T_{i\bar{j}}\|\chi_\varepsilon(w-z)\alpha^m(w)$$
$$\le \frac{m}{\sqrt{2}}\int_{w\in B(z;\varepsilon)}\chi_\varepsilon(w-z)T\wedge\alpha^{m-1}(w)$$
$$\le \frac{m}{\sqrt{2}}\varepsilon^{-2m}(\max\chi)\int_{w\in B(z;\varepsilon)}T\wedge\alpha^{m-1}$$
$$\le \frac{m}{\sqrt{2}}r^{-2m}(\max\chi)\int_{B(2r)}T\wedge\alpha^{m-1}$$
$$= 2^{2m-5/2}m(\max\chi)r^{-2}n(2r,T).$$

We have by (3.17) and (3.18)

$$(3.19)\qquad |H_{i\bar{j}}(z)| \le C'r^{-2}\{M(2r)+n(2r,T)\}$$

for $\|z\| \le r$, where C' is a positive constant independent of r and T. We infer from the assumption, Lemma (3.2), (3.19) and Lemma (3.11) that

$$|H_{i\bar{j}}(z)| \le O(r_\nu^{q-2}) \le O\left(\left(\frac{r_\nu}{2}\right)^{q-2}\right),\quad \|z\| \le \frac{r_\nu}{2}.$$

If $0 \le q \le 1$, then $H_{i\bar{j}} \equiv 0$. In the case of $q \ge 2$, we see by Lemma (1.4) that $H_{i\bar{j}}$ are polynomials of degree $\le q-2$. Then it follows from (3.15) that $H_{i\bar{j}} = 0$. Therefore $dd^c[U(q;z,T)] = T$. *Q.E.D.*

(3.20) Theorem (Stoll). *Let D be an effective divisor on $\mathbf{C}^m$ such that $O \notin \operatorname{supp} D$. Assume that there are $q \in \mathbf{Z}^+$ and a positive increasing sequence $r_\nu \uparrow \infty$ ($r \to \infty$) such that*

$$\lim_{\nu\to\infty}\int_1^{r_\nu} t^{-q}\,dn(t,D) < \infty.$$

Then there exists a unique holomorphic function F satisfying the conditions:

(i) $(F)=D$.

(ii) $F(O)=1$ *and all partial derivatives of order* $\leq q$ *vanish at* O.

(iii) *There is a positive constant* $A(q)$ *depending only on* q *such that*

$$\log|F(z)| \leq A(q)\|z\|^q \left\{ \int_0^{\|z\|} \frac{n(t, T)}{t^{q+1}} dt + \|z\| \lim_{\nu\to\infty} \int_{\|z\|}^{r_\nu} \frac{n(t, T)}{t^{q+2}} dt \right\}.$$

Proof. Take a locally finite open covering $\{W_\lambda\}_{\lambda=1}^\infty$ of $\mathbf{C}^m$ so that W_λ are balls and there are holomorphic functions F_λ on W_λ satisfying $(F_\lambda)=D|W_\lambda$. By Theorem (3.14) we have the canonical potential $U(q; z, D)$ of D satisfying

$$dd^c[U(q; z, D)] = D.$$

Put

$$h_\lambda = \frac{1}{2}U(q; z, D) - \log|F_\lambda|.$$

Then $dd^c h_\lambda = 0$; i.e., h_λ are pluriharmonic. Therefore there are holomorphic functions g_λ such that the real parts $\mathrm{Re}\, g_\lambda$ of g_λ coincide with h_λ. Hence we get

$$\frac{1}{2}U(q; z, D) = \mathrm{Re}\,(g_\lambda + \log F_\lambda)$$

When $W_\lambda \cap W_\mu \neq \varnothing$,

$$g_\lambda + \log F_\lambda = g_\mu + \log F_\mu + i a_{\lambda\mu},$$

where $a_{\lambda\mu} \in \mathbf{R}$ satisfying the cocycle condition

$$a_{\lambda\mu} = -a_{\mu\lambda}, \quad a_{\lambda\mu} + a_{\mu\nu} + a_{\nu\lambda} = 0.$$

Hence we have a Cech cohomology class $(a_{\lambda\mu}) \in H^1(\mathbf{C}^m, \mathbf{R})$. Since $H^1(\mathbf{C}^m, \mathbf{R}) = \{0\}$, there are constants $b_\lambda \in \mathbf{R}$ such that

$$a_{\lambda\mu} = b_\mu - b_\lambda.$$

Then we have

$$g_\lambda + i b_\lambda + \log F_\lambda = g_\mu + i b_\mu + \log F_\mu \quad \text{on } W_\lambda \cap W_\mu \neq \varnothing.$$

Then we define a holomorphic function F on $\mathbf{C}^m$ by

$$F = F_\lambda \exp(g_\lambda + i b_\lambda) \quad \text{on } W_\lambda.$$

By Theorem (3.14) this F satisfies the required properties.

Let G be a holomorphic function on $\mathbf{C}^m$ satisfying (i), (ii) and (iii). Then $A = F/G$ is a nowhere vanishing holomorphic function on $\mathbf{C}^m$. By the same method as in the proof of Corollary (3.13) and conditions, (ii) and (iii), we see that

$$\max\{|\log|F(z)||;\ \|z\|\le r\}=O(r^q),$$

$$\max\{|\log|G(z)||;\ \|z\|\le r\}=O(r^q),$$

Taking $\log A(z)$ so that $\log A(O)=\log 1=0$, we get

$$|\mathrm{Re}\log A(z)|=O(r^q).$$

Since $\mathrm{Re}\log A(z)$ is harmonic in $\mathbf{C}^m$, Lemma (1.4) implies that $\mathrm{Re}\log A(z)$ is a polynomial of degree $\le q$ in z^i and $\bar{z}^i$. Then condition (ii) implies that $\mathrm{Re}\log A(z)=0$ and hence $A(z)\equiv 1$. *Q.E.D.*

The unique holomorphic function F for a given divisor D in Theorem (3.20) is called the **Weierstrass-Stoll canonical function** of D.

We give a criterion of the algebraicity of a divisor on $\mathbf{C}^m$.

(3.21) Lemma. *Let P be a polynomial on $\mathbf{C}^m$. Then*

$$\textit{degree of } P=p \iff N(r,(P))=(p+o(1))\log r.$$

Proof. Remark that $N(r,(P))$ is a convex increasing function in $\log r$. Hence it suffices to show that for any $p\in\mathbf{R}$ with $p\ge 0$

$$\text{degree of } P\le p \iff N(r,(P))\le(p+o(1))\log r.$$

Suppose that the degree of $P\le p$. Then by Proposition (5.3.14)

$$N(r,(P))\le T(r,P)\le(p+o(1))\log r.$$

Conversely, suppose that $N(r,(P))\le(p+o(1))\log r$. Let p' be a degree of P. Then Theorem (5.1.15) implies that

$$N(r,(P))=\int_{\Gamma(r)}\log|P|\eta-\int_{\Gamma(1)}\log|P|\eta. \tag{3.22}$$

Let $\rho\colon \mathbf{C}^m-\{O\}\to\mathbf{P}^{m-1}(\mathbf{C})$ be the Hopf fibering and ω_0 the Fubini-Study Kähler form. Then

$$\eta=d^c\log\|z\|\wedge\rho^*\omega^{m-1}. \tag{3.23}$$

For every complex line $l\in\mathbf{P}^{m-1}(\mathbf{C})$ through O, we take a point $a_l\in l\cap\Gamma(1)$. Then it follows from (3.22) and (3.23) that

$$N(r,(P))=\int_{l\in\mathbf{P}^{m-1}(\mathbf{C})}\left\{\log|P(re^{i\theta}a_l)|d\theta-\log|P(e^{i\theta}a_l)|d\theta\right\}\omega_0^{m-1} \tag{3.24}$$

$$=\int_{l\in\mathbf{P}^{m-1}(\mathbf{C})}N(r,(P|l))\omega_0^{m-1}=\int_1^r\frac{dt}{t}\int_{l\in\mathbf{P}^{m-1}(\mathbf{C})}n(r,(P|l))\omega_0^{m-1}.$$

Since P is a polynomial of degree p', $n(r, (P|l)) \uparrow p'$ as $r \uparrow \infty$. Therefore we obtain

$$N(r, (P)) = (p' + o(1))\log r \leq (p + o(1))\log r,$$

so that $p' \leq p$. *Q.E.D.*

(3.25) Theorem (Stoll). *Let D be an effective divisor on $\mathbf{C}^m$. Then D is an algebraic divisor defined by a polynomial of degree p if and only if*

$$N(r, D) = (p + o(1))\log r.$$

Proof. By Lemma (3.21) it suffices to show that if $N(r, D) = O(\log r)$, then D is algebraic. Suppose that $N(r, D) = O(\log r)$. Then

$$n(r, D) = O(1). \tag{3.26}$$

We may assume that $O \notin \operatorname{supp} D$. Let $q = 0$ and take any positive increasing sequence $r_\nu \uparrow \infty$ in Theorem (3.20). Then we have the Weierstrass-Stoll canonical function F satisfying

$$\log |F(z)| \leq A(0)\left\{ \int_0^r \frac{n(t, D)}{t} dt + r \int_r^\infty \frac{n(t, D)}{t^2} dt \right\}$$

for $z \in B(r)$. It follows from this and (3.26) that

$$T(r, F) = m(r, F) \leq O(\log r).$$

By Proposition (5.3.14) F is a polynomial. *Q.E.D.*

Notes

In the case of $m = 1$, the Weierstrass canonical product of a given effective divisor on $\mathbf{C}$, of which order is finite, is well known (cf., e.g., Hayman [46]). Stoll [110] proved its higher dimensional version, Theorem (3.20). The proof given here is due to Lelong [63]. By the uniqueness, the function constructed here must coincide with that of Stoll. Let $l \in \mathbf{P}^{m-1}(\mathbf{C})$ denote a complex line of $\mathbf{C}^m$ through O. Then the restriction $F|l$ of the Weierstrass-Stoll canonical function of an effective divisor D on $\mathbf{C}^m$ are the Weierstrass canonical product of the intersections $D \cdot l$ for all $l \in \mathbf{P}^{m-1}(\mathbf{C})$. This is trivial by Stoll's construction. Stoll [114] is a nice survey on his method. There is an application for meromorphic mappings $f : \mathbf{C}^m \to \mathbf{P}^N(\mathbf{C})$ of finite order due to Noguchi [75]. Mok-Siu-Yau [70] applied the method of the canonical potential of Theorem (3.14) to the characterization problem of $\mathbf{C}^m$.

Bibliography

1. L. V. Ahlfors, "Über die Anwendung differentialgeometrischer Methoden zur Untersuchung von Überlagerungsflächen," *Acta Soc. Sci. Fennicae Nova Ser. A* **2**, pp. 1-17 (1937).
2. L. V. Ahlfors, "The theory of meromorphic curves," *Acta Soc. Sci. Fennicae, Nova Ser. A.* **3**, pp. 3-31 (1941).
3. S. Ju. Arakelov, "Families of algebraic curves with fixed degeneracies," *Izv. Akad. Nauk SSSR Ser. Mat.* **35**, pp. 1277-1302 (1971).
4. J. Ax, "Some topics in differential algebraic geometry II," *Amer. J. Math.* **94**, pp. 1205-1213 (1972).
5. E. Bedford and B. A. Taylor, "The Dirichlet problem for a complex Monge-Ampère equation," *Invent. Math.* **37**, pp. 1-44 (1976).
6. E. Bedford and B. A. Taylor, "A new capacity for plurisubharmonic functions," *Acta Math.* **149**, pp. 1-41 (1982).
7. E. Bishop, "Conditions for the analyticity of certain sets," *Michigan Math. J.* **11**, pp. 289-304 (1964).
8. A. Bloch, "Sur les système de fonctions holomorphes à variétés linaires," *Ann. Sci. École Norm. Sup.* **43**, pp. 309-362 (1926).
9. A. Bloch, "Sur les systèmes de fonctions uniformes satisfaisant à l'équation d'une variété algébrique dont l'irrégularité dépasse la dimension," *J. Math. Pures Appl.* **5**, pp. 9-66 (1926).
10. S. Bochner and D. Montgomery, "Groups on analytic manifolds," *Ann. Math.* **48**, pp. 659-669 (1947).
11. A. Borel, "Some metric properties of arithmetic quotients of symmetric spaces and an extension theorem," *J. Differential Geometry* **6**, pp. 543-560 (1972).
12. E. Borel, "Sur les zéros des fonctions entières," *Acta Math.* **20**, pp. 357-396 (1897).
13. R. Brody, "Compact manifolds and hyperbolicity," *Trans. Amer. Math. Soc.* **235**, pp. 213-219 (1978).

14. R. Brody and M. Green, "A family of smooth hyperbolic hypersurfaces in $\mathbf{P}_3$," *Duke Math. J.* **44**, pp. 873-874 (1977).

15. J. Carlson, "Some degeneracy theorems for entire functions with values in an algebraic variety," *Trans. Amer. Math. Soc.* **168**, pp. 273-301 (1972).

16. J. Carlson and P. Griffiths, "A defect relation for equidimensional holomorphic mappings between algebraic varieties," *Ann. Math.* **95**, pp. 557-584 (1972).

17. H. Cartan, "Sur les zéros des combinaisons linéaires de p fonctions holomorphes données," *Mathematica* **7**, pp. 5-31 (1933).

18. H. Cartan, "Idéaux de fonctions analytiques de n variables complexes," *Ann. Ecole Normal Sup.* **61**, pp. 149-197 (1944).

19. S. S. Chern, "An elementary proof of the existence of isothermal parameters on a surface," *Proc. Amer. Math. Soc.* **6**, pp. 771-782 (1955).

20. S. S. Chern, "The integrated form of the first main theorem for complex analytic mappings in several variables," *Ann. Math.* **71**, pp. 536-551 (1960).

21. S. S. Chern, "Holomorphic curves in the plane," pp. 73-94 in *Differential Geometry in Honor of K. Yano*, Kinokuniya, Tokyo (1972), pp. 73-94.

22. S. S. Chern and R. Osserman, "Complete minimal surfaces in euclidean n-space," *J. d'Analyse Math.* **19**, pp. 15-34 (1967).

23. M. Cowen and P. Griffiths, "Holomorphic curves and metrics of negative curvature," *J. d'Analyse Math.* **29**, pp. 93-153 (1976).

24. J.-P. Demailly, "Nombres de Lelong généralisès, théorèmes d'intègralité et d'analyticité," *Acta Math.* **159**, pp. 153-169 (1987).

25. D. A. Eisenman, *Intrinsic measures on complex manifolds and holomorphic mapping*, Memoires, Vol. 96, Amer. Math. Soc., Providence, Rhode Island (1970).

26. G. Faltings, "Endlichkeitssätze für abelsche Varietäten über Zahlköpern," *Invent. Math.* **73**, pp. 349-366 (1983).

27. G. Faltings, "Arakelov's theorem for Abelian varieties," *Invent. Math.* **73**, pp. 337-347 (1983).

28. H. Fujimoto, "Extension of the big Picard's theorem," *Tohoku Math. J.* **24**, pp. 415-422 (1972).

29. H. Fujimoto, "The defect relations for the derived curves of a holomorphic curves in $\mathbf{P}^n(\mathbf{C})$," *Tohoku Math. J.* **34**, pp. 141-160 (1982).

30. H. Fujimoto, "On the number of exceptional values of the Gauss maps of minimal surfaces," *J. Math. Soc. Japan* **40**, pp. 235-247 (1988).

31. H. Fukushima and N. Okada, "Probability theory and complex Monge-Ampère operator," *Acta Math.* **300**, pp. 1-50 (1988).

32. D. Gilbarg and N. S. Trudinger, *Elliptic Partial Differential Equations of Second Order*, Springer-Verlag, Berlin-Heidelberg-New York (1977).

33. I. Graham and H. Wu, "Some remarks on the intrinsic measures of Eisenman," *Trans. Amer. Math. Soc.* **288**, pp. 625-660 (1985).

34. I. Graham and H. Wu, "Characterization of the unit ball B^n in complex euclidean

space,'' *Math. Z.* **189**, pp. 449-456 (1985).

35. H. Grauert, ''Mordells Vermutung über rationale Punkte auf Algebraischen Kurven und Funktionenköper,'' *Publ. Math. I.H.E.S.* **25**, pp. 131-149 (1965).

36. H. Grauert and R. Remmert, ''Plurisubharmonische Funktionen in komplexen Räumen,'' *Math. Z.* **65**, pp. 175-194 (1956).

37. M. Green, ''Holomorphic maps into complex projective space,'' *Trans. Amer. Math. Soc.* **169**, pp. 89-103 (1972).

38. M. Green, ''Some examples and counter-examples in value distribution theory for several complex variables,'' *Compositio Math.* **30**, pp. 317-322 (1975).

39. M. Green, ''The hyperbolicity of the complement of $2n+1$ hyperplanes in general position in $\mathbf{P}_n$, and related results,'' *Proc. Amer. Math. Soc.* **66**, pp. 109-113 (1977).

40. M. Green, ''Holomorphic maps to complex tori,'' *Amer. J. Math.* **100**, pp. 615-620 (1978).

41. M. Green and P. Griffiths, ''Two applications of algebraic geometry to entire holomorphic mappings,'' pp. 41-74 in *The Chern Symposium 1979*, Springer-Verlag, New York-Heidelberg-Berlin (1980), pp. 41-74.

42. R. Greene and H. Wu, ''Function Theory on Manifolds Which Possess a Pole,'' in *Lecture Notes in Math.*, Springer-Verlag, Berlin-Heidelberg-New York (1979).

43. P. Griffiths, ''Holomorphic mappings into canonical algebraic varieties,'' *Ann. Math.* **93**, pp. 439-458 (1971).

44. P. Griffiths and J. King, ''Nevanlinna theory and holomorphic mappings between algebraic varieties,'' *Acta Math.* **130**, pp. 145-220 (1973).

45. R. C. Gunning and H. Rossi, *Analytic Functions of Several Complex Variables,* Prentice-Hall Inc., Englewood Cliffs, N. J. (1965).

46. W. K. Hayman, *Meromorphic Functions,* Oxford Univ. Press, London (1964).

47. H. Hironaka, ''Resolution of singularities of an algebraic variety over a field of characteristic zero: I; II,'' *Ann. Math.* **79**, pp. 109-203; 205-326 (1964).

48. L. Hörmander, *Introduction to Complex Analysis in Several Variables,* van Nostrand, Princeton (1966).

49. C. Horst, ''Compact varieties of surjective holomorphic mappings,'' *Math. Z.* **196**, pp. 259-269 (1987).

50. M. Kalka, B. Shiffman, and B. Wong, ''Finiteness and rigidity theorems for holomorphic mappings,'' *Michigan Math. J.* **28**, pp. 289-295 (1981).

51. K. Kawamata, ''On Bloch's conjecture,'' *Invent. Math.* **57**, pp. 97-100 (1980).

52. P. Kiernan, ''Extensions of holomorphic maps,'' *Trans. Amer. Math. Soc.* **172**, pp. 347-355 (1972).

53. H. Kneser, ''Zur Theorie der gebrochenen Funktionen mehrerer Veränderlicher,'' *Jber. Deutsch. Math. Verein.* **48**, pp. 1-38 (1938).

54. S. Kobayashi, *Hyperbolic Manifolds and Holomorphic Mappings,* Marcel Dekker, New York (1970).

55. S. Kobayashi, ''Intrinsic distances, measures, and geometric function theory,'' *Bull.*

Amer. Math. Soc. **82**, pp. 357-416 (1976).

56. S. Kobayashi and T. Ochiai, "Satake compactification and the great Picard theorem," *J. Math. Soc. Japan* **23**, pp. 340-350 (1971).

57. S. Kobayashi and T. Ochiai, "Meromorphic mappings onto compact complex spaces of general type," *Invent. Math.* **31**, pp. 7-16 (1975).

58. K. Kodaira, "Holomorphic mappings of polydiscs into compact complex manifolds," *J. Diff. Geometry* **6**, pp. 33-46 (1971).

59. K. Kodaira, "Nevanlinna Theory," in *Seminar Notes*, Univ. of Tokyo, Tokyo (1974).

60. M. H. Kwack, "Generalization of the big Picard theorem," *Ann. Math.* **90**, pp. 9-22 (1969).

61. S. Lang, "Higher dimensional Diophantine problems," *Bull. Amer. Math. Soc.* **80**, pp. 779-787 (1974).

62. S. Lang, "Hyperbolic and Diophantine analysis," *Bull. Amer. Math. Soc.* **14**, pp. 159-205 (1986).

63. P. Lelong, "Fonctions entières (n variables) et fonctions plurisousharmoniques d'ordre fini dans $\mathbf{C}^n$," *J. d'Analyse Math.* **12**, pp. 365-407 (1964).

64. P. Lelong, *Fonctions plurisousharmoniques et Formes différentielles positives,* Gordon and Breach, Paris (1968).

65. L. Lempert, "La métrique de Kobayashi et la representation des domaines sur la boule," *Bull. Soc. Math. France* **109**, pp. 427-474 (1981).

66. Ju. Manin, "Rational points of algebraic curves over function fields," *Izv. Akad. Nauk. SSSR. Ser. Mat.* **27**, pp. 1395-1440 (1963).

67. J. Milnor, "Morse Theory," in *Annals of Math. Studies*, Princeton Univ. Press, Princeton (1963).

68. J. Milnor, "On deciding whether a surface is parabolic or hyperbolic," *Amer. Math. Monthly* **84**, pp. 43-46 (1977).

69. H. E. Mir, "Sur le prolongement des courants positifs fermés," *Acta Math.* **153**, pp. 1-45 (1984).

70. N. Mok, Y.-T. Siu, and S. T. Yau, "The Poincaré-Lelong equation on complete Kähler manifolds," *Compositio Math.* **44**, pp. 183-218 (1981).

71. S. Mori and S. Mukai, "The uniruledness of the moduli space of curves of genus 11," in *Algebraic Geometry, Proc. Japan-France Conf. Tokyo and Kyoto 1982, Lecture Notes in Math.*, Springer-Verlag, Berlin-Heidelberg-New York (1983).

72. R. Narasimhan, "Introduction to the Theory of Analytic Spaces," in *Lecture Notes in Math.*, Springer-Verlag, Berlin-Heidelberg-New York (1966).

73. R. Nevanlinna, "Zur Theorie der meromorphen Funktionen," *Acta Math* **46**, pp. 1-99 (1925).

74. R. Nevanlinna, *Le Théorème de Picard-Borel et la théorie des fonctions méromorphes,* Gauthier-Villars, Paris (1939).

75. J. Noguchi, "A relation between order and defects of meromorphic mappings of $\mathbf{C}^m$ into $\mathbf{P}^N(\mathbf{C})$," *Nagoya Math. J.* **59**, pp. 97-106 (1975).

76. J. Noguchi, ''Meromorphic mappings of a covering space over $\mathbf{C}^m$ into a projective variety and defect relations,'' *Hiroshima Math. J.* **6**, pp. 265-280 (1976).

77. J. Noguchi, ''Holomorphic curves in algebraic varieties,'' *Hiroshima Math. J.* **7**, pp. 833-853 (1977).

78. J. Noguchi, ''Meromorphic mappings into a compact complex space,'' *Hiroshima Math. J.* **7**, pp. 411-425 (1977).

79. J. Noguchi, ''Open Problems in Geometric Function Theory,'' pp. 6-9 in *Proc. Conf. on Geometric Function Theory, Katata*, ed. S. Murakami, etc. (1978), pp. 6-9.

80. J. Noguchi, ''Supplement to ''Holomorphic curves in algebraic varieties'','' *Hiroshima Math. J.* **10**, pp. 229-231 (1980).

81. J. Noguchi, ''A higher dimensional analogue of Mordell's conjecture over function fields,'' *Math. Ann.* **258**, pp. 207-212 (1981).

82. J. Noguchi, ''Lemma on logarithmic derivatives and holomorphic curves in algebraic varieties,'' *Nagoya Math. J.* **83**, pp. 213-233 (1981).

83. J. Noguchi, ''Hyperbolic fibre spaces and Mordell's conjecture over function fields,'' *Publ. RIMS, Kyoto University* **21**, pp. 27-46 (1985).

84. J. Noguchi, ''On the value distribution of meromorphic mappings of covering spaces over $\mathbf{C}^m$ into algebraic varieties,'' *J. Math. Soc. Japan* **37**, pp. 295-313 (1985).

85. J. Noguchi, ''Logarithmic jet spaces and extensions of de Franchis' theorem,'' pp. 227-249 in *Contributions to Several Complex Variables, Aspects Math.*, Vieweg, Braunschweig (1986), pp. 227-249.

86. J. Noguchi, ''Moduli spaces of holomorphic mappings into hyperbolically imbedded complex spaces and locally symmetric spaces,'' *Invent. Math.* **93**, pp. 15-34 (1988).

87. J. Noguchi and T. Sunada, ''Finiteness of the family of rational and meromorphic mappings into algebraic varieties,'' *Amer. J. Math.* **104**, pp. 887-900 (1982).

88. T. Ochiai, ''On holomorphic curves in algebraic varieties with ample irregularity,'' *Invent. Math.* **43** , pp. 83-96 (1977).

89. T. Ochiai and J. Noguchi, *Geometric Function Theory in Several Complex Variables,* Iwanami, Tokyo (1984). (in Japanese)

90. K. Oka, *Collected Papers,* Springer-Verlag, Berlin-Heidelberg-New York-Tokyo (1984).

91. M. Okada, ''Espaces de Dirichlet generaux en analyse complexe,'' *J. Functional Analy.* **46**, pp. 396-410 (1982).

92. A. N. Parshin, ''Algebraic curves over function fields. I,'' *Izv. Akad. Nauk SSSR Ser. Mat.* **32**, pp. 1145-1170 (1968).

93. R. Remmert, ''Projecktionen analytischer Mengen,'' *Math. Ann.* **130**, pp. 410-441 (1956).

94. R. Remmert, ''Holomorphe und meromorphe Abbildungen komplexer Räume,'' *Math. Ann.* **133**, pp. 328-378 (1957).

95. G. de Rham, *Variétés différentiables,* Hermann, Paris (1973).

96. D. Riebesehl, ''Hyperbolische komplexe Räume und die Vermutung von Mordell,''

Math. Ann. **257**, pp. 99-110 (1981).

97. H. L. Royden, "Remarks on the Kobayashi metric," pp. 125-137 in *Several Complex Variables II Maryland 1970, Lecture Notes in Math.*, Springer-Verlag, Berlin-Heidelberg-New York (1971), pp. 125-137.

98. H. L. Royden, "The extension of regular holomorphic maps," *Proc. Amer. Math. Soc.* **43**, pp. 306-310 (1974).

99. F. Sakai, "Defect relations for equidimensional holomorphic maps," *J. Faculty of Sci., Univ. Tokyo Sec. IA* **23**, pp. 561-580 (1976).

100. H. L. Selberg, "Algebroide Funktionen und Umkehlfunktionen Abelscher Integrale," *Avh. Norske Vid. Akad. Oslo* **8**, pp. 1-72 (1934).

101. I. R. Shafarevich, *Basic Algebraic Geometry,* Springer-Verlag, Berlin-Heidelberg-New York (1974).

102. B. Shiffman, "Nevanlinna defect relations for singular divisors," *Invent. Math.* **31**, pp. 155-182 (1975).

103. N. Sibony, "A class of hyperbolic manifolds," pp. 357-372 in *Annals of Math. Studies*, Princeton Univ. Press, Princeton (1981), pp. 357-372.

104. N. Sibony, "Quelques problèmes de prolongement de courants en analyse complexe," *Duke Math. J.* **52**, pp. 157-197 (1985).

105. Y.-T. Siu, "Analyticity of sets associated to Lelong numbers and the extension of closed positive currents," *Invent. Math.* **27**, pp. 53-156 (1974).

106. Y.-T. Siu, "Extension of meromorphic maps into Kähler manifolds," *Ann. Math.* **102**, pp. 421-462 (1975).

107. Y.-T. Siu, "Defect relations for holomorphic maps between spaces of different dimensions," *Duke Math. J.* **55**, pp. 213-251 (1987).

108. S. Skoda, "Prolongement des courants positifs, fermés de masse finie," *Invent. Math.* **66**, pp. 361-376 (1982).

109. W. Stoll, "Die beiden Hauptsätze der Wertverteilungstheorie bei Funktionen mehrerer komplexer Veränderlichen (I)," *Acta Math.* **90**, pp. 1-115 (1953).

110. W. Stoll, "Ganze Funktionen endlicher Ordnung mit gegebenen Nullstellenflächen," *Math. Z.* **57**, pp. 211-237 (1953).

111. W. Stoll, "Die beiden Hauptsätze der Wertverteilungstheorie bei Funktionen mehrerer komplexer Veränderlichen (II)," *Acta Math.* **92**, pp. 55-169 (1954).

112. W. Stoll, "Einige Bemerkungen zur Fortsetzbarkeit analytischer Mengen," *Math. Z.* **60**, pp. 287-304 (1954).

113. W. Stoll, "The growth of the area of a transcendental analytic set. I; II," *Math. Ann.* **156**, pp. 47-78; 144-170 (1964).

114. W. Stoll, *Holomorphic Functions of Finite Order in Several Complex Variables*, Amer. Math. Soc., Providence, Rhode Island (1974).

115. W. Stoll, "Value Distribution on Parabolic Spaces," in *Lecture Notes in Math.*, Springer-Verlag, Berlin-Heidelberg-New York (1977).

116. G. Stolzenberg, "Volumes, Limits, and Extensions of Analytic Varieties," in *Lecture*

Notes in Math., Springer-Verlag, Berlin-Heidelberg-New York (1966).

117. P. R. Thie, "The Lelong number of a point of a complex analytic set," *Math. Ann.* **172**, pp. 269-312 (1967).

118. T. Urata, "Holomorphic mappings into a certain compact complex analytic space," *Tohoku Math. J.* **33**, pp. 573-585 (1981).

119. A. L. Vitter, "The lemma of the logarithmic derivative in several complex variables," *Duke Math. J.* **44**, pp. 89-104 (1977).

120. A. Weil, *Introduction à l'Etude des Variétés kähleriennes*, Hermann, Paris (1958).

121. R. O. Wells, Jr., "Differential Analysis on Complex Manifolds," in *Graduate Texts in Math.*, Springer-Verlag, New York-Heidelberg-Berlin (1980).

122. J. Weyl, "Meromorphic curves," *Ann. Math.* **39**, pp. 516-538 (1938).

123. H. Wu, "Remarks on the first main theorem in equidistribution theory III," *J. Diff. Geometry* **3**, pp. 83-94 (1969).

124. H. Wu, "A remark on holomorphic sectional curvature," *Indiana Univ. Math. J.* **22**, pp. 1103-1108 (1973).

125. H. Wu, Some open problems in the study of noncompact Kähler manifolds, Geometric Theory of Several Complex Variables, Lecture Notes Ser. RIMS, **340**, Kyoto, 1978, 12-25.

Index

Symbols